T0334212

GROUNDWATER ENVIRONMENT IN ASIAN CITIES

Concepts, Methods and Case Studies

GROUNDWATER ENVIRONMENT IN ASIAN CITIES

Concepts, Methods and Case Studies

Edited by

SANGAM SHRESTHA
Water Engineering and Management,
Asian Institute of Technology (AIT), Thailand

VISHNU PRASAD PANDEY
Water Engineering and Management,
Asian Institute of Technology (AIT), Thailand;
Department of Civil Engineering,
Asian Institute of Technology
and Management (AITM), Nepal

BINAYA RAJ SHIVAKOTI
Water Resources Management Team, Natural Resources
and Ecosystem Area, Institute for Global Environmental
Strategies (IGES), Japan

SHASHIDHAR THATIKONDA
Department of Civil Engineering,
Indian Institute of Technology, Hyderabad, India

Amsterdam • Boston • Heidelberg • London
New York • Oxford • Paris • San Diego
San Francisco • Singapore • Sydney • Tokyo

ELSEVIER Butterworth-Heinemann is an imprint of Elsevier

Butterworth-Heinemann is an imprint of Elsevier
The Boulevard, Langford Lane, Kidlington, Oxford OX5 1GB, UK
50 Hampshire Street, 5th Floor, Cambridge, MA 02139, USA

Copyright © 2016 Elsevier Inc. All rights reserved.

No part of this publication may be reproduced or transmitted in any form or by any means, electronic
or mechanical, including photocopying, recording, or any information storage and retrieval system,
without permission in writing from the publisher. Details on how to seek permission, further information about the Publisher's permissions policies and our arrangements with organizations such as
the Copyright Clearance Center and the Copyright Licensing Agency, can be found at our website:
www.elsevier.com/permissions.

This book and the individual contributions contained in it are protected under copyright by the Publisher
(other than as may be noted herein).

Notices
Knowledge and best practice in this field are constantly changing. As new research and experience
broaden our understanding, changes in research methods, professional practices, or medical treatment
may become necessary.

Practitioners and researchers must always rely on their own experience and knowledge in evaluating and
using any information, methods, compounds, or experiments described herein. In using such information
or methods they should be mindful of their own safety and the safety of others, including parties for whom
they have a professional responsibility.

To the fullest extent of the law, neither the Publisher nor the authors, contributors, or editors, assume
any liability for any injury and/or damage to persons or property as a matter of products liability,
negligence or otherwise, or from any use or operation of any methods, products, instructions, or ideas
contained in the material herein.

British Library Cataloguing-in-Publication Data
A catalogue record for this book is available from the British Library

Library of Congress Cataloging-in-Publication Data
A catalog record for this book is available from the Library of Congress

ISBN: 978-0-12-803166-7

For information on all Butterworth-Heinemann publications
visit our website at http://store.elsevier.com/

Working together
to grow libraries in
developing countries

www.elsevier.com • www.bookaid.org

LIST OF CONTRIBUTORS

Aditya Sarkar
Department of Geology, University of Delhi, India

Anirut Ladawadee
Bureau of Groundwater Conservation and Restoration, Department of Groundwater Resources, Thailand

Binaya Raj Shivakoti
Water Resources Management Team, Natural Resources and Ecosystem Area, Institute for Global Environmental Strategies (IGES), Japan

Bui Tran Vuong
Division of Water Resources Planning and Investigation for the South of Vietnam (DWRPIS), Ministry of Natural Resources and Environment of Vietnam

Domingos Pinto
Water Engineering and Management at the Asian Institute of Technology (AIT), Thailand

Ekaterina I. Nemaltseva
Laboratory rational ground water use, Institute of Water Problems and Hydropower, National Academy of Science, Kyrgyz Republic

Gennady M. Tolstikhin
Kyrgyz Hydrogeological Survey, State Agency on Geology and Mineral Resources of the Kyrgyz Republic

Haryadi Tirtomihardjo
Center of Groundwater Resources and Environmental Geology (CGREG), Ministry of Energy and Mineral Resources (MEMR), Indonesia

Heejung Kim
School of Earth and Environmental Sciences, Seoul National University, Seoul, Republic of Korea

Jin-Yong Lee
Department of Geology, Kangwon National University, Chuncheon, Republic of Korea

Jingli Shao
School of Water Resources and Environment, China University of Geosciences, China

Jiurong Liu
Beijing Institute of Hydrogeology and Engineering Geology, China

Kabita Karki
Department of Mines and Geology, Kathmandu, Nepal

Kang-Kun Lee
School of Earth and Environmental Sciences, Seoul National University, Seoul, Republic of Korea

Khin Kay Khaing
Department of Geography, University of Yangon, Myanmar

Le Hoai Nam
Division of Water Resources Planning and Investigation for the South of Vietnam (DWRPIS), Ministry of Natural Resources and Environment of Vietnam

Liya Wang
Beijing Institute of Hydrogeology and Engineering Geology, China

Muhammad Basharat
International Waterlogging and Salinity Research Institute (IWASRI), Pakistan Water and Power Development Authority (WAPDA), Lahore, Pakistan

Oranuj Lorphensri
Department of Groundwater Resources, Thailand

Phan Nam Long
Division of Water Resources Planning and Investigation for the South of Vietnam (DWRPIS), Ministry of Natural Resources and Environment of Vietnam

Qing Yang
Beijing Institute of Hydrogeology and Engineering Geology, China

Qiulan Zhang
School of Water Resources and Environment, China University of Geosciences (Beijing)

Rabin Malla
Center of Research for Environment, Energy and Water (CREEW), Kathmandu, Nepal

Rafael G. Litvak
Ground Water Modelling Laboratory, Kyrgyz Research Institute of Irrigation

Rong Wang
Land Subsidence Research Institute of Beijing Institute of Hydrogeology and Engineering Geology, China

Sangam Shrestha
Water Engineering and Management, Asian Institute of Technology (AIT), Thailand

Shakir Ali
Department of Geology, University of Delhi, India

Shashank Shekhar
Department of Geology, University of Delhi, India

Shashidhar Thatikonda
Department of Civil Engineering, Indian Institute of Technology, Hyderabad, India

Suman Kumar
Department of Geology, University of Delhi, India

SVN Rao
WAPCOS, Regional office, Hyderabad, India

Tussanee Nettasana
Bureau of Groundwater Conservation and Restoration, Department of Groundwater Resources, Thailand

Vishnu Prasad Pandey
Water Engineering and Management, Asian Institute of Technology (AIT), Thailand;
Department of Civil Engineering, Asian Institute of Technology and Management
(AITM), Nepal

Woo-Hyun Jeon
Department of Geology, Kangwon National University, Chuncheon, Republic of Korea

Zhiping Li
Beijing Institute of Hydrogeology and Engineering Geology, China

PREFACE

Groundwater contributes to the sustainable development of many Asian cities by providing water for domestic, industrial, and agricultural uses and by regulating ecosystem flows. In many Asian cities more than half the potable water supply comes from groundwater. In some cities, groundwater abstraction for industrial use is even higher than for potable use. Industrial use in total groundwater abstraction is 80% in Bandung and 60% in Bangkok. In India 60% of irrigated areas are served by groundwater. In Pakistan groundwater provides over 40% of total crop water requirements in the highly populous province of Punjab and 70% of the farmers receive 80–100% of their irrigation water from wells and tubewells. However, groundwater has not always been properly managed, which often has resulted in depletion and degradation of the resource. Much emphasis has been given to groundwater resource development without paying careful attention to its management despite its strategic role in sustainable development. Many cities are already suffering from water insecurity as a result of rapid population growth and economic development. Without proactive groundwater governance, the detrimental effects of poor management will nullify (or even surpass) the social gains made so far. Many cities are already suffering from water insecurity as a result of rapid population growth and economic development. To maintain the advantage of groundwater as an important resource for sustainable development, groundwater management should be more strategic and proactive to cope with increased demand from rapid industrialization and urbanization including the potential impacts of climate change.

This book presents what is currently scientifically known about the groundwater environment of certain Asian cities using the driver-pressure-state–impact–response (DPSIR) framework. The book presents in detail facts and figures on groundwater dependence, problems related to groundwater overexploitation, implementation of various policy instruments and management practices, and their results in 14 carefully selected Asian cities: Bandung (Indonesia), Bangkok (Thailand), Beijing (China), Bishkek (Kyrgyzstan), Chitwan (Nepal), Delhi (India), Dili (Timor-Leste), Ho Chi Minh City (Vietnam), Hyderabad (India), Khulna (Bangladesh), Lahore (Pakistan), Seoul (South Korea), Tokyo (Japan), and Yangon (Myanmar).

The book will be helpful to a wide range of readers – local, regional, and global – working directly or indirectly in the groundwater sector. It is specifically targeted at policy makers, decision makers, researchers and research/academic institutes, implementing agencies, and governments. The book's collection of maps, tables, and references will be invaluable to those who would otherwise have to search elsewhere for basic information on the groundwater of Asian cities.

The theoretical background of the topics discussed along with the case studies should help readers understand the similarities and differences about the status of groundwater development and use in each city. In addition, the information in the book will serve as a baseline for further research such as the mitigation of groundwater-related problems (e.g., land subsidence), impact of climate change on groundwater, and the importance of groundwater in implementing sustainable development goals in the future.

Sangam Shrestha
Vishnu Prasad Pandey
Binaya Raj Shivakoti
Shashidhar Thatikonda

ACKNOWLEDGMENTS

We would like to express our sincere gratitude to the many people who helped bring this book to fruition; to all those who provided support, talked things over, read, wrote, offered comments, allowed us to quote their remarks, and assisted in the editing, proofreading, and design.

The writing of this book was conceived as a result of feedback from a project entitled "Enhancing Groundwater Management Capacity in Asian Cities through Development and Application of Groundwater Sustainability Index in the Context of Global Change," which was supported by a scientific capacity-building project under the Asia Pacific Network (APN) for Global Climate Change's core capacity development program, CAPaBLE. We would like to thank the APN for providing the financial and technical support necessary for groundwater managers and researchers from eight cities across Asia to collaborate in this work. The cities were Hyderabad, Bandung, Vientiane, Yangon, Chitwan, Lahore, Bangkok, and Ho Chi Minh City.

Our sincere thanks go to all the contributing authors. They prepared their chapters despite their busy schedules and despite our frequent reminders. We would also like to thank all the reviewers for their valuable feedback.

Special thanks go to Smriti Malla for her tireless efforts in communicating with the contributing authors and in compiling and formatting all the chapters throughout the publication process.

FOREWORD

In April 2013 the Asia-Pacific Network for Global Change Research (APN) awarded a grant to Dr Sangam Shrestha of the Asian Institute of Technology, Thailand for a scientific capacity-building project under APN's core capacity development program, CAPaBLE. The title of the project was "Enhancing Groundwater Management Capacity in Asian Cities through Development and Application of Groundwater Sustainability Index in the Context of Global Change." The work of the project was accomplished in January 2015 through strong collaborative efforts that brought together key scholars and stakeholders from an array of reputable research institutions, government sectors, and other organizations across Asia.

Project activity focused specifically on increasing the understanding of policy makers and other relevant stakeholders and end users in eight selected cities across Asia: Hyderabad, Bandung, Vientiane, Yangon, Chitwan, Lahore, Bangkok, and Ho Chi Minh City. The specific aim of the activity was to develop the capacity of cities to assess their current situation and understanding of groundwater management. More details of specific project activity that has culminated in the writing of this book can be found by accessing the metadata link on APN's E-Library.[1] Dr Shrestha and his team added more cities through their strong regional network to cover various subregions in Asia and came up with this excellent publication.

Taking this opportunity, APN would like to congratulate the dedication and collaborative efforts of the authors of the book under the leadership of Dr Sangam Shrestha. The book will be a valuable resource for groundwater adversaries in the scientific, decision-making, and end user communities, particularly for understanding and assessing the state

---

[1] Enhancing the Groundwater Management Capacity in Asian Cities through the Development and Application of Groundwater Sustainability Index in the Context of Global Change, APN E-Lib, accessed June 18, 2015, http://www.apn-gcr.org/resources/items/show/1901

of groundwater resources in the region as well as learning from the responses practiced so far.

This book will also ensure the continuity and sustainability of APN's core mission, which is to support regional cooperation, strengthen interactions among scientists and decision makers, and improve the scientific and technical capabilities of countries in the Asia-Pacific region.

Dr Linda Anne Stevenson

Division Head
Communication and Scientific Affairs
APN Secretariat, Kobe, Japan

ASIA-PACIFIC NETWORK FOR
GLOBAL CHANGE RESEARCH

FOREWORD

Groundwater is a precious but invisible resource; to assess, monitor, use, or protect something that we cannot see is not an easy task, especially in urban areas characterized by numerous (known and unknown) human impacts on groundwater environment. Additionally, groundwater specialists face the challenge of explaining the importance of the invisible resource in a vivid way. Try to imagine groundwater in (or rather under) urbanized areas; when we walk the streets of our cities do we think about the groundwater beneath? Not really, despite probably using it every day at home and at work.

Groundwater plays an important role in the sustainable development of cities, especially those cities that are growing rapidly. The urban population in these cities in 1960 accounted for about 34% of total global population; last year (2014) it reached 54% and still continues to grow. Demand for groundwater is growing accordingly. So, how can we ensure sustainable use of limited groundwater resources in this situation? How can we prevent (or at least mitigate) depletion, pollution, land subsidence, or the intrusion of sea water in coastal cities? The book in your hands gives a clear answer: only by informed management, by learning about the groundwater environment in the cities, and sharing that knowledge for reuse elsewhere.

The fast-growing cities of Asia have many common features and specifics in terms of groundwater availability and management practice. That makes reading this book even more interesting, and comparative analysis even more beneficial. Books like this one make the invisible visible.

igrac
International Groundwater Resources Assessment Centre

Dr Neno Kukuric

Director
International Groundwater Resources Assessment Centre (IGRAC)
Delft, The Netherlands

FOREWORD

The UNESCO International Hydrological Programme (IHP) is an intergovernmental program with the mandate in the United Nations system to devote all possible efforts to water sciences research, water resources management, education and capacity building. Since its inception in 1975, IHP has evolved from an internationally coordinated hydrological research program into an encompassing, holistic program to facilitate education and capacity building, and enhance water resources management and governance. IHP facilitates an interdisciplinary and integrated approach to water-

shed and aquifer management, which incorporates the social dimension of water resources, and promotes and develops international research in hydrological and freshwater sciences. The overarching focus of the UNESCO IHP's Eighth Phase (2014–2021) is on *"Water security: Responses to local, regional, and global challenges"* and it comprises six thematic areas; two are respectively devoted to studies on "Groundwater in a changing environment (Theme 2)" and "Water and human settlements of the future (Theme 4)".

Considering that groundwater in aquifers ensures drinking water supplies for nearly half of the world's population (including most of the urban areas and megacities) and that one of the major challenges of the twenty-first century is to provide access to improved water sources to cities all over the world, the IHP supports the development of up-to-date scientific knowledge on groundwater resources management in urban areas.

According to recent studies (WWDR, 2014) the rate of groundwater abstraction globally is increasing by 1–2% per year with an estimated 20% of the world's aquifers being overexploited (WWAP, 2012). Furthermore, the world's urban population is projected to grow to 6.3 billion by 2050 (UNDESA ST/ESA/SER.A/354) and cities are facing a range of pressures, such as climate change, population growth, and deterioration of urban infrastructures. Due to these pressures, cities of the future will experience difficulties in efficiently managing scarcer and less reliable water resources and in providing sufficient sanitation. These challenges call for comprehensive research and studies, and the implementation of new science-based methodologies and endorsement of principles for groundwater resources management and cities.

The book "*Groundwater Environment in Asian Cities: Concepts, Methods and Case Studies*" responds to this call with concrete examples and lessons for readers that are eager to understand urban groundwater management challenges in Asia.

The book presents up-to-date scientific knowledge on the groundwater environment in selected Asian cities that are among the ones most impacted by the global changes. The comparative analyses on policy and management responses adopted – including their strengths and weaknesses – are clearly presented along with discussions of relevant policy recommendations. For researchers and practitioners alike the book offers important insights on the inherently political nature of groundwater and cities, and can be considered a context for wide-ranging dialog between experts in the relevant disciplines. Given its strong science-policy focus, it also represents a valuable contribution to the achievements of the Eighth Phase of UNESCO-IHP.

Conscious of the need to raise political awareness globally on the urgency to improve groundwater governance and construct new paradigms for sustainable cities, I sincerely hope that this book will prompt discussions at high levels on urban groundwater challenges and opportunities and will moreover inspire new actions in this domain in Asia and beyond.

Dr Aureli Alice

Chief of Section
Groundwater System and Human Settlements
Division of Water Sciences
International Hydrological Programme (IHP)
United Nations Education, Scientific and Cultural
Organization (UNESCO)
Paris, France

United Nations Educational, Scientific and Cultural Organization

International Hydrological Programme

FOREWORD

Groundwater is an important source of water supply for many Asian cities. Rapid urban transition in Asia is largely supported by the easy and adequate access to groundwater. However, benefits of groundwater use have come at the cost of resource depletion and degradation. Urbanization, population growth, industrial development and the impacts of climate change are exerting huge pressure on groundwater resources in Asia. In many cities the resource sustainability is already threatened due to unwise development and use of groundwater and increasing pollution. Appropriate response strategies for improving the groundwater governance in Asian cities are still inadequate and needs further attention in terms of filling information gap, identifying solution through exchange of experiences and policy interventions.

The Institute of Global Environmental Strategies (IGES) has been trying to find ways to address groundwater management problems in Asia since 2004 with its vision to improve groundwater governance through research, capacity building and knowledge networking. This book is an excellent outcome of such a collaborative effort. The book by analyzing systematically the status of groundwater environment in Asian cities under the driver-pressure-state-impact-response (DPSIR) framework has generated very useful knowledgebase of groundwater in Asian cities. As a "regional hub for groundwater management in the Asia-Pacific region", IGES finds this book as a very much useful reference for knowledge hub partners, groundwater managers, academic institutions, research scholars, and international organizations working in the areas of groundwater in Asia and beyond. I am highly hopeful that it induces policy responses for the sustainable management of groundwater in this region.

Hideyuki Mori

President
Institute of Global Environmental
Strategies (IGES)
Hayama, Kanagawa, Japan

IGES

Institute for Global
Environmental Strategies

CONTENTS

ACRONYMS AND ABBREVIATIONS

ADB	Asian Development Bank
APN	Asia-Pacific Network for Global Change Research
BCM	Billion cubic meters
BWSMB	Bharatpur Water Supply Management Board
CBDC	Central Bari Doab canal command
cfs	Cubic feet per second
CGWB	Central Ground Water Board
cusec	Cubic meters per second
DEG	Directorate of environmental geology
DGB	Dili groundwater basin
DGR	Department of Groundwater Resources, Thailand
DJB	Delhi jal board
DONRE	Department of Natural Resources and Environment
DPSIR	Driver-pressure-state-impact-response
DPSWR	Driver-pressure-state-welfare-response
DTW	Deep tubewell
EC	Electrical conductivity
ET_0	Reference crop evapotranspiration
FY	Fiscal year
GB	Groundwater basin
GDP	Gross domestic product
GHMC	Gram panchayats with the Municipal Corporation of Hyderabad
GIS	Geographic information system
GW	Groundwater
GWRDB	Groundwater Resources Development Board
ha-m	hectare-meter
HCMC	Ho Chi Minh City
HMWSSB	Hyderabad Metropolitan Water Supply and Sewerage Board
HUA	Hyderabad urban agglomeration
HUDA	Hyderabad Urban Development Authority
IDA	Industrial development area
IT	Information technology
ITB	Institute of Technology Bandung
IWT	Indus Water Treaty
JICA	Japan International Cooperation Agency
JNU	Jawaharlal Nehru University
km	Kilometers
KWSP	Krishna Water Supply Project
LCP	Lower Central Plain in Thailand
lpcd	Liters per capita per day
lpm	Liters per minute
lps	Liters per second
LULC	Land use/land cover

m	Meters
MAF	Million acre feet
masl	Meters above mean sea level
mbgl	Meters below ground level
MCH	Municipal Corporation of Hyderabad
MCM	Million cubic meters
mg/l	Milligrams per liter
MGD	Million gallons a day
MLD	Million liters a day
MONRE	Ministry of Natural Resources and Environment
MPC	Maximum permissible concentration
MPL	Maximum permissible limit
MWA	Metropolitan Waterworks Authority
PWA	Provincial Waterworks Authority
NCT	National capital territory
NERI	National Environment Research Institute
NGOs	Non-governmental organizations
NRW	Non-revenue water
NWSC	Nepal Water Supply Corporation
P	Precipitation
PCE	Tetrachloroethylene
ppm	Parts per million
PPP	Public private partnership
SMG	Seoul metropolitan government
STW	Shallow tubewell
SWL	Static water level
S_y	Specific yield
T	Transmissivity
TCE	Trichloroethylene
TDS	Total dissolved solids
TRGWR	Total renewable groundwater resources
TRWR	Total renewable water resources
UN	United Nations
WASID	Water and soils investigation division
WHO	World Health Organization
YCDC	Yangon City Development Committee

SECTION I

Concepts and Methods

CHAPTER 1

Groundwater as an Environmental Issue in Asian Cities

Sangam Shrestha* and Vishnu Prasad Pandey*,**
*Water Engineering and Management, Asian Institute of Technology (AIT), Thailand
**Department of Civil Engineering, Asian Institute of Technology and Management (AITM), Nepal

1.1 INTRODUCTION

Groundwater represents by far the largest storage of unfrozen ubiquitous high-quality fresh water on Earth. It is more widely accessible and less vulnerable to quality degradation and drought than surface water (Foster and Chilton, 2003; Schwartz and Ibaraki, 2011). These characteristics promote its widespread development, which can be scaled and localised according to demand, obviating the need for substantial infrastructure (Giordano, 2009). Globally, groundwater is the source of one third of all freshwater withdrawals, supplying an estimated 36, 42, and 27% of water used for domestic, agricultural, and industrial purposes, respectively (Döll et al., 2012). In many environments, natural groundwater discharge sustains baseflow to rivers, lakes, and wetlands during periods of low or no rainfall.

While irrigation dominates groundwater use globally and contributes to food security, the traditional role of groundwater as a basic habitation need is equally important. Around 2 billion of the rural/urban population obtain drinking water from groundwater, which accounts for around 32% of total drinking water supply (Morris et al., 2003). Today, 54% of the world's population live in urban areas, a proportion that is expected to increase to 66% by 2050. Projections show that urbanisation, combined with overall global growth could add another 2.5 billion people to urban populations by 2050, with close to 90% of the increase concentrated in Asia and Africa. The urban population of the world has grown rapidly from 746 million in 1950 to 3.9 billion in 2014. Asia, despite its lower level of urbanisation, is home to 53% of the world's urban population, followed by Europe with 14% and Latin America and the Caribbean with 13%. The world's urban population is expected to surpass 6 billion by 2045 (UN, 2014). It has been estimated that about one third of Asia's population, some 1000 to 1200 million people, and some

Groundwater Environment in Asian Cities
http://dx.doi.org/10.1016/B978-0-12-803166-7.00001-5

Copyright © 2016 Elsevier Inc.
All rights reserved.

150 million Latin Americans are groundwater reliant (BGS-ODA-UNEP-WHO, 1996). Half of the world's 23 megacities – cities like Bangkok, Beijing, Buenos Aires, Calcutta, Cairo, Dhaka, Jakarta, Mexico City, Lagos, Lima, London, Manila, Paris, Shanghai, Teheran, and Tianjin – as well as hundreds of smaller towns and cities are also groundwater dependent (UN, 2001).

Dependence on groundwater is most evident in Asia, especially throughout South Asia and China where irrigation dominates groundwater withdrawal. Large numbers of people living in rural areas, small towns, and cities are dependent on groundwater for domestic use; over 85% of the rural drinking water supply in India comes from groundwater, while groundwater supplies about half of the water consumed in cities like Delhi (World Bank, 2010). Major cities and municipalities in the region rely either fully or partially on groundwater as part of the water supply network, where it is also used by small-scale rural or town water supply systems. In the case of Jakarta, only 30% of water is supplied from surface sources; the remainder is harvested from groundwater (Delinom, 2012).

Industrial groundwater is usually excluded from discussions on groundwater management, partly due to its insignificance when compared with agriculture. However, the industrial sector is a major user of groundwater in urban areas (Figure 1.1). Industrial use in total groundwater abstraction is 80% in Bandung and 60% in Bangkok. There is a strong correlation between groundwater use and gross domestic product (GDP) in these cities (Figure 1.2). Along with population increase and socioeconomic development, continued growth in groundwater use is therefore expected.

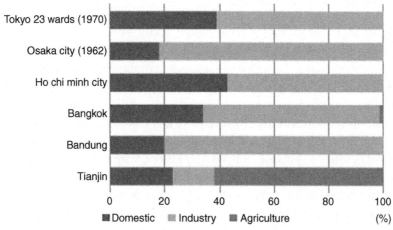

Figure 1.1 *Groundwater Use for Selected Cities in Asia (IGES, 2007).*

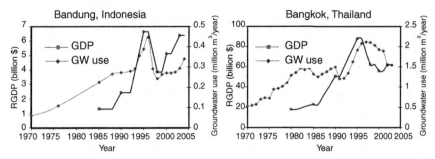

Figure 1.2 *Groundwater Abstraction and Correlation with City-Level GDP. Source: IGES (2007).*

Despite the significance of groundwater for sustainable development, it has not always been properly managed, which has often resulted in depletion and degradation of the resource. Due to various pollution sources and climate change in urban areas, the quality and quantity of groundwater have become important issues for urban groundwater environments (Collin and Melloul, 2003). Much emphasis has been placed on groundwater resource development without giving careful attention to its management despite its strategic role in sustainable development. Uncontrolled groundwater development and indiscriminate waste disposal often accompany urban expansion, resulting in growing water scarcity and deteriorating water quality. This degradation occurs within urban areas on periurban fringes and downstream and is a contributory cause of spiralling water supply costs. The substitution of degraded urban groundwater by alternative out-of-town supplies is expensive, with unit water costs often two to three times higher than current costs (Calow et al., 1999).

To maintain the advantages groundwater has as an important resource for sustainable development and as a reserve freshwater resource for current and future generations, groundwater management should be more strategic and proactive to cope with increased demand from rapid industrialisation and urbanisation, including the potential impact of climate change. A comprehensive understanding of the groundwater environment is the very first step to formulating and implementing effective solutions to reduce degradation of the groundwater resource and maintain it as a vital resource for safe water supply in the face of competing political, societal, and economic issues, as well as limited financial resources for technological development and essential infrastructure.

Therefore, this book aims to present the state of the groundwater environment in terms of facts and figures on groundwater dependence, problems related to groundwater overexploitation, and the existence and implementation

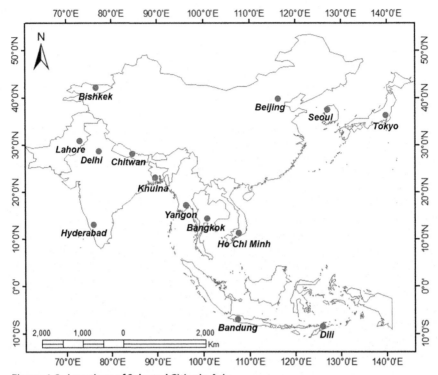

Figure 1.3 *Locations of Selected Cities in Asia.*

of various policy instruments, management practices, and results. Fourteen Asian cities have been selected: Bandung (Indonesia), Bangkok (Thailand), Beijing (China), Bishkek (Kyrgyzstan), Chitwan (Nepal), Delhi (India), Dili (Timor-Leste), Ho Chi Minh (Vietnam), Hyderabad (India), Khulna (Bangladesh), Lahore (Pakistan), Seoul (South Korea), Tokyo (Japan), and Yangon (Myanmar) (Figure 1.3). The driver–pressure–state–impact–response (DPSIR) framework, described in Chapter 2, is applied to evaluate the state of the groundwater environment. DPSIR is a causal framework for describing the interactions between society and the environment, and serves as a communication tool between researchers from different disciplines, policy makers, and all stakeholders involved in aquifer resources.

1.2 CASE STUDY CITIES

The groundwater environment in 14 cities (see Figure 1.3 for locations and Table 1.1 for key characteristics) in Asia is analysed and presented in this book. Some cities consider the city core as well as its vicinity as the study

Table 1.1 A brief profile of the 14 case study cities (see respective chapters for details)

City (country)	Key characteristics
Bandung (Indonesia)	Bandung, the capital of West Java, is the country's third largest city and second largest metropolitan area. Located at 768 masl and 140 km southeast of Jakarta, the city has a population of about 2.38 million and a density of 15,640 persons/km^2 (January, 2012). Bandung has a tropical monsoon climate characterised by a rainy season from October to May and a low rainfall season from June to September. Mean annual precipitation in the Bandung Basin ranges from 1,500 mm/y to 3,500 mm/y. The city has a cooler temperature year-round compared to other Indonesian cities. Water supply relies both on surface and groundwater sources. Some 70% of domestic and 60% of industrial water demand was met by groundwater in 2007. The transmissivity of highly productive aquifers is in a range of 500–1,500 m^2/d and optimum well yields in the deep aquifer are from 2.0 L/s to 9.5 L/s. Urbanization and industrial development has had a profound influence on both the quantity and quality of water. Some 43% of groundwater reserves in the deep aquifer are already exploited. Groundwater has dropped by over 50 m from its original level, forming a cone of depression and creating critical zones, especially in industrial areas. Major groundwater-related issues are groundwater overexploitation and subsequent depletion in groundwater levels, deterioration of groundwater quality, and land subsidence.
Bangkok (Thailand)	Bangkok is the capital and most populated city in Thailand. The city is located along the bank of the Chao Phraya River and has a population of 11.5 million and a density (in the city core) of 3600 persons/km^2 (in 2013). Bangkok lies in the humid tropics and is hot throughout the year with a tropical wet-and-dry climate, which is under the influence of the South Asian monsoon system. The water supply comes from both surface and groundwater sources. In the vicinity of Bangkok, water supply systems are mainly based on groundwater. Bangkok has eight confined aquifers and the most abstracted are Phra Pradaeng, Nakhon Luang, and Nonthaburi due to their higher productivity, accessibility, and good water quality. Major groundwater-related issues are land subsidence (rate = 1.0 cm/y during 2006–2012), depletion and recovery of groundwater levels, and degradation in groundwater quality.

(Continued)

Table 1.1 A brief profile of the 14 case study cities (see respective chapters for details) *(cont.)*

City (country)	Key characteristics
Beijing (China)	Beijing ($39°28'$–$41°05'$N, $115°25$–$117°30'$E) covers an area of 16,410 km² and lies on an alluvial plain whose topography is high in the northwest and low in the southeast. It is the centre for politics, culture, and the economy of China. Beijing's population is growing rapidly and reached some 21.15 million in 2013 with a density of 3240 persons/km². The climate is characterised as typically warm temperate, semihumid, and semiarid. Precipitation in normal years varies from 300 mm/y to 800 mm/y. Between June and September 80% of annual precipitation is received. Water demand in the city has redoubled in nearly half a century as a result of rapid urbanisation and economic growth. Available water resources in the city have decreased recently to below 200 m³/capita/y. Water scarcity is predominant and water resources have become a major bottleneck to sustainable development of the city. Groundwater accounts for over 70% of total water supply in the city. Successive overexploitation of groundwater since the 1970s and overseeing its protection has resulted in a series of groundwater environmental issues such as depletion in groundwater levels (rate = 2.0 m/y from 2000 to 2011) and storage, deterioration of groundwater quality, land subsidence (cumulative subsidence in 2013 was over 100 mm; 3.2 times deeper than in 1999), formation of groundwater fissures, and degradation of wetlands.
Bishkek (Kyrgyzstan)	Bishkek ($42°50'$N–$74°35'$E), originally Pishpek and currently the capital of Kyrgyzstan with a population of about one million and a density of nearly 5000 persons/km², lies in the northern part of the republic. It is the most economically developed and densely populated city in the country. Altitude varies from 550 masl to 900 masl. It is a manufacturing centre that produces about half of Kyrgyzstan's output. Climate in the city is arid. Monthly average air temperature drops to $-10°C$ during December to February and the maximum reaches $+25.8°C$ during July. Long-term (1978–2008) average precipitation in the city is 462 mm/y. The city is 100% aquifer dependent for various water uses, which are provided by both intraurban and periurban wellfields. The geology in the north comprises a multilayered system with numerous low-permeability silt and clay horizons intermingled with high-permeability sands and fine gravels. More complex semiconfined aquifer conditions occur in the flatter northern part of the city. The possibility exists for significant pumping-induced vertical leakage of potentially contaminated urban recharge to deeper parts of aquifer systems. Unconsolidated alluvial and fluvioglacial deposits have high transmissivity n a range of 900–6,000 m²/d.

Table 1.1 A brief profile of the 14 case study cities (see respective chapters for details) *(cont.)*

City (country)	Key characteristics
	There are recharge zones in the south and discharge zones in the north. Major groundwater-related issues are groundwater flooding in the north, depletion of groundwater reserves in the south, and degradation in groundwater quality due to discharge of contaminated water from industry and untreated sewage from parts of the city not connected to the central sewerage system.
Chitwan (Nepal)	Chitwan (27°21′45″–27°52′30″N, 83°54′45″–84°48′15″E), the fifth largest city in Nepal, covers an area of 2218 km² in the central development region. It is the commercial and service centre of central-south Nepal and a major destination for higher education, healthcare, and transportation in the region. Like other cities in Terai, the southern belt of Nepal, Chitwan also relies entirely on groundwater from alluvial aquifers to meet most of its water demands. Intensive groundwater irrigation is a predominant practice to support the livelihood and food security of some 0.58 million people (2011 estimate) residing in Chitwan. The climate varies from lower tropical to subtropical, with very cold winters and extremely hot summers. Average maximum and minimum temperatures are 27.3°C (in May) and 4°C (in December), respectively. Mean annual rainfall (1998–2012 average) is 2156 mm. Groundwater depletion is not currently an issue, which implies it is still in the early stages of groundwater development. However, with the growing population and urbanisation, there has been a severe increase in the amount of solid waste generation, and its untreated disposal is polluting both surface water and groundwater resources.
Delhi (India)	Delhi, the capital city of India, is the third largest city in the country by area and the second largest by population. It supports a population of 16.7 million. The climate is humid subtropical to semiarid. Temperature varies from 25–45°C from April to July and 5–22°C from December to January. The Delhi National Capital Territory (NCT) (28°24′17″–28°53′00″N; 77°50′24″–77°20′37″E) has traditionally faced water scarcity as is evident from centuries-old water storage and recharge structures. The deficit between water supplied by the Delhi Jal Board and water demand is 250 million gallons a day. Groundwater is widely abstracted to meet various water demands. It is in a state of groundwater overexploitation. The alluvial floodplain of the Yamuna River (area = 97 km²) is bounded by bunds (earthen dikes built to restrain water movement) and is the most proficient area for groundwater resources to the city. Transmissivity and specific yield of the floodplain aquifer are 1371 and 0.24 m²/d, respectively. Major groundwater-related issues are groundwater-level decline and degradation of quality.

(Continued)

Table 1.1 A brief profile of the 14 case study cities (see respective chapters for details) *(cont.)*

City (country)	Key characteristics
Dili (Timor-Leste)	Dili (8°340S, 125°340E), the capital city of Timor-Leste, covers an area of 48 km². It is located downstream of the Comoro River. Topography varies from 0 masl to 60 masl. The population in Dili is 0.2 million (2010 estimate) with a density of 1402 persons/km². Dili has a tropical hot climate into dry and wet seasons. Highest and lowest temperatures in a year are 36 and 14°C, respectively. Average annual precipitation ranges from 940 mm to 1761 mm. With high population growth and urbanisation, the city experienc water shortage during the dry season. Of the total water demand of 17.9 MCMy (in 2010), only 14.53 MCM was fulfilled from the water supply. Groundwater abstracted from alluvial aquifers of Quaternary sediments contributes around 60% to total water supply in the city. Total volume of groundwater abstraction is very close to annual recharge, and therefore the aquifer is the verge of overexploitation. Recent development activities such as the construction of riverbanks, wetlands, and surfacesubsurface drainage without considering the health of the groundwater environment is stress on the aquifer by limiting recharge in the wet season. Major groundwater-related issues are depletion in groundwater levelsdecline in production capacity of wellsdecline in recharge due to urbanizationand salt-water intrusion in deeper parts of the aquifer.
Ho Chi Minh City (Vietnam)	Ho Chi Minh City (HCMC) (10°10′–10°38′N, 106°22′–106°54′E) is the largest city in Vietnam located in the southeast region 1760 km south of Hanoi. The city covers an area of 2095 km² and is home to over 7.95 million people (in 2014), accounting for approximately 0.6% of total area and 8.79% of total population of Vietnam. It is the most important economic centre, contributing 20% of the country's GDP. The average elevation of the entire city ranges from 0.5 masl to 1.0 masl. The climate is tropical, specifically tropical wet and dry, with an average humidity of 75% and an average temperature of 27°C. It receives average annual rainfall of 1946 mm, of which 1130 mm is lost in evaporation. Water supply is derived from both surface and groundwater sources. The rapid increase in groundwater withdrawal started when economic policies in Vietnam were instigated in 1990. The share of groundwater in total water supply, which was 44.7% in 2012, is planned to reduce to 15.5% in 2015 and 2.7% in 2020. Major groundwater-related issues are overexploitation of groundwater resources as reflected in the groundwater environment by a decline in groundwater levels (rate: 0.5–1.0 m/y), degradation of groundwater quality (NO_3 content is four times the permissible value; iron content is 100 times the permissible value), saltwater intrusion in some areas, and land subsidence of a few centimetres per year in heavily pumped areas.

Table 1.1 A brief profile of the 14 case study cities (see respective chapters for details) *(cont.)*

City (country)	Key characteristics
Hyderabad (India)	Hyderabad (17.3700°N, 78.4800°E), the capital of Andhra Pradesh, occupies an area of 650 km² along the banks of the Musi River and supports a 7.75 million population (as of 2012) at a density of about 18,500 persons/km². It is one of the fastest growing metropolitan cities and ranks as the sixth largest urban agglomeration in India. Elevation ranges from 460 masl to 560 masl. Hyderabad receives total rainfall of 780 mm/y, 75% of which is received during the southwest monsoon (January–September). The city experiences a warm temperate climate under semiarid tropical conditions. Annual mean temperature is 26°C. Only 30% of the city's water demand is met by groundwater. However, the aquifer is already in a state of overexploitation with total abstraction exceeding recharge. Common groundwater abstraction structures are borewells, but their yields mainly depend on recharge conditions in the area. The entire district is underlain by consolidated rocks. Due to an increase in the number of borewells, yields have fallen drastically, leading to the failure of wells. Though land subsidence is less likely due to the presence of consolidated rocks, major groundwater-related issues are limiting of groundwater recharge due to land cover changes, groundwater overexploitation, decline in groundwater levels (in a range of 0.23–9.5 m), and degradation of groundwater quality.
Khulna (Bangladesh)	Located in the southwest, Khulna is the third largest city in Bangladesh after Dhaka and Chittagong. It covers an area of 8000 km² along the banks of the Rupsha and Bhairab rivers. It has a population of 1.045 million (in 2011). The climate is humid during summer and pleasant in winter. Only 17% of citizens have access to a piped water supply. In 2011, of total water demand of 240 million litres per day (MLD) only 35 MLD were met through piped water supply. Khulna depends entirely on groundwater to meet water demand. There are a large number of privately owned tubewells in and around the city with a depth greater than 1000 ft, as a result of which there is significant depletion in the groundwater table. Major groundwater-related issues are water pollution from urban solid waste and wastewater, salinity ingress in surface and groundwater, arsenic and excessive iron in groundwater, and flooding and water logging.

(Continued)

Table 1.1 A brief profile of the 14 case study cities (see respective chapters for details) *(cont.)*

City (country)	Key characteristics
Lahore (Pakistan)	Lahore (31°15–31°45N, 74°01–74°39E), the provincial capital of Punjab (meaning "five waters" or "five rivers") occupies an area of 404 km² is the largest metropolitan city in the country, the largest in South Asia, and the largest city in the world. It has a population of 8.83 million with a density of 4983 persons/km². The climate in Lahore is subtropical semiarid with average rainfall of 575 mm/y, which varies from 300 mm/y to 1,200 mm/y. Mean annual temperature is about 24.3°C ranging from 33.9°C (in June) to 12.8°C (in January). Groundwater is the only source of water in the city. Around 89% of the total area of Lahore uses the tubewell facility mostly provided by the Water and Sanitation Agency (WASA). These tubewells are of different capacities, installed at depths of 150–200 m roughly pump 800 cubicfeet per second (cusec) of water around the clock. As there is no municipal act, laws or rules on water rights for groundwater use, in the city are water at rate of about 375 cusec. Due to excessive use of groundwater, water levels during 2003–2011 are plummeting at a rate of 0.57–1.35 m/y and the production capacity of wells is also declining. Flow in the Ravi River, which was once the major source of recharge to groundwater in Lahore, has since 2000 due to construction of the Thein in India. About 728.8 t/d of untreated sewage is into the River Ravi, severely affecting groundwater quality.
Seoul (South Korea)	Seoul (37°25′–37°41′N, 126°45′–127°11′E), the capital city of Korea, is located in the midwest of the Korean Peninsula covering a total area of 605.5 km². The population has grown rapidly since the Korean War to about 10.5 million in 2010 with a density of 17,473 persons/km² (in 2010), one of the highest of all Asian cities. The climate shows four distinct seasons and obvious seasonal variations in air temperature, and is largely influenced by the North Pacific high-pressure system. Long-term average minimum and maximum temperatures and precipitation are 38.4°C, −16.4°C, and 1532 mm, respectively. Human activity such as land cover change, urbanisation, and expansion of subway lines is influencing and disturbing the water environment, making the urban groundwater system vulnerable to depletion and pollution. Due to various pollution sources and climate in urban areas, the quality and quantity of groundwater have become important issues.

Table 1.1 A brief profile of the 14 case study cities (see respective chapters for details) *(cont.)*

City (country)	Key characteristics
	Groundwater abstracted from unconsolidated alluvium aquifers (hydraulic conductivity = 0.01–0.1 cm/s) and fractured bedrock (hydraulic conductivity = 2.3 × 10^{-5} to 7.4 × 10^{-3} cm/s) contribute only 10% to the city's total water supply system. Total groundwater abstraction in Seoul is about 0.1 m^3/y, which is about 23% of estimated sustainable yield. Major groundwater-related issues are depletion in groundwater levels (0.8–38.9 m from 2010 to 2014), degradation in groundwater quality, rising groundwater temperature, and land subsidence.
Tokyo (Japan)	Tokyo, the capital of Japan and the most populated metropolitan city in the world, lies in the Kanto region on the southeastern side of the main island, Honshu, and includes Izu and Ogasawara islands. With an area of 2188 km^2, it has a population of 13.19 million (August 2011), which is about 10% of the national population. Tokyo lies in the humid subtropical climate zone with hot humid summers and generally mild winters with cool spells. There are five primary rivers flowing into the Tokyo Basin: the Tonegawa, Arakawa, Tamagawa, Sagamigawa, and Tsurumigawa. This area is flood prone and sits at a low elevation. Approximately 1.5 million of Tokyo's residents reside below sea level. Groundwater is an important source of water supply in Tokyo, although it has not been managed systematically as in other Japanese cities. Issues related to water table depletion, land subsidence, and sea water intrusions were major groundwater problems during the 1950s and 1960s. However, due to enforcement of several laws such as the Industrial Water Law and Building Water Law, the land subsidence problem is by and large under control. Contamination from domestic wastewater or fertilisers was also identified, but the introduction of water purification plants and sewerage plants has been effective in this case. Major groundwater-related issues are contamination from volatile organic carbons (VOCs), such as trichloroethylene and tetrachloroethylene, produced before the 1980s. Their prevention has been successful to some extent; however, once contaminated it takes a long time to recover.

(Continued)

Table 1.1 A brief profile of the 14 case study cities (see respective chapters for details) *(cont.)*

City (country)	Key characteristics
Yangon (Myanmar)	Yangon (16°44′–17°2′N, 96°0′–96°21′E) covers an area of 948 km² and is the most populated and urbanised city in Myanmar. Yangon is located at the confluence of the Yangon and Bago rivers on the eastern margin of the Ayeyarwady Delta, about 30 km from the Gulf of Martaban. Total population in the city is 4.93 million (in 2013), with a density of over 6000 persons/km². Most parts of the city are within an elevation of 3–6 masl. The climate is tropical and average annual rainfall, maximum temperature, and minimum temperature are 2500 mm, 38.1°C, and 16°C, respectively. Water supplies in the city are derived from two main sources: surface (from Hlawga, Gyobu, Pugyi, and Ngamoeyeik reservoirs) and groundwater. A significant amount of water also comes from rainwater, lakes, and ponds. Groundwater contributes some 53% to the water supply system. Groundwater overexploitation is linked to population growth, urbanisation, industrialisation, and changing lifestyles of the people. A water and environmental crisis in the city is predicted if the city's population doubles to 10 million by 2020. Major groundwater-related issues are decrease in recharge areas and rates due to urbanisation, depletion in groundwater reserves due to overexploitation, and potential land subsidence even though no records exist to date.

GDP, gross domestic product; masl, meters above sea level.

area. The largest and smallest cities by area as well as population are Beijing (area = 16,410 km², population = 21.15 million) and Dili (area = 48 km², population = 0.2 million), respectively. The study cities are scattered within a geographical window between 70–140°E and 10°S–50°N.

1.3 STRUCTURE OF THE BOOK

This book is divided into four parts: (i) Concepts and Methods; (ii) Groundwater Environment in South Asia; (iii) Groundwater Environment in Southeast Asia; and (iv) Groundwater Environment in Central and East Asia. The first part, comprising two chapters, contextualises the study and presents the analysis framework. The second part consists of six chapters, one of which introduces the water environment of the region and the remainder present case studies of five cities: Chitwan (Nepal), Delhi (India), Hyderabad (India), Khulna (Bangladesh), and Lahore (Pakistan). Each case study is divided into eight sections; the first two provide a general introduction and outline the hydrogeologic characteristics of the city; the following five describe and analyse indicators of the DPSIR framework; and the last provides a summary of the chapter. The third part also consists of six chapters; one introducing the region and the remaining five presenting case studies of five cities: Bandung (Indonesia), Bangkok (Thailand), Dili (Timor-Leste), Ho Chi Minh City (Vietnam), and Yangon (Myanmar). The final part, comprising five chapters, describes the region and presents the groundwater environment of four cities: Beijing (China), Bishkek (Kyrgyzstan), Seoul (South Korea), and Tokyo (Japan).

REFERENCES

BGS-ODA-UNEP-WHO, 1996. Characterization and assessment of groundwater quality concerns in Asia Pacific region. UNEP/WHO Report UNEP/DEIA/AR/96-1, Nairobi, Kenya, 102 pp.

Calow, R.C., Morris, B.L., Macdonald, D.M., Talbot, J.C., Lawrence, A.R., 1999. Tools for assessing and managing groundwater pollution threats in urban areas. BGS Technical Report WC/99/18, Keyworth, UK.

Collin, M.L., Melloul, A.J., 2003. Assessing groundwater vulnerability to pollution to promote sustainable urban and rural development. J. Clean. Prod. 11 (7), 727–736.

Delinom, R.M., 2012. Groundwater pricing and regulation in Greater Jakarta area (PowerPoint presentation), Groundwater Governance: A Global Framework for Fourth Regional Consultation: Asia and Pacific Region, December 3–5, 2012, Shinizhuang, China.

Döll, P., Hoffmann-Dobrey, H., Portmann, F.T., Siebert, S., Eicker, A., Rodell, M., Strassberg, G., Scanlon, B.R., 2012. Impact of water withdrawals from groundwater and surface water on continental water storage variations. J. Geodyn., 59–60, 143–156.

Foster, S.S.D., Chilton, P.J., 2003. Groundwater: The processes and global significance of aquifer degradation. Philos. Trans. R. Soc. Lond. B 358, 1957–1972.

Giordano, M., 2009. Global groundwater? Issues and solutions. Annu. Rev. Env. Resour. 34, 153–178.

Institute for Global Environmental Strategies (IGES), 2007. Sustainable groundwater management in Asian cities. Hayama, Japan.

Morris, B.L., Lawrence, A.R., Chilton, P.J., Adams, B., Calow, R., and Klinck, B.A., 2003. Groundwater and its susceptibility to degradation: A global assessment of the problems and options for management. Early Warning and Assessment Report Series, RS, 03-3. United Nations Environment Programme, Nairobi, Kenya.

Schwartz, F.W., Ibaraki, M., 2011. Groundwater: a resource in decline. Elements 7, 175–179.

The World Bank, 2010. Deep Wells and Prudence: Towards Pragmatic Action for Addressing Groundwater overexploitation in India. The World Bank, Washington, DC, (2010) p. 97.

UN, 2014. 2014 Revision of World Urbanization Prospects.

CHAPTER 2

DPSIR Framework for Evaluating Groundwater Environment

Vishnu Prasad Pandey*,** and Sangam Shrestha**
*Department of Civil Engineering, Asian Institute of Technology and Management (AITM), Nepal
**Water Engineering and Management, Asian Institute of Technology (AIT), Thailand

2.1 INTRODUCTION

The driver–pressure–state–impact–response (DPSIR) framework has evolved into a widely used tool to provide and communicate knowledge on the state and causal factors regarding environmental issues. Due to the interdisciplinary nature of environmental issues and their links with certain sociotechnical, political, and economic factors, holistic analysis is required in the collection, assessment, and evaluation of various data/information. Indicator-based frameworks like DPSIR, which can recognize the complexity of environmental interactions and provide a means of analyzing them, are therefore frequently used for the purpose.

Cooper (2013) and Svarstad et al. (2008) provide a comprehensive review on the theoretical basis of the DPSIR framework. It is rooted in the stress–response framework developed by Statistics Canada in the late 1970s (Rapport and Friend, 1979). In the 1990s this approach saw further development by, among others, the OECD (1991, 1993) and UN (1996, 1999, 2001). The DPSIR framework was first elaborated in its present form in two studies by the European Environmental Agency (EEA, 1995; Holten-Andersen et al., 1995) and subsequently modified by EEA (2012) in its glossary. Table 2.1 provides a summary of various versions of causal chain frameworks along with their focus and content.

The DPSIR framework and its earlier incarnations are commonly used for interdisciplinary indicator development, system and model conceptualization, and the structuring of integrated research programs and assessments (e.g., EEA, 2005; OECD, 2003; UNEP, 2002; Walmsley, 2002). The framework serves as a communication tool between researchers of different disciplines, as well as policy makers and stakeholders sharing the environment. The five components in the framework are related by logical relations: drivers

Groundwater Environment in Asian Cities
http://dx.doi.org/10.1016/B978-0-12-803166-7.00002-7
Copyright © 2016 Elsevier Inc.
All rights reserved.

Table 2.1 Summary of earlier incarnations of the DPSIR framework, information categories, and their contents

Information categories

Framework	Driver	Pressure	State	Impact	Response
DPSIR	*Driving force* (EEA, 1999) Social, demographic, and economic development in societies and corresponding changes in lifestyles, overall levels of consumption, and production patterns	Pressure (EEA, 1999) Developments in (the) release of substances (emissions), physical and biological agents, the use of resources, and the use of land	State (indicators) (EEA, 2012) Indicator of condition of different environmental compartments and systems in physical, chemical, or biological variables	(Environmental) impact (EEA, 2012) Impacts on human beings, ecosystems, and man-made capital resulting from changes in environmental quality	Response (EEA, 1999) Responses by groups (and individuals) in society, as well as government attempts to prevent, compensate, ameliorate, or adapt to changes in the state of the environment Policy response options
PSIR (Turner, 2000; Turner et al., 1998)	Socioeconomic drivers Urbanization and transport/ trade, agricultural intensification/land use change, tourism, and recreation demand, etc.	Environmental pressures Land conversions and reclamation, dredging, aggregates, oil and gas abstraction, waste disposal, etc.	Environmental state changes Changes in ... fluxes across coastal zones, loss of habitats and biodiversity, etc.	Impacts Consequential impacts on human welfare *via* produc– tivity, health, amenity, and existence value changes	

				State	Human health and welfare	Societal response
PSR/E (Schulze and Colby, 1994)	Pressures underlying social and techno-logical forces that ... drive economic activity	Indirect★ Human activi-ties related to ... im-provement of human welfare	Direct pressures Biophysical inputs and outputs that may exert immedi-ate stress on ecosystems	State Ambient conditions and trends Valued environmental attributes	Human health and welfare Longevity, morbidity, value of ecological goods and services, other nonuse values	Societal response Purposeful actions to address ... ecological, human health, or welfare changes or impacts that are considered undesirable
PSR (OECD, 1993, 2003)	Effects Relationship between two or more variables within any of the pressure, state, and response categories					
	Pressure Indirect Human activities that lead to proximate pressures		Proximate Pressures directly exerted on the environment (e.g., emissions)	State Quality of the environment and quality and quantity of natural resources		Response Actions to mitigate, adapt to, or prevent human-induced negative impacts on the environment
FDES (UN, 1984,1991)	Action Social and economic activities, natural events			Impact Environmental impacts of activities/ events		Reaction Responses to environmental impacts

(Continued)

Table 2.1 Summary of earlier incarnations of the DPSIR framework, information categories, and their contents *(cont.)*

Information categories

Framework	Driver	Pressure	State	Impact	Response
S–RESS (Friend, 1979)	Stressor Activities with the potential to degrade the of the natural environment, to effect si/the health of man to threaten the survival of species, to place pressure on nonrenewable resources, and to deteriorate the quality of human settlement	Stress Elements that place pressures on and contribute to the breakdown of, natural and humanmade environment	Stock Natural resources Environmental response Observed effects of stress upon natural and humanmade environments Stock resource (cf. flows in other measures)		Collective and individual responses Mankind's reaction to environmental changes

* Also includes "natural processes and factors" that may act alone or together with human actions to create biophysical pressures.
Source: Cooper, 2013.

generate pressure; pressures influence/modify state; states provoke or cause impacts; impacts stimulate or ask for responses; and responses modify or substitute drivers, eliminate/reduce/prevent pressures, restore/influence states, and compensate or mitigate impacts.

One of the presumed strengths of the DPSIR framework is that it captures, in a simple manner, the key relationships between factors in society and the environment, and can therefore be used as a communication tool between researchers, policy makers, and stakeholders. The key features of the framework, resulting in its wider application, are as follows:

- Transparency and simplicity using five concepts that are readily obvious to both scientists and stakeholders.
- Ability to simplify complex interactions between humans and the environment to enhance communication between scientists and stakeholders.
- Ability to isolate particular linkages or interactions while retaining conceptual relevance to the larger system.
- Appealing to the public and decision makers, and therefore inherently human centric.
- Appealing to policy makers as it links political objectives to environmental problems and implies causal relationships among the factors.
- Having components and elements that can be mapped onto other frameworks, including those of the Millennium Ecosystem Assessment and Long Term Ecological Research Program.

The framework has also been criticized, partly as a consequence of its simplicity. It has been argued that the framework cannot take into account the dynamics of its own system model, cannot handle cause-and-consequence relationships, suggests linear unidirectional causal chains, and ignores key non-human drivers of environmental change (Berger and Hodge, 1998; Rapport et al., 1998; Rekolainen et al., 2003). Recently, the framework has been further modified and taken on a slightly different format: driver–pressure–state–welfare–response (DPSWR) (Cooper, 2013). Nevertheless, in a relatively short time, the DPSIR framework has become popular among researchers and policy makers alike as a conceptual framework for structuring and communicating policy-relevant research regarding various environmental issues such as air pollution, solid waste, and biodiversity. It is increasingly used as a framework for analyzing landscape and seascape issues (Elliott, 2002; La Jeunesse et al., 2003; Odermatt, 2004; Scheren et al., 2004; Holman et al., 2005), biodiversity (Delbaere, 2002; Maxim et al., 2009), sustainable development of natural heritage sites (Wei et al., 2007), water resource management in a river basin (Benini et al., 2010), soil erosion (Gobin et al., 2004), and marine and coastal systems (Bowen and Riley, 2003; Cave et al., 2003), among others. A few studies have also applied the framework to evaluating subsurface environments (Danielopol et al., 2003; Jago-on et al., 2009; Pandey et al., 2010) by considering groundwater as an "environmental issue."

This chapter describes the DPSIR framework, its application in various sectors, and provides a generic set of DPSIR indicators applicable to evaluating the groundwater environment. Subsequent chapters apply the DPSIR framework to selected cities in South Asia, Central and East Asia, and Southeast Asia to develop a systematic knowledge base of the groundwater environment in these cities. From the groundwater environment perspective, the DPSIR framework provides a holistic approach to separating various factors that influence the groundwater environment under DPSIR. It enables analysis of logical interlinks between the components, as well as visualizing the extent and interrelationships of their representative indicators. The results will be useful to all stakeholders, including policy makers, in helping to understand the extent of the situation, and can serve as a basis in the decision-making process for sustainable management of groundwater resources in cities and aquifers.

2.2 STRUCTURE OF THE FRAMEWORK

The DPSIR framework embodies a systems perspective, implying the demarcation of a particular system of interest, with explicit or implicit boundaries. The system is bounded in two ways. First, it is bounded in terms of the scale on which impacts are defined (e.g., from a single river to the entire world). Second, it is bounded in terms of the scale of the responses and driving forces affecting this system (e.g., from local economic changes to global environmental agreements).

The boundaries will not necessarily coincide; responses and drivers that act on different scales will often determine impacts individually. The drawing of these boundaries depends on the particular issue of interest and its conceptualization, which are strongly influenced by the perspective of those using the framework.

By considering the groundwater environment as a system, one can evaluate its state by collecting published or unpublished reports, papers, data, and information relating to the groundwater environment from a large number of sources, and analyzing them under an established indicator-based framework. The framework in its most basic form consists of five components with certain logical relations as mapped in Figure 2.1.

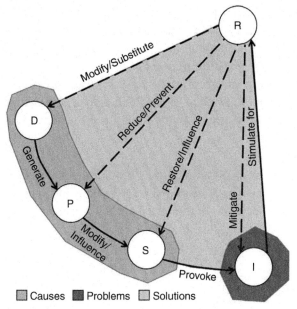

Figure 2.1 *Structure of the DPSIR Framework.*

Driving forces, in the form of social, economic, or environmental development, exert pressures; as a consequence, the state of the environment changes. This leads to impacts that may elicit societal responses that feed back to the driving forces, pressures, states, or impacts (EEA, 2001). As an example, an increased demand for food (driving force) can lead to the intensification of agriculture via increased fertilizer use, resulting in an increase in nitrate runoff into nearby streams (pressure), leading to eutrophication of downstream water bodies (state), and subsequent changes in aquatic life and biodiversity (impact). To address this situation (response), taxes on fertilizer could be increased or changes made to land management practices to reduce nitrate leaching.

2.2.1 Driver

Drivers refer to fundamental processes in a society, which drive activities having a direct impact on the groundwater environment. A "driving force" is a need (e.g., agricultural land, energy, industry, transport, housing). It is the underlying cause that leads to environmental pressures. Examples of primary driving forces for individuals are the need for shelter, food, and water, while examples of secondary driving forces are the need for mobility, entertainment, and culture. For the industrial sector, a driving force could be the need to be profitable and to produce at low costs, while for a nation a driving force could be the need to keep unemployment levels low. In a macroeconomic context, production or consumption processes are structured according to economic sectors (e.g., agriculture, energy, industry, transport, households). In the case of water, population (number, growth rate, age structure, education level, etc.) is of course a primary driver for water-intensive activities. In addition, urbanization, tourism, agricultural activities, industrialization, etc. may be other drivers that generate pressure on the water environment.

2.2.2 Pressure

Pressures are referred to as direct stress brought about by expansion in an anthropogenic system and associated interventions in the natural environment. Generally, depletion of the available resource, degradation of its quality and overexploitation, climate change, etc., can be considered "pressures" on the system.

2.2.3 State

State describes conditions and trends in the groundwater environment induced mainly by human activities. These include, but are not limited to, physical state, chemical state, and biological state.

2.2.4 Impact

Impacts deal with effects on the anthropogenic system, which result from changes in the state of the natural environment, contributing to the vulnerability of both natural and social systems. However, vulnerability to change varies between different systems depending on their geographic, economic, and social conditions, exposure to change, and capacity to mitigate or adapt to change.

2.2.5 Response

Responses consist of the actions of society and/or decision makers to modify/substitute drivers, to reduce/prevent pressures, to restore/influence states, and mitigate/reduce impacts. Responses address vulnerability issues for both people and the environment, and provide opportunities for enhancing human wellbeing by sustainable use of available resources. The level of responses has to be related to the magnitude of impacts. These various responses need different planning processes and the involvement of certain decision makers. Planning levels may include formulating policies, plans, programs, and designing projects from the macro to micro level. Authorities at the central and local level could be involved in formulating and executing responses.

2.3 SELECTION OF INDICATORS

Indicators of the drivers, pressures, states, impacts, and responses depend on the environmental issue under consideration. A thorough review of the DPSIR framework for application to various environmental issues reveals that the DPSIR indicator varies in accordance with the location, scale, and issues. To provide a broader perspective on application of the framework, a set of indicators applied in selected case studies are presented in the following subsections. This helps identify a set of potential indicators for evaluating the groundwater environment of a city under the DPSIR framework. The indicators for a particular city can then be selected based on certain criteria, such as suitability to the study area, availability of data, and ease of interpretation.

2.3.1 A Case of Integrated Environment Assessment

Figure 2.2 shows the DPSIR indicators proposed by the National Environment Research Institute (NERI) in Denmark to analyze the state and causal factors for integrated environment assessment. These indicators have been

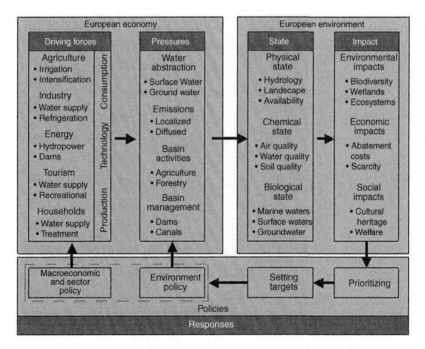

Figure 2.2 *Integrated Environment Assessment in a DPSIR Framework. Source: Adapted from NERI (1997).*

applied in a European context. The driving forces were selected in relation to production, consumption, and technology, and exert pressure on water abstraction, emissions, basin activities, and basin management. The drivers and pressures are considered factors related to the economy. State and impact, on the other hand, are considered factors related to the environment. Macroeconomic policies, sectoral policies, and environmental policies are considered responses, and may therefore be both structural (i.e., hardcore engineering) and nonstructural (i.e., softcore engineering).

2.3.2 Land and Soil Degradation Cases

Two cases are presented that use different sets of indicators for land and soil degradation analysis under the DPSIR framework (Table 2.2). The first outlines a set of indicators used for La Barreta (Queretaro, Mexico), whereas the second is for the entire dryland area of the world.

Table 2.2 DPSIR indicators for land and soil degradation analysis

DPSIR components	Indicators	
	Porta and Poch (2011)	FAO (2013)
Drivers	• Land ownership • Human population growth • Agriculture • Land use • Growth of demand for agricultural and forest products	• Incidence of poverty/wealth • Access rights/tenure • Population density • Labor availability • Inputs and infrastructure • Occurrence of conflicts • Education, knowledge of and access to support service • Protected areas
Pressures	• Deforestation • Exposure of bare soil • Overexploitation • Changes in water infiltration and flow	• Trends in land use and its intensity • Crop management level • Deforestation • Overexploitation of vegetation • Overgrazing • Industrial activities • Urbanization • Natural causes • Discharge of effluents • Washing out of pollutants • Airborne pollutants

Table 2.2 DPSIR indicators for land and soil degradation analysis *(cont.)*

DPSIR components	Indicators	
	Porta and Poch (2011)	FAO (2013)
State	• Soil degradation • Physical deterioration • Biological deterioration • Nutrient depletion • Soil erosion and loss • Increase in rock outcrops	• Type of land degradation (soil, biological, water) • Degree of land degradation • Rate of land degradation
Impacts	• Deterioration of soil quality • Habitat destruction • Loss of biodiversity • Water stress • Decline in agriculture and forestry production • Decline in agricultural income • Change in human population	• Degradation of ecosystem services • Production decline • Carbon storage loss • Water availability decline • Degradation of water quality • Decline in biodiversity • Decline in tourism
Responses	• Conservation of biodiversity • Formulation of rehabilitation strategies • Valuing landscape • Making recreational parks	• Macroeconomic policies • Land policies and policy instruments • Conservation and rehabilitation • Monitoring and early-warning systems • Commitments to international conventions • Investment in land–water resources

2.3.3 Air Quality Case

Table 2.3 presents the indicators used to analyze urban air quality in Kathmandu Valley. It can be seen that they are entirely different than for land and soil degradation, and integrated environment assessment.

2.3.4 Water Issue Cases

The state of water is determined by natural factors, such as geology and climate, and pressures stemming from human activity. Many of the pressures and the underlying driving forces are common to all or some of the issues. For example, agriculture is a significant driving force in terms of ecological quality, nutrient and organic pollution, hazardous substances, and water

Table 2.3 DPSIR indicators for urban air quality in Kathmandu Valley, Nepal

Components	Indicators
Drivers	• Kathmandu–centric development • Weak institutions • Increased affluence and modernization
Pressures	• Rapid urbanization • Haphazard growth • Unplanned settlements • Increased emissions (from vehicles, industry, mismanagement of solid waste) • Increase in number of vehicles
State	• Particular matters • Gaseous pollutants • Air toxins
Impacts	• Health impacts • Economic impacts
Response	• Identifying and mapping key stakeholders • International support for air quality management • Setting up an institutional framework • Setting up a policy and legal framework

Adapted from ICIMOD (2007).

quantity. Table 2.4 provides an outline of how indicators vary with water issues such as quantity and quality.

Water quantity issues prevail when demand exceeds availability in a certain period. This may occur frequently in areas with low rainfall and high population density, and in areas with intensive agricultural or industrial activity. Large spatial and temporal differences in water availability may indicate the water state, which is expected to change due to climate change. Other pressures on water quantity arise from the main sectoral users of water such as agriculture, households, energy production, and industry. Seasonal demand from tourism is a significant pressure. However, when we consider water quality issues – such as organic pollution and eutrophication of the aquatic environment, in particular; increased industrial and agricultural production; and more of the population being connected to sewerage systems – the result will be increased discharge of nutrients and organic loads into surface water bodies. The overloading of nutrients (nitrogen and phosphorus) into seas, coastal waters, lakes, and rivers can result in a series of adverse effects known as eutrophication. Impacts and responses differ when it comes to tackling water quality and quantity issues.

Table 2.4 DPSIR indicators for analyzing water issues

	Indicators for		
DPSIR components	**Water in general**	**Water quantity issues**	**Water quality (organic pollution and eutrophication)**
Drivers	• Industry • Energy • Agriculture • Aquaculture • Households • Tourism • Climate • Geology	• Industry • Energy • Agriculture • Aquaculture • Households • Tourism • Climate	• Industry • Agriculture – livestock density and fertilizer use • Households – wastewater treatment
Pressures	• Climate change • Point source pollution • Diffused source pollution • Water abstraction • Physical intrusions	• Climate change • Total abstraction • Sectoral water use: household, industry, agriculture, etc	• Discharge from point source • Atmospheric deposition • Loads to coastal water • Nitrogen balance
State	• Water quantity • Groundwater status • Ecological status • Chemical • Physical • Biological	• Available water	• Nitrogen, phosphorus in lakes, rivers, and marine waters • Nitrates in groundwater • Organic matter in rivers • Chlorophyll in lakes and marine waters
Impacts	• Loss of habitats/species • Ill health • Droughts/floods • Desertification • Salinization • Loss of amenity • Coastal erosion • Nonindigenous species • Eutrophication • Acidification	• Freshwater shortage • Modification of stream flows • Saltwater intrusion • Groundwater levels	• Exceedance of standards for drinking water and bathing water • Secchi depth in lakes • Low oxygen in bottom layers of marine waters • Harmful phytoplankton in coastal waters

(Continued)

Table 2.4 DPSIR indicators for analyzing water issues *(cont.)*

	Indicators for		
DPSIR components	Water in general	Water quantity issues	Water quality (organic pollution and eutrophication)
Responses	• Water use restrictions • Alternative supplies • Subsidized water prices • Improved information • Demand-side management • Voluntary agreements • Regional cooperation • Wastewater treatment • Ban on products • Storage reservoirs	• Increasing reservoir stocks • Charging price for water • Increasing water use efficiency • Reducing water leakages	• Measures to reduce nonpoint sources (agriculture) • Wastewater treatment (households and industry)

Adapted from Kristensen (2004).

2.3.5 Subsurface Environment Case

Only a few studies have applied the framework to the subsurface environment. Jago-on et al. (2009) did so, and Pandey et al. (2010) further customized it for the Kathmandu Valley in Nepal. Figure 2.3 outlines the framework and indicators used for subsurface environment analysis. Pandey et al.'s study clearly distinguished the quantity and quality issues of groundwater in the subsurface.

2.3.6 Generic Set of Indicators for Evaluating the Groundwater Environment

A review of DPSIR applications on a wide range of issues has reflected commonalities and differences in the selection of indicators for those applications. The Asia Pacific Network for Global Climate Change Research (APN) funded a project entitled "Enhancing the Groundwater Management Capacities in Asian Cities through the Development and Application of Groundwater Sustainability Index in the Context of Climate Change",

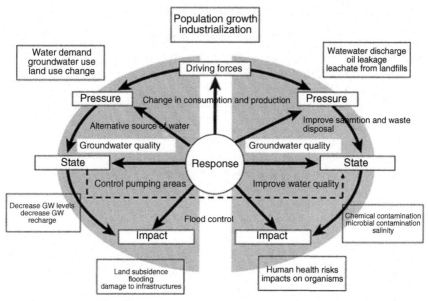

Figure 2.3 *DPSIR Framework for Subsurface Environment Problems. Source: Jago-on et al. (2009).*

which was held at the Asian Institute of Technology (AIT) in January 2014. After consultation with stakeholders, a generic set of DPSIR indicators were considered to be applicable for evaluating the groundwater environment (Table 2.5). The indicators are by no means comprehensive and prescriptive, and some are more suitable to one case than another. However, this is expected to provide a generic guideline to select a suitable set of indicators for the case study of interest.

2.4 INTERPRETING THE RESULTS

The relevant set of DPSIR indicators for a city can be identified by undertaking a literature review. After considering the suitability of the study area and the availability of data, a set of indicators for a particular city can be finalized. The indicator values for different time periods can be quantified from secondary and primary data. Each indicator can be discussed and interpreted in terms of spatial and temporal trends, their ranges, and interlinks with other indicators. This will ultimately result in generation of a knowledge base for the status of the groundwater environment. Additionally, a comprehensive summary table for easy visualization and comparison can be prepared as shown in Table 2.6.

Table 2.5 A generic set of indicators for evaluating groundwater environments

Components	Indicators
Drivers	• Population growth (density and growth rate) • Migration or urbanization • Tourism • Economic growth (industrial and agricultural development) • Climate change
Pressures	• Increase in water demand • Inadequate surface water supply • Increase in groundwater abstraction • Land use changes • Occupational shift • Wastewater discharge • Leachage from landfill sites
States	• Well statistics • Groundwater abstraction (volume or quantity) • Chemical and microbial contamination • Groundwater level • Groundwater recharge • Legal framework and institutions
Impacts	• Deterioration of groundwater quality • Decline in groundwater level • Land subsidence • Human health risk • Decline in production capacity of wells • Modification of stream flows • Saltwater intrusion
Responses	• Groundwater monitoring (quality, level) • Alternative water resources • Change in consumption and production • Improved sanitation and waste disposal • Control in pumping areas • Increasing water use efficiency • Application of advance technology • Environmental standards and guidelines

Modified after Danielopol et al., 2003; Jago-on et al., 2009; Pandey et al., 2010, and the Stakeholders Consultation Workshop held in January, 2014 at the Asian Institute of Technology, Thailand.

2.5 SUMMARY

DPSIR analysis may show whether anthropogenic systems (population growth, urbanization, and tourism) or natural systems are the predominant driving force exerting pressure on the groundwater environment of a city by various means. As a result, various impacts (specified in quantitative

Table 2.6 Summary of DPSIR indicator values for Kathmandu Valley, Nepal

	Indicators	1991	2001
Drivers	Population growth within GW basin (area: 327 km²)	Population: 1.03×10^6 Population density: 3150 persons/km² Annual growth rate during 1991–2001	1.53×10^6 4680 persons/km² 3.96%
	Urbanization within GW basin	Urban area: 31.0 km², in 1984 Urban population: 0.61×10^6	97.5 km², in 2000 1.29×10^6
	Increase in tourism	Tourist arrivals: 0.29×10^6 (in 1991) Number of hotel wells: 23 (until 1991)	0.56×10^6 (in 2007) 62 (1991–2000)
Pressures	Inadequate surface water resources	Expansion in social system, urbanization, rising economic activities, and quality of life are water intensive, which has reduced per capita surface water availability.	
	Land cover change (1984–2000)	Urban area: 31.0 km² Nonagricultural area: 35.7 km²	Urban area: 97.5 km² Nonagricultural area: 184.1 km²
	Overabstraction of GW resources	Abstraction: 40 MLD (or 14.6 MCM/y) Recharge: 9.6 MCM/y	59.06 MLD (or 21.56 MCM/y) 9.6 MCM/y
State	Well statistics	Total wells: 115	Total: 386, production well: 249, abandoned wells >100
	GW abstraction	Abstraction: 40 MLD (or 14.6 MCM/y) Recharge: 9.6 MCM/y Rainfall: 1755 mm/y = 150.93 MCM/y (@86 km² recharge area)	59.06 MLD (or 21.56 MCM/y) 9.6 MCM/y 1755 mm/y = 150.93 MCM/y (@86 km² recharge area)
	GW level	Shallower toward southern part, decreasing toward NW–SE and W–E direction. In general, water level rises from June to July and declines from November to December.	
	GW quality	Shallows are characterized by coliform bacteria, high manganese, iron, locally nitrate and BOD; and deeps by high iron (1–3 mg/L), manganese (0.02–0.42 mg/L), BOD, arsenic and dissolved gases like ammonia (1.05–6.5 mg/L) and methane. Electrical conductivity, NH_4–N, and arsenic increases from north toward the central area.	
	Recharge	4.61–14.6 MCM/y; average value is = 9.6 MCM/y	4.61–14.6 MCM/y; average value is = 9.6 MCM/y

(Continued)

Table 2.6 Summary of DPSIR indicator values for Kathmandu Valley, Nepal *(cont.)*

	Indicators	1991	2001
Impacts	Decline in GW level	13–33 m during mid-1980s–2000 in northern area	1.38–7.5 m during 2000–2008 in NWSC wellfields.
	Decline in production capacity of wells (L/s)	Decline during mid-1980s–2000: 7.89 in Dhobikhola, 4.97–36.17 in Bansbari, 19.10 in Gokarna, 31.31 in Bhaktapur/Bode, and 15.17–27.81 in Manohara.	
	Land subsidence	No monitoring and no evidence of land subsidence so far, but few studies have cautioned the possibility of land subsidence in areas with significant decline in water level and high percent of compressible clay and silt in the subsoil.	
	Public health	GW fulfills 50% of the valley's water demand. However, its quality is under threat because of solid waste disposal and effluent discharge into rivers and open ground. Bacterial indicators, which cause waterborne diseases, have already been detected in the groundwater samples. Moreover, waterborne diseases have been the cause for 16.5% of total deaths in the Teku Hospital in the valley.	
Responses	GW monitoring	Monitoring began in the early 1970s, but was limited to the duration of a particular project only.	Continuous monitoring since 1999 under the initiation of Metcalf and Eddy and later continued by the GWRDP.
	Environmental standards and guidelines	The Ministry of Environment, Science & Technology developed environmental standard guidelines through the Nepal Bureau of Standards & Metrology (NBSN, 2001, 2003); however, implementation remains poor in practice.	
	Melamchi Drinking Water Project	The project will bring 510 MLD water from off-the-valley sources in three stages, to fulfill the valley's water demand until 2030 and reduce pressure on GW.	

MCM, Million cubic meters; mm, millimeter; y; year, GW, groundwater; EC, electrical conductivity; MoEST, ministry of environment science and technology; exponential growth rate calculated using $r = (1/t) \times \ln(P_t/P_0)$; where P_t = population after t years from the base period, P_0 = base year population.
Source: Pandey et al., 2010.

terms) on the groundwater environment can be observed. The various attempts to maintain a healthy groundwater environment may be structural (e.g., construction of a monitoring network, underground dams, recharge ponds, remediation measures, and development of information and decision systems), institutional (establishing new institutions related to groundwater or reforms in the existing institutional setup), or political and regulatory. The response might have come from different actors such as government, NGOs, academia, individuals, and communities. An overview of indicators used in selected DPSIR applications and a generic set of DPSIR indicators for groundwater environment studies would be very helpful in evaluating the groundwater environment of other cities both in Asia and beyond.

ACKNOWLEDGMENT

The authors would like to acknowledge SEA-EU-NET II Project (http://www.sea-eu.net/about/aims_results) for supporting the development of this chapter.

REFERENCES

Benini, L., Bandini, V., Marazza, D., Contin, A., 2010. Assessment of land use changes through an indicator-based approach: a case study from the Lamone river basin in Northern Italy. Ecol. Indic. 10, 4–14.

Berger, A.R., Hodge, R.A., 1998. Natural change in the environment: a challenge to the pressure–state–response concept. Soc. Indic. Res. 44, 255–265.

Bowen, R.E., Riley, C., 2003. Socio-economic indicators and integrated coastal management. Ocean Coast. Manage. 46, 299–312.

Cave, R.R., Ledoux, L., Turner, K., Jickells, T., Andrews, J.E., Davies, H., 2003. The Humber catchment and its coastal area: from UK to European perspectives. Sci. Total Environ. 314, 31–52.

Cooper, P., 2013. Socio-ecological accounting: DPSWR, a modified DPSIR framework, and its application to marine ecosystems. Ecol. Econ. 94, 106–115.

Danielopol, D., Griebler, C., Gunatilaka, A., Notenboom, J., 2003. Present state and future prospects for groundwater ecosystems. Environ. Conserv. 30 (2), 104–130.

Delbaere, B., 2002. An inventory of biodiversity indicators in Europe. European Environmental Agency Technical Report 92.

EEA, 1999. Environmental indicators: typology and overview. Technical Report No.25. European Environment Agency (EEA), Copenhagen.

EEA, 2001. Environmental Signals 2001. European Environment Agency, Copenhagen.

EEA, 2012. Environmental terminology and discovery service. Available from: http://glossary.eea.europa.eu/ (accessed 30.08.12.).

EEA, 1995. Europe's Environment: The Debris Assessment. European Environment Agency, Copenhagen.

EEA, 2005. European Environmental Outlook. European Environment Agency, Copenhagen.

Elliott, M., 2002. The role of the DPSIR approach and conceptual models in marine environmental management: an example for offshore wind power. Mar. Pollut. Bull. 44 (6) .

FAO, 2013. Land Degradation Assessment in Dry Lands: Methodology and Results. Food and Agricultural Organization (FAO) of the United Nations, Rome, Italy.

Gobin, A., Jones, R., Kirkby, M., Campling, P., Govers, G., Kosmas, C., Gentile, A.R., 2004. Indicators for pan-European assessment and monitoring of soil erosion by water. Environ. Sci. Policy 7 (1), 25–38.

Holman, I.P., Rounsevell, M.D.A., Shackley, S., Harrison, P.A., Nicholls, R.J., Berry, P.M., Audsley, E., 2005. A regional, multi-sectoral and integrated assessment of the impacts of climate and socio-economic change in the UK. Climate Change 71, 9–41.

Holten-Andersen, J., Paalby, H., Christensen, N., Wier, M., Andersen, F.M., 1995. Recommendations on strategies for integrated assessment of broad environmental problems. Report submitted to the European Environment Agency (EEA) by the National Environmental Research Institute (NERI), Denmark.

ICIMOD, 2007. Kathmandu Valley Environment Outlook. International Center for Integrated Mountain Development (ICIMOD), Kathmandu, Nepal.

Jago-on, K.A.B., Kaneko, S., Fujikura, R., Fujiwara, A., Imai, T., Matsumoto, T., Zhang, J., Tanikawa, H., Tanaka, K., Lee, B., Taniguchi, M., 2009. Urbanization and subsurface environmental issues: an attempt at DPSIR model application in Asian cities. Sci. Total Environ. 407 (9), 3089–3104.

Kristensen, P., 2004. The DPSIR Framework. Paper presented at the September 27–29, 2004 Workshop on a Comprehensive Assessment of the Vulnerability of Water Resources to Environmental Change in Africa using the River Basin Approach. UNEP Headquarters, Nairobi, Kenya.

La Jeunesse, I., Rounsevell, M., Vanclooster, M., 2003. Delivering a decision support system tool to a river contract: a way to implement the participatory approach principle at the catchment scale? Phys. Chem. Earth 28 (12–13), 547–554.

Maxim, L., Spangenberg, J.H., O'Connor, M., 2009. An analysis of risks for biodiversity under the DPSIR framework. Ecol. Econ. 69, 12–23.

National Environmental Research Institute (NERI), 1997. Integrated environmental assessment on eutrophication. Technical Report No. 207. Report submitted to the European Environment Agency (EEA) by the National Environmental Research Institute (NERI), Denmark.

NBSN, 2001. Industrial Effluent Standards. HMG/NBSM, Kathmandu.

NBSN, 2003. Wastewater, Combined Sewerage Treatment Plant and Industrial Wastewater Standard. HMG/NBSM, Kathmandu.

Odermatt, S., 2004. Evaluation of mountain case studies by means of sustainability variables. A DPSIR model as an evaluation tool in the context of the North–South discussion. Mt. Res. Dev. 24 (4), 336–341.

OECD, 1991. Environmental Indicators, a Preliminary Set. OECD, Paris.

OECD, 1993. OECD Core Set of Indicators for Environmental Performance Reviews, Environment Monographs No. 83. Organization for Economic Co-operation and Development (OECD), Paris.

OECD, 2003. OECD Environmental Indicators: Development, Measurement and Use, Reference Paper. Organization for Economic Co-operation and Development (OECD), Paris.

Pandey, V.P., Chapagain, S.K., Kazama, F., 2010. Evaluation of groundwater environment of Kathmandu Valley. Environ. Earth Sci. 60 (6), 1329–1342.

Porta, J., Poch, R.M., 2011. DPSIR analysis of land and soil degradation in response to changes in land use. Span. J. Soil Sci. (SJSS) 1 (1), 100–115.

Rapport, D., Friend, A., 1979. Towards a comprehensive framework for environmental statistics: a stress–response approach. Statistics Canada Catalogue 11-510. Minister of Supply and Services Canada, Ottawa.

Rapport, D.J., Gaudet, C., Karr, J.R., Baron, J.S., Bohlen, C., Jackson, W., Jones, B., Naiman, R., Norton, B., Pollock, M.M., 1998. Evaluating landscape health: integrating societal goals and biophysical process. J. Environ. Manage. 53 (1), 1–15.

Rekolainen, S., Kämäri, J., Hiltunen, M., 2003. A conceptual framework for identifying the need and role of models in the implementation of the water framework directive. Int. J. River Basin Manage. 1 (4), 347–352.

Scheren, P.A.G.M., Kroeze, C., Janssen, J.J.J.G., Hordijk, L., Ptasinski, K.J., 2004. Integrated water pollution assessment of the Ébrié Lagoon, Ivory Coast, West Africa. J. Mar. Syst. 44 (1–2), 1–17.

Schulze, I., Colby, M., 1994. A Conceptual Framework to Support the Development and Use of Environmental Information, EPA 230 R 94 012. Environmental Statistics and Information Division. Environmental Protection Agency, Washington, DC.

Svarstad, H., Petersen, L.K., Rothman, D., Siepel, H., Watzold, F., 2008. Discursive biases of the environmental research framework DPSIR. Land Use Policy 25, 116–125.

Turner, R.K., Lorenzoni, I., Beaumont, N., Bateman, I.J., Langford, I.H., McDonald, A.L., 1998. Coastal management for sustainable development: analysing environmental and socio-economic changes on the UK coast. Geogr. J. 164 (3), 269–281.

Turner, R.K., 2000. Integrating natural and socio-economic science in coastal management. J. Mar. Syst. 25, 447–460.

UN, 1984. A Framework for the Development of Environment Statistics, Statistical Papers Series M, No. 78. United Nations (UN) Department of International Economic and Social Affairs, Statistical Office, New York.

UN, 1991. Concepts and Methods of Environment Statistics: Statistics of the Natural Environment, Studies in Methods Series F, No. 57. United Nations (UN) Department of International Economic and Social Affairs, Statistical Office, New York.

United Nations Environment Program, 2002. Global Environmental Outlook 3. Earthscan, London.

UN, 1996. Indicators of Sustainable Development. Division for Sustainable Development, New York.

UN, 2001. Indicators of Sustainable Development: Framework and Methodologies. Background Paper No. 3, Commission on Sustainable Development, Division for Sustainable Development, Department of Economic and Social Affairs, New York.

UN, 1999. Work Programme on Indicators of Sustainable Development of the Commission on Sustainable Development. Division for Sustainable Development, Department of Economic and Social Affairs, New York.

Walmsley, J.J., 2002. Framework for measuring sustainable development in catchment systems. Environ. Manage. 29 (2), 195–206.

Wei, J., Yountao, Z., Houqin, X., Hui, Y., 2007. A framework for selecting indicators to assess the sustainable development of the natural heritage site. J. Mt. Sci. 4, 321–330.

SECTION II

Groundwater Environment in South Asia

CHAPTER 3

Water Environment in South Asia: An Introduction

Vishnu Prasad Pandey*,**, Sangam Shrestha**, and Shashidhar Thatikonda†

*Department of Civil Engineering, Asian Institute of Technology and Management (AITM), Nepal
**Water Engineering and Management, Asian Institute of Technology (AIT), Thailand
†Department of Civil Engineering, Indian Institute of Technology, Hyderabad, India

3.1 PHYSIOGRAPHY AND CLIMATE

South Asia covers an area of nearly 4.78 million km² (WB, 2015) and extends from near the Equator to 38°26′N and 60°34′E–97°25′E (Figure 3.1). It is surrounded by (from west to east) Western Asia, Central Asia, Eastern Asia, and Southeast Asia. The subregion comprises eight countries: Afghanistan, Bangladesh, Bhutan, India, Maldives, Nepal, Pakistan, and Sri Lanka. The physiography is dominated by the Himalayan Mountains, which form a physical and cultural barrier separating the subregion from China. The subregion is home to a variety of geographical features such as glaciers, rainforests, valleys, deserts, and grasslands. Two major rivers flow out of the Himalayas: the Ganges, which flows across northern India into Bangladesh, and the Indus, which flows across Pakistan. Home to the highest mountains and largest mangrove forests in the world, lush jungles, the tenth-largest desert on Earth, deep river valleys, and many other unique landscape features, this subregion boasts a most diverse assortment of geographic features (Babel and Wahid, 2008).

Climate in South Asia varies acutely from tropical monsoon in the south to temperate in the north. The variety is influenced not only by altitude but also by factors such as proximity to the seacoast and seasonal impact of monsoons (Weiss and Mughal, 2012). The weather of this subregion is dominated by the monsoon or seasonal reversal of wind flow due to changing low and high-pressure patterns. The large Eurasian (Europe and Asia) landmass and the equally large Indian Ocean produce differences between the heating capacities of the former and the latter. Because land heats and cools down faster than water, seasonal reversal of winds occurs: the monsoon. As discussed in Babel and Wahid (2008), southwesterly winds blow toward the

Groundwater Environment in Asian Cities
http://dx.doi.org/10.1016/B978-0-12-803166-7.00003-9

Copyright © 2016 Elsevier Inc.
All rights reserved.

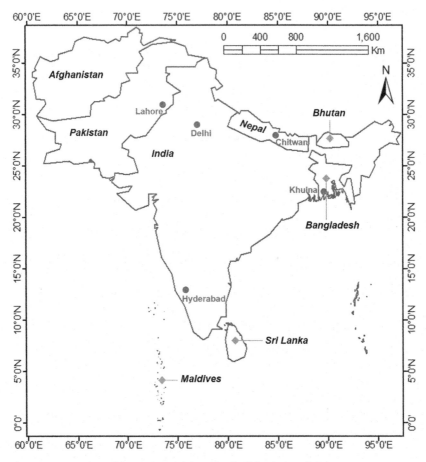

Figure 3.1 *Location of Countries and Case Study Cities in South Asia.*

shore in South Asia during the Northern Hemisphere summer. In contrast, northeasterly winds blow offshore during the Northern Hemisphere winter. During the dry season, the winds from the northeast are dry because they lose their moisture on the Asian landmass. As the winds approach the southern tip of India, specifically the state of Tamil Nadu, they pass over the Bay of Bengal and pick up moisture. Tamil Nadu then receives most of its precipitation during these months. Toward late spring and early summer, however, the weather is hot and dry over most of the subcontinent. During the summer, air is drawn in from the Indian Ocean and Arabian Sea as land surface temperature increases. The winds pick up large volumes of moisture, with resulting rains falling first along India's western coast. The winds later flow around the southern tip of India, as a result of being funneled up the

Bay of Bengal into the delta area of the Ganges and Brahmaputra rivers. The rains later reach the upper Ganges Valley, with the Indian capital of New Delhi receiving less moisture than the other areas mentioned earlier since the winds arrive there later in June and July. The Deccan Plateau to the east of the Western Ghats (mountains) receives significantly less precipitation than coastal areas. As the summer (wet) monsoons approach the west coast of India, they rise over the Western Ghats and the air subsequently cools. The cool air is less able to hold moisture, which is then released as precipitation, a process known as orographic precipitation.

3.2 SOCIOECONOMICS AND ENVIRONMENTAL ISSUES

Nearly 1.65 billion people (in 2012) live in this subregion, with density varying from 20 (Bhutan) to 1203 persons/km^2 (Bangladesh) (FAO, 2015). Population is heavily concentrated in the fertile river valleys of the Ganges and the Indus. Agriculture is a crucial sector of the economy contributing approximately 25% to its gross domestic product (GDP). The contribution ranges from 4.2% (in the Maldives) to 36.5% (in Nepal) (in 2012) (FAO, 2015), which is very high compared with the worldwide average.

This subregion is not only one of the fastest growing in the world, but also one of the poorest. This reality puts water and energy at the very heart of the subregion's development process. The subregion's huge population number, coupled with expanding economies, has resulted in ever-increasing demands for water and energy. Inadequate drinking water and sanitation services are responsible for poor environmental conditions and water-related diseases. The water quality and environment in South Asia are generally much degraded. The major environmental issues associated with population factors in this region include (Babel and Wahid, 2008): (i) increasing demands on natural resources including water; (ii) salt water intrusion in coastal aquifers; (iii) spread of waterborne diseases; (iv) water and soil pollution; and (v) decline in fishery resources. Indeed, the issues and challenges in this subregion regarding the water sector are major in scale, diversity, and complexity.

3.3 WATER AVAILABILITY AND WITHDRAWAL

South Asia is endowed with considerable water resources; however, its uneven distribution hinders its access. The subregion is home to about one quarter of the global population, but has less than 5% of the world's annual renewable water resources. Precipitation and total renewable water

Table 3.1 Water resource availability in South Asian Countries in 2012

Country	P (BCM/y)	TRWR BCM/y	TRWR m³/persons/y	TRGWR (BCM/y)
Afghanistan	213.5	65.33	2,190	10.65
Bangladesh	395.8	1227.00	7,932	21.12
Bhutan	84.5	78.00	105,121	8.10
India	3,560.0	1911.00	1,545	432.00
Maldives	0.6	0.03	89	0.03
Nepal	220.8	210.20	7,651	20.00
Pakistan	393.3	246.80	1,378	55.00
Sri Lanka	112.3	52.80	2,503	7.80
South Asia	4,981	3,791	2,229	555

BCM, billion cubic meters; P, precipitation; TRWR, total renewable water resources; TRGWR, total renewable groundwater resources.
Source: FAO, 2015.

resources vary across countries (Table 3.1), with regional values of 4981 billion cubic meters (BCM)/y and 3791 BCM/y, respectively. Total renewable groundwater resources in the subregion are 555 BCM/y, which varies among countries from 0.03 BCM/y (in the Maldives) to 432.0 BCM/y (in India). Low per capita water availability (2229 m³/persons/y) coupled with a very high relative level of water use (dominated by irrigation) make this subregion one of the most water-scarce regions of the world, which seriously impacts economic development (SAWI, 2015).

Inadequate drinking water supplies (quantity and quality) and poor sanitation are ubiquitous in South Asia. More than 91% of total water withdrawals in the subregion are for agriculture (Table 3.2), which is higher than the global average. Groundwater is the primary source of drinking water. Irrigation in South Asia accounts for 344.12 BCM/y or 33.6% of total fresh water withdrawals, which is nearly half total global withdrawal. Total withdrawal in terms of percent total renewable water resource availability, an indicator of pressure on fresh water resources, is about 27%. Groundwater quality issues are widespread caused by ingress of untreated urban wastewater and chemical-laden irrigation drainage. Natural contamination of groundwater with arsenic and fluoride is very common in Bangladesh, Nepal, and Pakistan, and fluoride contamination is widespread in India.

There are a number of key water resource management challenges in this subregion: floods and droughts, inadequate availability of water and its supply, poor sanitation, groundwater depletion and pollution, food and

Table 3.2 Annual water withdrawals by sector, and sources in South Asian countries

Country	Agriculture sector Volume (BCM)	% of total	Total withdrawal Volume (BCM)	m³/ person	With-drawal as % of TRWR	Groundwater Volume (BCM/y)	% of total
Afghanistan	20.00	98.62	20.28	913.4	31.0	3.04	15.0
Bangladesh	31.50	87.82	35.87	231.9	2.9	28.48	79.4
Bhutan	0.32	94.08	0.34	455.5	0.4	0.00	0.0
India	688.00	90.41	761.00	615.4	33.9	251.00	33.0
Maldives	0.00	0.00	0.01	17.5	15.7	NA	NA
Nepal	9.32	98.14	9.50	366.0	4.5	NA	NA
Pakistan	172.40	93.95	183.50	1024.0	74.4	61.60	33.6
Sri Lanka	11.31	87.34	12.95	637.9	24.5	NA	NA
South Asia	932.85	91.15	1023.44	620.5	27.0	344.12	33.6

BCM, billion cubic meters; NA, not applicable; TRWR, total renewable water resources.
Source: FAO, 2015.

energy insecurity, environmental degradation, inadequate water institutions, and poor sharing of information with stakeholders (SAWI, 2015).

3.4 CASE STUDY CITIES

Groundwater is key to supporting socioeconomic development of the subregion. Since abstraction is uncontrolled in many cases, groundwater environmental issues – such as aquifer degradation, land subsidence, depletion of groundwater reserves, decline in well yields, and salinity intrusions in coastal aquifers – are widespread in aquifers in South Asia. In this section of the book, five cities and their aquifer systems have been selected randomly, based on the availability of contributors and as diverse a cross-section as possible. The cities in alphabetical order are Chitwan (Nepal), Delhi (India), Hyderabad (India), Khulna (Bangladesh), and Lahore (Pakistan). They are detailed in Chapters 4–8, respectively. The general characteristics of the cities are provided in Table 1.1.

ACKNOWLEDGMENT

The authors would like to acknowledge SEA-EU-NET II Project (http://www.sea-eu.net/about/aims_results) for supporting the development of this chapter.

REFERENCES

Babel, M.S., Wahid, S.M., 2008. Freshwater Under Threat: South Asia, Vulnerability Assessment of Freshwater Resources to Environmental Change. United Nations Environment Programme and Asian Institute of Technology, Bangkok.

FAO, 2015. AQUASTAT database, Food and Agriculture Organization of the United Nations (FAO). Available from: http://www.fao.org/nr/water/aquastat/data/query/index.html?lang=en (accessed 12.06.15).

South Asia Water Initiative (SAWI), 2015. Available from: http://www.southasiawaterinitiative.org/SAWIAbout (Accessed 12.06.15).

Weiss, A.M., Mughal, M.A.Z., 2012. Pakistan. Kotzé, L., Morse, S. (Eds.), Berkshire Encyclopedia of Sustainability, 9, Great Barrington, MA, Berkshire, pp. 236–240.

WB, 2015. World Development Indicators, World Bank (WB). Available from: http://data.worldbank.org/indicator (accessed 14.04.15).

CHAPTER 4

Groundwater Environment in Chitwan, Nepal

Rabin Malla* and Kabita Karki**
*Center of Research for Environment, Energy and Water (CREEW), Kathmandu, Nepal
**Department of Mines and Geology, Kathmandu, Nepal

4.1 INTRODUCTION

Almost all the people in the Terai, the southern plain of Nepal, rely on groundwater to meet their domestic water demands. In recent decades, groundwater irrigation practices have intensified and become crucial for the livelihood and food security of millions of people living in the area.

Chitwan District occupies 2218 km² of Nepal in the western part of Narayani Zone (Figure 4.1) and has a population of about 0.58 million. Bharatpur, the district headquarter, is the fifth largest city in Nepal and home to 0.15 million people. It is a commercial and service center in central–south Nepal and a major destination for higher education, health-care, and transportation in the region. The district takes its name from the Chitwan Valley, one of Nepal's Inner Terai valleys between the Mahabharat and Siwalik ranges, foothills of the Himalayas.

Both natives and migrants from the hills inhabit the district. About 77.5% of them live in villages and 22.5% in municipalities. Agriculture is the main occupation. Groundwater is the major source of water for industry and irrigation, while surface water is mainly made available by using the Narayani lift irrigation system. Installation of deep tubewells (DTWs) and shallow tubewells (STWs) has increased over the years. Groundwater use for irrigation purposes is making significant contributions socioeconomically to the lives of the people. The effects of pesticide and fertilizer use on groundwater are readily visible.

The Bharatpur Water Supply Management Board (BWSMB), a separate entity providing water supply services to the Bharatpur Municipality, was established in 2011. Water supply in the municipality before the BWSMB was established fell within the jurisdiction of the Nepal Water Supply Corporation (NWSC) based in Chitwan. The BWSMB has a

Copyright © 2016 Elsevier Inc.
All rights reserved.

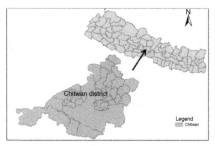

Figure 4.1 *Location of Chitwan District in Nepal.*

mandate to manage, monitor, regulate, and control the use of groundwater resources within the municipality. Groundwater is the major source of drinking water supply for the BWSMB and all other parts of the district. The BWSMB monitors the water quality of 19 deep borings used for water supply monthly. The water quality of the deep borings monitored are reported as safe for drinking as long as they meet national drinking water quality standards. Furthermore, there are no reports of interruptions in the supply of water for drinking, industrial use, and irrigation from groundwater and surface water sources.

The availability of water has so far not been a problem. Nevertheless, major concerns relating to aquifers in the area include initiating management at an early stage, addressing deficiencies in policies and knowledge, and monitoring the quality and quantity of groundwater at adequate intervals. Therefore, understanding and protecting the groundwater environment of the area is important and requires adequate knowledge of driver–pressure–state–impact–response (DPSIR), and their causal interlinks. This chapter aims at highlighting these components by analyzing the status of relevant indicators.

4.2 ABOUT THE STUDY AREA

4.2.1 Physiography, Climate, and River Systems

Chitwan is a NNW–SSE-trending dun valley (an alluvial basin) surrounded by the Mahabharat and Churia hills. It is also called Bhitrimadesh (inner plain land of the Inner Terai) and lies in the Narayani Zone of the Central Development Region. The district center, Bharatpur, is situated on the East–West Highway. Chitwan District lies between 83°54′45″ to 84°48′15″E and 27°21′45″ to 27°52′30″N.

Table 4.1 Ecological zones, geographical areas, and groundwater potential in Chitwan District

Ecological zones	Geographical area and elevation	Area covered by the zones (%)	Groundwater potential
Lower tropical	Terai and inner Terai below 300 m	58.2	These are potential groundwater recharge and storage basins
Upper tropical	Siwalik range 300–1000 m	32.6	
Subtropical	Mahabharat Range 1000–2000 m	6.7	—
Water body zone (only river deltas and some lakes)	—	2.5	—

Source: Ecological zones and their coverage are from Lillesø et al. (2005).

The climate of the district varies from lower tropical to subtropical (Table 4.1). Average maximum and minimum temperatures are 37.3°C in May and 4°C in December, respectively. High humidity is prevalent throughout the year except in winter. Mean annual rainfall in the 15 years from 1998 to 2012 was 2156 mm, 85% of which was received in the monsoon season, which lasts from June to September. The major climatic zones, geographical areas, and their coverage are shown in Table 4.1.

The Narayani is the largest first-grade river in the district; it forms the western border with Nawalparasi District. The second largest river is the Rapti, which flows westward through the middle of the district. There are a number of streams – such as the Lothar, Kayar, Rigdi, and Rieu – in the drainage system in the district (Figures 4.2 and 4.3). All the streams flow from north to south. Many swamps, wetlands, and depressions are present in the area.

4.2.2 Economic Activities

Agriculture is the most important sector of the economy in Chitwan, and 251,000 ha of agricultural land are dedicated to it. In terms of area planted, value of sales, and significance as a staple food, rice is the most important crop grown in the district followed by mustard, maize, wheat, pulses, buckwheat, and sesame. As human pressures on arable land increase, so the utilization of land resources intensify. Improved technologies and extended irrigation facilities have increased cropping intensity and cereal grain production to such an extent that three crops a year can be produced on

Figure 4.2 *Major Rivers in Chitwan, Nepal.*

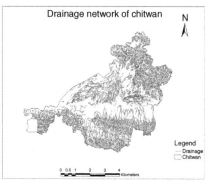

Figure 4.3 *Drainage Network Density in Chitwan, Nepal.*

Table 4.2 Irrigation status of agriculture land

Total area (ha)	Cultivable land (ha)	Cultivated land (ha)	Irrigable area (ha)	Surface and groundwater irrigation (ha)	Balance area for year-round irrigation (ha)
251,000	99,510	52,410	86,550	25,831	33,719

good-quality land. Two important irrigation systems in Chitwan – the Khageri Canal System and the Narayani Lift System – play vital roles in providing irrigation for agriculture. Adequately developed water resources will provide much of the water needed to irrigate the land. The detailed irrigation status of agricultural land in the area is presented in Table 4.2.

In addition to agriculture, tourism is another important sector of the economy in Chitwan. The Chitwan National Park is a popular tourist destination in the valley. Increasing tourism is providing a steady source of income for the residents of the area. On the other hand, it is exerting pressure on existing groundwater resources as well.

4.2.3 Geological Setting

Chitwan is a NNW–SSE-trending synclinal dun valley (about 140 km long and 60 km wide) situated within the Siwalik Hills, a sub-Himalayan range of the Nepal Himalaya. It is a river valley transporting sediment to the west via the Narayani River and to the east by the Rapti River. Within this valley there are two dun valleys formed by neotectonic movement: the Rapti and Reu valleys (Neupane, 2009, unpublished data). Based on the physiography, the valley is divided into the Mahabharata Range, Siwalik, and the Terai.

The Siwalik is a highly populated region. It is made up of soft sandstone and mudstone that encircle the valley, whereas the Terai region is low-lying land in the central part of the valley, which consists of sand, silt, and gravel (Dangol and Poudel, 2004; Tamrakar et al., 2008).

The Siwalik belt consists of molassic sediments, mainly conglomerates, sandstones, and clays dating to the Neogene (upper Miocene to Early Pleistocene). Hagen (1969) divided the Siwalik into the upper, middle, and lower Siwalik and described their composition as well. The upper Siwalik consists of well sorted round to subrounded conglomerates and less indurate coarse to fine-grained sandstones. The lower part of the upper Siwalik consists of pebble conglomerates and the upper part consists of boulder conglomerates. The Middle Siwalik consists of thick-bedded, less indurate, medium to coarse-grained "salt-and-pepper" sandstone with a smaller proportion of mudstone beds. The salt-and-pepper texture in the sandstone is due to the presence of biotite and feldspar. The lower Siwalik consists of thick-bedded variegated mudstone with subordinate fine to very fine–grained sandstone beds and pseudoconglomerate. The lower and middle Siwalik make a continuous belt all along the Himalaya, but the upper Siwalik only occurs intermittently in some areas. Sediment size increases upward and becomes conglomeratic in the upper Siwalik.

The broad valleys in the Himalayan foothills, the so-called "duns," represent geological depressions (synclinals) in the Siwalik. A geological map of the area and its cross-section are shown in Figures 4.4 and 4.5, respectively.

The Siwalik range in the east consists of conglomerates of the upper Siwalik strata, while formations from the lower Siwalik predominate in the west (Hagen, 1969). To the north of the Siwalik is the Rapti Valley, which steadily widens from east to west. The eastern valley floor is covered by the broad alluvial fans of tributary rivers emerging from the Mahabharat Range. To the west the floor of the basin consists of alluvial deposits from the Narayani River (Haffner, 1979).

Fluvial deposits in Chitwan's dun valley are considered to be Quaternary in age and include alluvial fan deposits, channel deposits, and floodplains that date to the Pleistocene and Holocene. They are still being accumulated to the present day (Table 4.3).

4.2.4 Hydrogeology

The large aquifer system underlying Chitwan's dun valley is predominantly filled with highly porous, permeable, and unconsolidated to poorly consolidated alluvial or fan deposits that date from the Late Pleistocene to

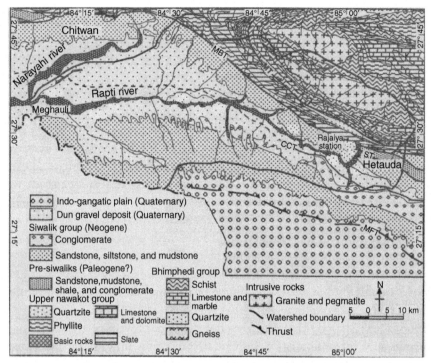

Figure 4.4 *Geological Map of Chitwan Dun Valley (Tamrakar et al., 2008).*

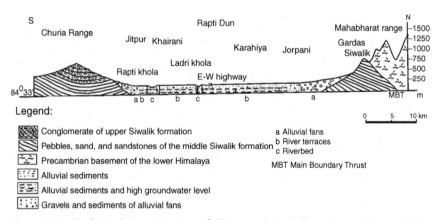

Figure 4.5 *Geological Cross-Section of Chitwan Dun Valley. Source: Hagen, 1969, Fig 85,91; Haffner, 1970, Fig 56; GoN map NO. 35, 1:25000; GoN geological map No. 72 AB 1: 125000; Graphics: K. Wegner.*

Table 4.3 Stratigraphic profile of the study area

Strati- graphic age	Stage	Stratigraphic name	Lithology	Depth (m)	Thick- ness (m)	Function in aquifer system
Holocene	Post glacial stage	Narayani alluvium	Swamp, levee, and riverbed sediments	0–5	5	Upper aquitard
		Narayanghat sand	Fine to coarse sand	5–10	5	Aquifer shallow
		Bharatpur sand	Unsorted sediments, boulder/ cobble/ pebble	10–20	10	Aquifer shallow
Pleistocene	Last glacial stage	Devghat gravel	Boulders/ cobbles/ pebbles	20–120	100	Deep aquifer

Source: Morris et al. (2003).

Holocene. These unconsolidated valley fill deposits consist of thickly bed-ded conglomerates with pebble to boulder clasts in a fine-grained matrix. They are locally called dun fan gravels or dun gravels. The sediments are finer toward the confluence of the Rapti and Narayani rivers. Deposits here form the main aquifer system, which is characterized as being homoge-neous in nature as is evident from the hydraulic properties recorded in tube-well logs. Previous researchers named the aquifers Bhabar deposits. River fans and ancient river terraces are found mainly in the valley. The Bhabar Zone area in Chitwan District covers an area of about 280 km² while that of the dun valley is about 800 km².

The valley contains colluviums around the periphery and a thick pile of very coarse and poorly sorted fluvial deposits in the main valley. The sedi-ments in the valley are characterized by a widespread occurrence of hard, cemented conglomerate about 9–10 m thick with a 2–3 m thick clay layer overlying it. In the southeast part of the valley near the Rapti River, a 2–3 m thick clay layer overlies fine sands. Very thick gravel and boulder deposits subsequently underlie this. The porosity and permeability of these deposits are high and generally form good aquifer conditions. Both types of aquifer, unconfined/confined or semiconfined/leaky, are present in the area.

The hydrogeological characteristics of shallow aquifers in the Chitwan District as reported in Neupane (2009) are: average depth = 40 m; static water level = <5 m; discharge = 10 L/s; average percentage of aquifer = 40; transmissivity = 767–6423 m^2/d; dynamic reserve (Duba's estimate) = 421.6 million cubic meters (MCM).

4.2.5 Groundwater Potentials

Groundwater investigations in Nepal started in the 1960s. Many organizations – such as the US Agency for International Development (USAID), Groundwater Development Consultants (GDC), the World Bank, and the Government of Nepal (GoN) – have been involved in groundwater investigations in the Terai. These showed that coarse sediment deposits of a very permeable and unconfined nature represented a good hydrogeological setup for recharge and infiltration to aquifers in the Chitwan Valley. The water table of aquifers varies from springs down to depths of 20 m depending on the dimensions of individual deposits.

The first groundwater potential map of the whole Terai region was developed by GDC in 1993 from an irrigation development perspective. However, the Shallow Tubewell Investigation Project initiated by the GoN can be considered the most important study focused specifically on the Chitwan Valley. No updates have been made since then. As per the latest understanding, groundwater potential varies within the valley (Table 4.4 and Figure 4.6).

Table 4.4 Groundwater production from selected wells in the Chitwan Valley (GWRDB, 2001, 2005a, 2005b, 2006a, 2006b, 2013a, 2013b)

| Well location | Drilling depth, m | Groundwater production | | | |
		Static water level, meters below ground level (mbgl)	Dynamic water level (mbgl)	Discharge adopted for test: l/sec	Specific capacity, $m^3/d/m$
Chainpur-6	120	30	40	14	50.4
Jagatpur-7	90	3.10	10.03	40	498
Kesarbag	–	5.7	8	16	62.84
Padampur-3		1.50	31.85	30	6766.2
Padampur-4	116	18.08	19.84	15	736
Padampur-6	114	7.7	11.4	46	–
Padampur-8	115	22.15	25.16	20	574
Shukranagar-3	–	9.25	18.65	–	393

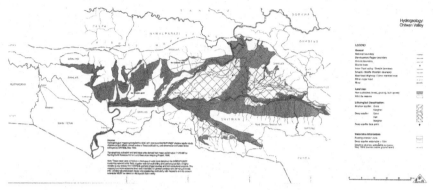

Figure 4.6 *Hydrogeology of Chitwan with Lithological Classification for Shallow Aquifer (as Good and Marginal) and Deep Aquifer (as Good, Fair, Marginal) (GWRDB, 2010).*

Depending on the history of sediment deposition and other related factors, groundwater potentials are classified as poor, marginal, and good.

The Village Development Committees (VDCs)/Municipalities demonstrating potential for shallow and deep aquifer development (Figure 4.6) are Jutpani, Bharatpur Municipality, Ratnanagar Municipality, Piple, Bhandara, Birendranagar, Khairahani Municipality, Kumroj, Bachhayauli, Kathar, Gitanagar, Shivanagar, Phulbari, Pithuwa, Mangalpur, Sharadanagar, Patihani, Divyanagar, Gunjanagar, Shukranagar, Meghauli, Jagatpur, Parvatipur, Gardi, and Bagauda.

Groundwater potential is fairly good in eastern and western parts of the valley for shallow, medium, and deep aquifers. It is low in western and northern parts for shallow aquifers. The primary source of groundwater in the valley is local precipitation. Groundwater storage potentials of shallow aquifers in the valley are estimated at 984 MCM.

4.3 DRIVERS

Population growth, urbanization, tourism, and agricultural intensification are the main drivers exerting pressure on the groundwater environment of the Chitwan Valley. The current status of indicator values for DPSIR are provided in Table 4.7.

4.3.1 Population Growth

Chitwan is known as the 76th district of Nepal where people from every district of the country have migrated. Migration can mainly be attributed

Table 4.5 Population in Chitwan District during 1971–2011 (Chitwan District Profile, 2013)

Census year	Number of households	Total population	Decadal increase in population	Population growth rate (%)	Population density (persons/km²)
1971	28,912	183,644	—	—	83
1981	41,414	259,571	75,927	3.46	117
1991	65,147	354,488	94,917	3.12	160
2001	92,863	472,048	117,560	2.86	213
2011	132,462	579,984	107,936	2.06	261

Table 4.6 Rural and urban population of Chitwan during 1981–2011 (Chitwan District Profile, 2013)

Year	Total population	Urban population	Urban population (%)	Rural population	Rural population (%)
1981	259,571	27,602	10.63	231,969	89.37
1991	354,488	54,670	15.42	299,818	84.58
2001	472,048	89,323	18.92	382,725	81.08
2011	579,984	194,144	33.47	385,840	66.53

to the fact that it is close to the capital of the country, transit to the capital is made easy (the same applies to other Terai districts), plenty of fertile agricultural land, and popular tourist destinations like Chitwan National Park. As a result, population in the district increases with every census year (Table 4.5). Population density in the district increased from 83 persons/km² in 1971 to 160 in 1991 and 261 in 2011.

4.3.2 Urbanization

Total urban area in Chitwan including the Bharatpur and Ratnanagar municipalities is 197.78 km². In recent decades the valley has started to develop as an urban center. It is now the second most developed in the country in terms of infrastructure and facilities. It is also clear (Table 4.6) that the urban population of the district is increasing compared with the rural population with every census year.

Deep groundwater constitutes the only single source of water supply in the Bharatpur Municipality. There are 19 deep borings with a production capacity of 27 million liters a day (MLD). Present abstraction stands at 15.2 MLD (BWSMB, 2013). However, there are no census data after 2001 to estimate the pressure on groundwater resources due to urbanization and migration. Apart from increased groundwater abstraction, unmanaged urbanization results in haphazard disposal of solid waste, and leachage from those sites further degrades groundwater quality.

4.3.3 Industrialization

The Chitwan Valley is gradually industrializing as indicated by the number of registered industries such as tourism (Table 4.7). There were 91 registered industries in FY 2005/06, which increased to 144 in FY 2012/13. Chitwan is the third largest tourist destination in Nepal. The Chitwan National Park established in 1973 has always been a popular destination for national and international tourists. The park was granted the status of Natural World Heritage Site in 1984. It is home to the rare one-horned rhino and Royal Bengal tiger. The park lies between two east–west-trending river valleys at the base of the Siwalik Hills of the outer Himalayas and covers an area of 932 km². Tourist numbers in the district are increasing (Figure 4.7), and accommodation is provided by 68 hotels. This increase in the number of tourists visiting the district puts added pressure on groundwater to meet the water demand of hotels and restaurants. Hotels depend on groundwater for run-of-the-mill daily purposes, hence the need to increase water abstraction.

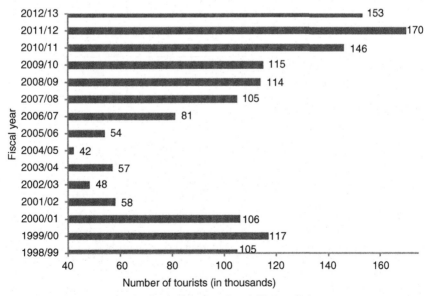

Figure 4.7 *Yearly Tourists in Chitwan National Park (CNP, 2012).*

Table 4.7 Status of DPSIR indicators in Chitwan Valley, Nepal

	Indicators	2001	2011
Drivers	Population growth		
	Population (million)	0.47	0.58
	Density (persons/km²)	213	261
	Annual growth rate (%) (CBS, 2013)	3.3 (2001–1991)	2.3 (2011–2001)
	Urbanization		
	Urban area (km²)	162.16 km² (7.31%; one municipality) (CBS, 2013)	197.78 km² (8.91%); two municipalities) (CBS, 2013)
	Urban population	89,323 (CBS, 2007)	194,384 (CBS, 2013)
	Industrialization		
	No. of tourist arrivals	58,000 (approx.)	170,000 (approx.)
	No. of registered industries	91 (in FY 2005/06) (Department of Industry, 2006).	144 (in FY 2012/13) (Department of Industry, 2013).
	Agricultural intensification	The commercialization of agriculture began a decade or so ago. The government is promoting usage of shallow groundwater for agriculture intensification. As of 2012/2013, 2903 shallow tubewells and 37 deep tubewells had been installed.	
Pressures	Land cover change (see Table 4.8)		
	Cultivated area (km²)	646.7 km² (28.9%) in 1992	638.6 km² (28.5%) in 2009
	Technological interventions in agriculture	The command area of the NLC has reduced to 3500 ha from 4700 ha resulting in increased dependence on groundwater. As of 2012/13, 37 deep tubewells and 2940 shallow tubewells had been installed.	
State	Well statistics and groundwater abstraction	Total numbers of wells in Chitwan has increased continuously from 397 in FY 2010/11 to 418 in 2011/12 and 2940 in FY 2013/14. Total groundwater abstraction for all uses is estimated as 135.7 MCM.	

Table 4.7 Status of DPSIR indicators in Chitwan Valley, Nepal *(cont.)*

	Indicators	2001	2011
	Groundwater level		SWL in the premonsoon season in 2013 varied from 4.8 mbgl at Bhimnagar to 11.2 mbgl at Prembasti (Figure 4.8), whereas the postmonsoon SWL remained almost constant.
	Groundwater quality		Chettri and Smith (1995) reported nitrate contamination in shallow groundwater in Chitwan. Chapagain et al. (2011) found 15% of shallow-groundwater samples were contaminated with nitrate in Mangalpur, Chitwan. Deep-groundwater quality in BWSMB-abstracted wells in 2014 was safe and potable.
	Groundwater recharge		284.7 MCM/y (Neupane, 2009, unpublished data)
Impacts	Decline in groundwater level		A gradually declining trend is observed in premonsoon SWL at Prembasti, Khaireni, and Bhimnagar (slight decline during 2005–2009 and over 2 m decline during 2007–2009). Postmonsoon SWL, in general, has remained constant over the years.
	Deterioration in groundwater quality		Deep groundwater in Chitwan is not contaminated so far (BWSMB, 2014); however, deterioration of the quality of shallow groundwater due to pesticides, chemical fertilizers, and organic manures is reported. Increase in the population of *Eichhornia crassipes*, a free-floating aquatic plant that grows in polluted water, lakes, and *Ghols*, indicates underlying contamination of groundwater.
Responses	Formulation of policies and laws		There are a number of policies, acts, and regulations related to groundwater such as Water Resources Act (1992), Local Self Governance Act (1999), Water Resources Strategy (2002), National Water Plan (2005), Bharatpur Water Supply Management Board Regulation (2009), Supreme Court Order (2010), Irrigation Policy (2013), Constitutional Provision (2007), and Groundwater Resources Act (under progress).

(Continued)

Table 4.7 Status of DPSIR indicators in Chitwan Valley, Nepal *(cont.)*

Indicators	2001	2011
Groundwater monitoring		The GWRDB began monthly monitoring of shallow-groundwater levels in a network of 17 wells in 2004. The BWSMB began monitoring groundwater quality in 2014, which they will monitor on a yearly basis.
Setting up environmental standards and guidelines		MoEST has set the maximum tolerance limits for three types of effluents: industrial waste into surface water, wastewater from treatment plants into surface water, and industrial effluents into public sewers (NBSM, 2003). These untreated waters may contaminate the shallow groundwater *via* the surface water system.
Organizational adjustments		The GoN has set up bodies in an attempt to get organizations to improve their management of groundwater resources in the area. They include the GWRDB (established 1976) and the BWSMB (established 2007).
Public awareness and interest		A rapid questionnaire of local people and interviews with personnel from the GWRDB revealed that local people and officials are aware of groundwater exploitation, decline in groundwater level, and its consequences. Conflicts between neighbors over the installation of deep tubewell have also surfaced as a result of concern that the sustainability of groundwater resources in the area could be affected. People want to contribute to conserving groundwater resources.

BWSMB, Bharatpur Water Supply Management Board; FY, fiscal year; GoN, Government of Nepal; GWRDB, Groundwater Resources Development Board; mbgl, meters below ground level; MCM, million cubic meters; MoEST, Ministry of Environment, Science and Technology; NLC, Narayani Lift Canal; SWL, static water level.

4.3.4 Agricultural Intensification

The commercialization of agriculture in Chitwan began a decade or so ago. Though monsoon rainfall is the major source of water for agriculture, the Agricultural Development Bank and the government have been promoting the use of shallow groundwater in the Terai for agriculture for decades. As of FY 2012/2013, 2903 shallow tubewells and 37 deep tubewells had been installed in the district. It is highly likely that more tubewells will be installed in coming years to abstract more groundwater and further pressurize the groundwater environment.

4.4 PRESSURES

Pressures in the case at hand can be defined as direct stresses brought about by expansion of the anthropogenic system and associated interventions in the natural environment. Pressure on the groundwater environment in Chitwan comes from land cover change and technological interventions in agriculture.

4.4.1 Land Cover Change

Chitwan attracts people because of its fertile land as well as its proximity to the Terai and the capital. Settlement areas in the valley have increased from 1.82 ha in 1978 to 3.78 ha in 1992 and 12 ha in 2009. However, sand and gravel, which remained relatively unchanged between 1978 and 1992, decreased by more than 33% between 1992 and 2009 (Table 4.8). The increase in settlement area and decrease in the sand and gravel area clearly suggest a reduction in aquifer recharge potential. The increased size of settlements in the northern part, which is similar to the Bhabar Zone of the Terai (Kharel, 2000, unpublished data) or the recharge zone, resulted in decreased discharge into the Khageri River (Shukla and Sada, 2012) and *Ghols* (artificial water bodies) (Karki, 2013, unpublished data). This is a good example of how the pressure on groundwater has been increasing in recent years from land cover changes.

4.4.2 Technological Interventions in Agriculture

The dependence on surface water for irrigation has decreased as a result of the increased number of DTWs and STWs constructed for irrigation. This is mainly because of the special focus of the GoN to enhance farmer access to irrigation as a strategy for poverty reduction. *Abhiyan, Garibiko Nidan* is a popular slogan meaning "shallow tubewell access to farmers is a means of reducing poverty." The slogan is widely used in the Terai/Inner Terai districts. This has helped improve the economy, on the one hand, but has increased pressure on the groundwater environment, on the other hand.

Table 4.8 Land cover in Chitwan from 1978 to 2009

Land use	1978		1992			2009		
	Area	%	Area	%	Change %	Area	%	Change %
Settlement	1.82	0.80	3.78	0.17	1.96	12.00	0.54	8.22
Airport area	—	—	0.53	0.02	—	0.53	0.02	—
Agricultural land	561.68	25.30	646.67	28.87	79.99	638.63	28.51	8.02
Forest	1440.31	64.53	1245.39	55.60	199.91	1216.82	54.32	28.53
Grassland	100.92	4.49	76.64	3.42	23.88	170.23	7.60	93.59
Gardens	—	—	1.11	0.05	19.69	0.54	0.02	0.57
Bush	20.68	0.93	112.44	5.02	—	94.94	4.21	18.20
Barren land	—	—	7.11	0.32	—	3.59	0.16	3.51
Forest area with sparse trees	—	—	1.10	0.05	—	0.24	0.01	0.87
Ponds and lakes	—	—	1.25	0.06	—	1.55	0.01	0.31
Water bodies	—	—	39.25	1.75	—	30.42	1.36	8.82
Sand and gravel	104.86	4.68	103.77	4.42	1.42	70.11	3.13	33.33
Erosion and river cutting area	—	—	1.22	0.05	—	1.02	0.05	0.19
Total	2239.93	100	2239.93	100	—	2239.93	100	

Source: DAO Chitwan (2012).

The command area of the Narayani Lift Canal (NLC) has now been reduced to 3500 ha from 4700 ha. KC (2012, unpublished data) reported that more respondents are in favor of an alternative to the NLC because the system is expensive, temporary, irregular, and inadequate. Technological interventions, especially in terms of improving access to groundwater through STW programs, have exerted more pressure on groundwater despite contributing to poverty reduction.

4.5 STATE

State in the case at hand describes the conditions, tendencies, and trends in the groundwater environment that have mainly been brought about by human activities. There are five indicators of the state of the groundwater environment: well statistics, groundwater abstraction, groundwater level, groundwater quality, and groundwater recharge.

4.5.1 Well Statistics and Groundwater Abstraction

The number of wells in Chitwan is constantly increasing and had reached 2940 (Table 4.9) as of FY 2012/13. These wells irrigate 5605 ha of land. Total groundwater abstraction in Chitwan for drinking, domestic use, industrial use, and irrigation is estimated as 135.7 MCM approximately (see Annex 4.1).

4.5.2 Groundwater Level

Groundwater levels are deduced from the static water level (SWL) observed at 17 monitoring wells. This has been done on a monthly basis since 2004. The SWL in the premonsoon season varies at different locations. For example, the SWL in the premonsoon season of 2013 was 4.8 meters–below ground level (mbgl) at Bhimnagar, 6.2 mbgl at Khaireni, and 11.2 mbgl at Prembasti (Figure 4.8). Data from the postmonsoon season of 2010 (Figure 4.10) also showed

Table 4.9 Number of tubewells and corresponding irrigated areas in Chitwan (Shakya et al., 2012)

	Well numbers			Total irrigated area (ha)		
Well type	in FY 2010/11	in FY 2011/12	Total until FY 2012/13	in FY 2010/11	in FY 2011/12	Total until FY 2012/13
Shallow	396	417	2903	990	1042.5	5165
Deep	1	1	37	—	80	440
Total	397	418	2940	990	1122.5	5605

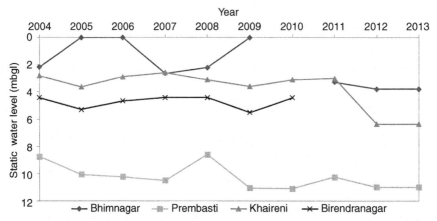

Figure 4.8 *Static Water Level (SWL) in Selected Shallow Wells in Premonsoon Season between 2004 and 2013. Source: GWRDB, Chitwan Branch.*

spatial variations: from 2.0 mbgl at Khaireni to 2.8 mbgl at Birendranagar and 6.9 mbgl at Prembasti. Water levels in the postmonsoon season are shallower than in the premonsoon season as groundwater infiltrates bedrock in the rainy season and increases the groundwater table.

4.5.3 Groundwater Quality

Improved disposal of organic waste and excessive fertilizer use on agricultural land in the hope of a good yield have resulted in nitrate contamination in shallow groundwater (WHO, 2004). Shallow groundwater in several parts of Nepal, including Chitwan, is contaminated with nitrate (Chettri and Smith, 1995). As shown in Figure 4.9, nitrate contamination in 15.6% of shallow-groundwater samples (based on samples from Mangalpur VDC, Chitwan) exceeded WHO guideline values (i.e., 50 mg/L) and 37.5% samples were within a range of 25–50 mg/L (Figure 4.9). Using δ^{15}N and δ^{18}O of nitrate as tracers suggests that soil manure is the primary source of nitrate contamination in shallow groundwater.

The BWSMB relies on deep groundwater to supply water to municipal residents. It has 19 deep wells. The water quality in these deep wells has been found to be safe and within the National Drinking Water Quality Standard. The water quality of some of the deep wells is provided in Table 4.10.

4.5.4 Groundwater Recharge

Groundwater recharge is a complicated phenomenon, especially when recharge into deeper parts of aquifers is considered. Precipitation amount is

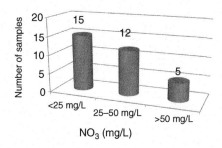

Figure 4.9 *Shallow Groundwater Samples in Chitwan (Chapagain et al., 2011).*

the primary factor; however, recharge depends on surface characteristics and hydrogeology as well (Driscoll, 1989).

There are many methods that can be used for groundwater recharge calculation such as Duba's estimation (1982) method (Neupane, 2009, unpublished data), the effective porosity method, and the conservative method. The conservative method was used for groundwater recharge analysis in this study, as it was the simplest. The method assumes only 10% of rainfall reaches the aquifer. Based on this calculation, groundwater recharge can be estimated as 284.7 MCM/y (Neupane, 2009, unpublished data), which is more than double the estimated abstraction of 135.7 MCM/y (Annex 4.1). It is therefore clear that Chitwan's aquifers are not overexploited.

4.6 IMPACTS

Impacts in the case at hand relate to effects on both the human system and the environment itself as a result of changes in the state of the environment. They contribute to the vulnerability of both natural and social systems. Visible impacts on the groundwater environment in Chitwan are decline in groundwater level and deterioration in groundwater quality.

4.6.1 Decline in Groundwater Level

As discussed in previous sections, almost every year Chitwan has witnessed increases in the urban population, arrival of tourists, registration of industries, and agricultural activity. Groundwater is the sole major resource meeting the water demands of all these sectors. In addition, increased settlement in the northern part of Chitwan (where aquifer systems are recharged) has resulted in decreased discharge into the Khageri River (Shukla and Sada, 2012). The impacts of groundwater use by all sectors on groundwater

Table 4.10 Water quality in selected deep wells of the BWSMB in 2014 (BWSMB, 2014)

Parameters	Unit	Observed value (Well 4, Saradpur)	Observed value (Well 5, Tri-chowk)	Observed value (Well 6, Aaptari)	Observed value (Well 15, Prembasti)	Observed value (Well 18, Anandpur)	National Drinking Water Quality Standards (2006)
Turbidity	NTU	1.0	0.5	1.0	1.0	0.5	5–10
pH	–	7.9	7.6	6.5	8.0	7.8	6.5–8.5
TDS	mg/L	150	152	107	107	200	1000
EC	µS/cm	300	304	214	213	400	1500
Iron (Fe)	mg/L	ND	0.001	0.002	ND	ND	0.3–3
Manganese (Mn)	mg/L	ND	ND	ND	ND	ND	0.2
Arsenic (As)	mg/L	0.010	0.007	0.0	0.015	0.01	0.05
Cadmium (Cd)	mg/L						0.003
Chromium (Cr)	mg/L	Negative	ND	ND	ND	ND	0.05
Cyanide (CN)	mg/L						0.07
Fluoride (F–)	mg/L	0.1	0.1	0.1	0.3	0.1	0.5–1.5
Lead (Pb)	mg/L						0.01
Ammonia (NH_3)	mg/L	<0.1	<0.1	<0.1	<0.1	<0.1	1.5
Chloride (Cl^-)	mg/L	9.9	14.2	16.3	6.0	7.5	250
Sulfate (SO_4^{2-})	mg/L						250
Nitrate (NO_3^-)	mg/L	<0.1	<0.1	0.01	<0.1	0.1	50
Copper (Cu)	mg/L						1
Total hardness	mg/L	170.0	130.0	96.0	142.0	156.0	500
Calcium (Ca)	mg/L	41.6	32.8	24.0	24.8	33.6	200
Zinc (Zn)	mg/L						3
Mercury (Hg)	mg/L						0.001
Aluminium (Al)	mg/L						0.2

BWSMB, Bharatpur Water Supply Management Board; EC, electrical conductivity; ND, not detectable; NTU, nephelometric turbidity unit; TDS, total dissolved solids.

level during the premonsoon season (April) and the postmonsoon season (October) are discussed in the following:

- *Premonsoon:* SWL in shallow observation wells during the premonsoon period is shown in Figure 4.8. The water level in Bhimnagar declined to some extent between 2005 and 2009 but has now become constant. SWL in Bhimnagar almost reached ground level (i.e., 0 mbgl) between 2004 and 2006, then dropped to more than 2 mbgl between 2007 and 2009, and further dropped to 4 mbgl between 2011 and 2013, clearly demonstrating a lowering trend in SWL. The gradually declining trend in premonsoon SWL is also observed at Prembasti and Khaireni.

- *Postmonsoon:* SWL in the postmonsoon season is shown in Figure 4.10. SWL in Bhimnagar lowered to about 3 mbgl in 2005 from its previous level of 2.5 mbgl, but regained its 2.5 mbgl level in subsequent years. SWL in Birendranagar raised in every postmonsoon season except 2007. SWL in Prembasti has declined gradually over the years, but has remained almost constant at about 2 mbgl in Khaireni. In general, though, postmonsoon SWL has remained more or less constant over the years.

4.6.2 Deterioration in Groundwater Quality

Deep-groundwater quality abstracted by the BWSMB to supply the Baharatpur Municipality has no reported case of water quality deterioration (see Table 4.10). However, shallow-groundwater quality has shown indications of deterioration. In lakes and *Ghols*, the population of *Eichhornia crassipes*, a

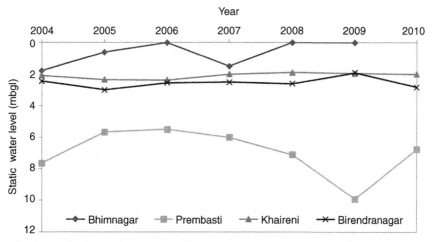

Figure 4.10 *Static Water Level (SWL) in Selected Shallow Wells in Postmonsoon Season between 2004 and 2010. Source: GWRDB, Chitwan Branch.*

free-floating aquatic plant that flourishes in polluted water, predominates throughout the year (Burlakoti and Karmacharya, 2004). The presence of *Eichhornia* is indicative of water pollution in lakes and *Ghols* and subsequent pollution of groundwater recharged from those water bodies. Chettri and Smith (1995) reported nitrate contamination in shallow groundwater in Chitwan. Similarly, Chapagain et al. (2011) found that 15% of shallow-groundwater samples were already contaminated with nitrate and 37% were more in line with contamination in Mangalpur, a VDC in Chitwan.

4.7 RESPONSES

The GoN, in close collaboration with interested stakeholders, has made several attempts to improve the water environment by means of a development program, organizational adjustments, and research activities. Other important initiatives include endorsement of the Water Resources Act (1992), development of standard guidelines for discharging industrial wastewater into the environment, attempts to recharge shallow groundwater through dugwells and ponds, and rainwater harvesting to reduce the pressure on groundwater resources.

4.7.1 Formulation of Policies and Laws

A number of policies, acts, and regulations relating to groundwater development and management have been formulated:

- *Water Resources Act (1992)*: This act deals with the rational utilization, conservation, management, and development of water resources available in Nepal in whatsoever form. The state owns all water resources, including groundwater, within its territory.
- *Local Self Governance Act (1999)*: According to this act, local bodies are tasked with making decisions on matters affecting the day-to-day needs and lives of people. They can levy tax on businesses commercially exploiting natural resources and heritage in their area. Local bodies play an important role in protecting the forests, vegetation, and other natural resources within the area.
- *Water Resources Strategy (2002)*: The objective of this strategy is sustainable use of water resources in which environmental conservation is insured in a holistic and systematic way. It envisages provision of adequate and safe drinking water (a declared commitment of government). Improvements in groundwater development and management subject to legislative provisions for groundwater use and management are also

included in the strategy. It emphasizes joint use of surface and groundwater resources to achieve year-round irrigation status.

- *National Water Plan (2005)*: The objectives of this plan include sustainable management of watersheds, adequate supply and access to potable water and sanitation, enhanced water-related information systems, and appropriate legal frameworks. It has put in place a regulatory mechanism to establish safe and sustainable groundwater abstraction and utilization for different uses. This regulatory mechanism will continue to consolidate and expand the groundwater-monitoring program and enhance regulatory measures. The plan further provides for regular monitoring of groundwater for water level fluctuations and surface/groundwater quality as well as setting up standards for irrigation use.
- *Bharatpur Water Supply Management Board Regulation (2009)*: This regulation includes the provision of regular monitoring and regulating of groundwater resources, issuing licenses for commercial extraction of groundwater if and only if deemed necessary, and controlling the overexploitation of groundwater using suitable instruments. The BWSMB monitored the water quality of its deep borings in 2014 as part of its mandate for better management of the supply of drinking water, something it plans to do once in year or at least every other year.
- *Supreme Court Order (2010)*: The GoN was issued a court order regarding the Kathmandu Valley in order to stop the groundwater level continuing to fall at such an alarming rate. The court ordered the GoN to put in place a Groundwater Action Plan that would restrict the number of tubewells, limit abstraction to maintain environmental balance, assess recharge potentials, strictly implement groundwater licensing, and promote groundwater recharge and reuse.
- *Irrigation Policy (2013)*: The objective of this policy is to highlight the importance of groundwater management and use by promoting joint use of surface water and groundwater, developing and utilizing groundwater resources, protecting groundwater, and imposing a special subsidy or tariff for electricity generated using groundwater.
- *Constitutional provision*: Access to groundwater for agriculture (irrigation), drinking water, and sanitation purposes are the fundamental rights of every citizen. The local community takes priority when it comes to water resources.
- *Groundwater Resources Act*: This act has been formulated and is under review by the Ministry of Law, Justice, Constituent Assembly, and Parliamentary Affairs of the GoN. After review by the ministry, it is expected

to be approved by the legislative parliament in 2016. This act also includes establishment of a Groundwater Authority tasked with studying, investigating, regulating, conserving, and controlling groundwater resources in the country.

4.7.2 Groundwater Monitoring

SWLs of 17 shallow observation tubewells have been monitored monthly since 2004 by the Chitwan Branch of the Groundwater Resources Development Board (GWRDB). As shown in Figure 4.8 and Figure 4.10 and discussed in Section 4.6.1, the trend in groundwater level declines in the premonsoon season (April) indicating increased use of shallow groundwater, whereas groundwater level in these shallow wells in the postmonsoon season remains more or less constant (October). Besides the GWRDB, the BWSMB began monitoring the groundwater quality of its deep borings in 2014, something it plans to do once a year or at least every other year.

4.7.3 Setting up Environmental Standards and Guidelines

The Ministry of Environment Science and Technology (MoEST), the agency responsible for determining environmental standard guidelines, has set the maximum tolerance limits for three types of effluents: industrial waste into surface water, wastewater from treatment plants into surface water, and industrial effluents into public sewers (NBSM, 2003). In addition, general standards have been determined as the tolerance limit of wastewater effluents discharged into surface water and public sewers for nine different industries: leather, wool processing, fermentation, vegetable ghee and oil, dairy products, sugar, cotton textiles, soap, and paper and pulp (NBSM, 2001). However, implementation of these standards remains poor in practice.

4.7.4 Organizational Adjustments

The GoN has set up a couple of boards in an attempt to make organizations improve their management of groundwater resources in the area. These boards are discussed in the following.

- *Ground Water Resource Development Board (GWRDB)*: This government agency was established in 1976. The GWRDB has nine branch offices in Nepal, including one in Chitwan, which was established in 2011. The GWRDB is tasked with the following:
  - Groundwater investigation, exploration, and study together with groundwater utilization for irrigation, drinking water, and other uses.

- Managing the operation of tubewells and distributing water for irrigation.
- Project implementation in accordance with agreements between the GoN and any foreign country or any agencies.
- Carrying out policy decision work on project formulation, operation, and implementation and monitoring such work.

- *Bharatpur Water Supply and Management Board (BWSMB)*: The BWSMB was established as a public statutory in Chitwan in 2007 as per the Water Supply Management Board Act (2006) of the GoN. Its objectives are to manage water supply and sanitation services in Bharatpur Municipality regularly and maintain high-quality standards and convenience.

4.7.5 Public Awareness and Interest

A rapid questionnaire of local people coupled with interviews with personnel from the Chitwan Branch Office of the GWRDB revealed that government officials and local people were aware of groundwater exploitation in their area, decline in groundwater levels, and concomitant consequences. The GWRDB also gets involved in conflicts between neighbors regarding deep-tubewell installations fearing that sustainability of groundwater resources in the area could be affected.

A survey conducted with users of the Panchakanya Irrigation System found that they wanted to contribute to conserving natural resources and water resources. Urban dwellers suggested local institutions be given greater authority to manage groundwater and seek alternatives for reducing stress on groundwater resources.

4.8 SUMMARY

DPSIR analysis shows anthropogenic systems — such as population growth, urbanization, agriculture, and tourism — exert pressure on the groundwater environment as a result of changing land use as well as changes in agricultural cropping patterns and practices. This has led to an increase in the number of wells drilled and the amount of groundwater abstracted, in addition to affecting the SWL and quality of groundwater. For instance, SWL monitoring in shallow wells (from 2004 to 2013) in the premonsoon season showed the groundwater level declining, whereas in the postmonsoon season (from 2004 to 2010) it was more or less constant. Currently, abstraction is less than recharge and therefore no adverse impact on the groundwater environment in terms of groundwater quantity has been observed. However, the quality of shallow groundwater is reported to be contaminated with nitrate, which is attributed to chemical fertilizers, organic manures, and human and livestock waste.

There are a number of legal frameworks that directly or indirectly relate to groundwater. The Groundwater Resources Act is the most important. It is currently under review but is expected to be approved by the legislative parliament in 2016. This act makes provision for the establishment of a Groundwater Authority with mandates to study, investigate, regulate, conserve, and control groundwater resources in Nepal. A rapid survey of people and interviews with government officials revealed that they were aware and concerned about groundwater resources. Increased participation of local institutions in protecting and conserving the resource was also suggested.

Chitwan is rich in groundwater resources. Several legal frameworks and appropriate institutions are in place for its proper management. Effective management of groundwater resources by strict implementation of the legal frameworks by the appropriate institutions is arguably the best way forward. Capacity building of government personnel may be needed as a result of changing legal frameworks and technologies such as water level–monitoring data loggers, designing water level and water quality–monitoring programs, water quality analysis, and database management and documentation. Many aspects of groundwater development deemed necessary for sustainable utilization of groundwater resources for irrigation in the area need to be investigated. A systematic study on the suitability of various drilling technologies for different geological setups in different parts of the valley will help enhance knowledge needed for the selection of appropriate drilling methods, tools, and affordability. Similarly, the review and critical study of the hydrogeological investigation previously carried out may need to be updated with the latest research findings and technologies for more scientifically enhanced management of groundwater resources.

REFERENCES

Bhandari, S., 2013. Hydro-geological study in western parts of the Chitwan Dun Valley. MSc dissertation, Central Department of Geology, Tribhuvan University, Nepal (unpublished).

Burlakoti, C., Karmacharya, S.B., 2004. Quantitative analysis of macrophytes of Beeshazar Tal, Chitwan. Nepal. Him. J. Sci. 2 (3), 37–41.

BWSMB, 2013. Annual report of Bharatpur Water Supply Management Board (BWSMB) for FY 2013/014.

BWSMB, 2014. Water quality test report. Bharatpur Water Supply Management Board (BWSMB).

CBS, 2007. Statistical Year Book of Nepal 2007, 1st ed. Central Bureau of Statistics, Government of Ramshahpath, Thapathali, Kathmandu, Nepal.

CBS, 2013. Statistical Year Book of Nepal, 14th ed. Central Bureau of Statistics, Government of Ramshahpath, Thapathali, Kathmandu, Nepal.

Chapagain, S.K., Nakamura, T., Malla, R., Kazama, F., 2011. Nitrate contamination of shallow groundwater in Mangalpur VDC of Chitwan District, Nepal. In: Proceedings of Ninth International Symposium on Southeast Asian Water Environment, Bangkok, December 1–3, 2011.

Chettri, M., Smith, G.D., 1995. Nitrate pollution in groundwater in selected districts of Nepal. Hydrogeol. J. 3, 71–76.

Chitwan District Profile, 2013. Statistics Branch Office, Chitwan, pp. 26–45.

CNP, 2012. Chitwan National Park (CNP) Office, Kasara, Chitwan. Annual Report 2012/013, pp. 73.

Dangol, V., Poudel, K., 2004. Channel shifting of Narayani River and its ramification in West Chitwan. J. Nepal Geol. Soc. 30, 153–156.

DAO Chitwan, 2012. District Agriculture Office (DAO) Profile, Chitwan District, Nepal.

Department of Industry, 2006. Industrial Statistics, Fiscal Year 2005/06, Government of Nepal, p. 19.

Department of Industry, 2013. Industrial Statistics, Fiscal Year 2012/13, Government of Nepal, pp. 37–38.

Driscoll, F.G., 1989. Groundwater and wells, Jhonson Filtration System Inc., St. Paul., Minnesota, 55112, 2nd Ed. pp. 1086.

Gleick, P.H., 1996. Basic water requirements for human activities: meeting basic needs. Water Int. 21, 83–92.

GWRDB, 2001. Deep tubewell reports, Shukranagar and Jagatpur; Groundwater Irrigation Project, Bharatpur, Chitwan. Groundwater Resource Development Board (GWRDB).

GWRDB, 2005a. Report on water well drilling at Padampur-4, Chitwan; Groundwater Irrigation Project, Bharatpur, Chitwan. Groundwater Resource Development Board (GWRDB).

GWRDB, 2005b. Report on water well drilling at Padampur-6, Chitwan; Groundwater Irrigation Project, Bharatpur, Chitwan. Groundwater Resource Development Board (GWRDB).

GWRDB, 2006a. Report on water well drilling at Padampur-3, Chitwan; Groundwater Irrigation Project, Bharatpur, Chitwan. Groundwater Resource Development Board (GWRDB).

GWRDB, 2006b. Report on water well drilling at Padampur-6, Chitwan; Groundwater Irrigation Project, Bharatpur, Chitwan. Groundwater Resource Development Board (GWRDB).

GWRDB, 2010. Souvenir. Groundwater Resource Development Board (GWRDB).

GWRDB, 2013a. Report on water well drilling at Kesarbag, Chitwan; Groundwater Irrigation Project, Bharatpur, Chitwan. Groundwater Resource Development Board (GWRDB).

GWRDB, 2013b. Report on water well drilling at Chainpur-6, Siddhipur, Chitwan. Groundwater Irrigation Development Committee, Parsakhaireni, Chitwan.

Haffner, W., 1979. Nepal Himalaya, Untersuchugenzumvertikalen Landschaftsaufbanzentral- und Ostnepals. Erdwissenschaftlicheforschung 12, Wiesbaden, Germany.

Hagen, T., 1969. Report on geological survey of Nepal, V. 1: preliminary reconnaissance. Denkschrifter der Schweizerischen Naturforschenden Gesellschaft, 86 (I) Jurich, pp. 185.

Lillesø, J-P.B., Shrestha, T.B., Dhakal, L.P., Nayaju, R.P., Shrestha, R., 2005. The Map of Potential Vegetation of Nepal – a forestry/agroecological/biodiversity classification system. Forest & Landscape Development and Environment Series 2-2005 and CFC-TIS Document Series No. 110.

Morris, B.L., Seddique, A.A., Ahmed, K.M., 2003. Response of the Dupi Tila aquifer to intensive pumping in Dhaka, Bangladesh. Hydrogeol. J. 11, 496–503.

NBSM, 2001. Industrial effluent standards. HMG/NBSM, Kathmandu.

NBSM, 2003. Wastewater, combined sewerage treatment plant and industrial wastewater standard. HMG/NBSM, Kathmandu.

Shakya, N., Shrestha, S.R., Aryal, B., Shrestha, I.K., Pokhrel, M., Raut, S.B., Poudel, A.K., 2012. Souvenir. Groundwater Resource Development Board.

Shukla, A., Sada, R., 2012. State of disjuncture, water conflict and approach to PIWRM: a narration on case for the field visit training course on Participatory Integrated Water Resources Management (PIWRM). Jointly organized by AIT and Nepal Engineering College in association with INPIM-Nepal, pp. 15.

Tamrakar, N.K., Maharjan, S., Shrestha, M.B., 2008. Petrology of Rapti River sand, Hetauda–Chitwan Dun Basin, Central Nepal, an Example of Recycle provenance. Bulletin of the Department, TU, Kathmandu, vol. 11, pp. 23–30.

WHO, 2004. Guidelines for Drinking Water Quality, vol. 1, 3rd ed. World Health Organization, Geneva, Switzerland.

ANNEX 4.1 CALCULATION OF TOTAL AMOUNT OF GROUNDWATER ABSTRACTION IN CHITWAN

In the absence of actual data on the amount of water needed by a person in a day, this study assumed the value as 45 liters/person/day (lpcd) (or 0.045 m^3/d). The assumed value is near to the general water requirement of water for drinking, sanitation, hygiene and cooking suggested by Gleick (1996).

Total groundwater abstraction is the sum of water used for irrigation through DTWs, STWs, and the water used by the population. For groundwater abstraction from tubewells, it is assumed that tubewells are used for 12 h for 120 d/y.

A. Total annual abstraction from deep tubewells for irrigation

The total number of deep tubewells as of FY 2012/13 was 37.

The discharge of water from deep tubewells ranged from 20 to 40 L/s (Bhandari, 2013) with average discharge of 30 L/s (108,000 L/h) = 108 m^3/h

Annual groundwater abstraction by deep tubewells = no. of wells × discharge × 12 × 120

= 37 × 108 × 12 × 120 = 5,754,240 m^3

B. Total annual abstraction from shallow tubewells for irrigation

The total number of shallow tubewells as of FY 2012/13 was 2903.

The discharge of water from shallow tubewells ranged from 6 to 10 L/s (Bhandari, 2013) with average discharge of 8 L/s (291,600 L/h) = 28.86 m^3/h)

Annual groundwater abstraction by shallow tubewells = number of wells × discharge × 12 × 120

= 2903 × 28.86 × 12 × 120 = 120,393,216 m^3

C. Total annual abstraction for domestic use

Total population of Chitwan (CBS, 2011) = 579,984

Per capita demand of water per day = 0.045 m^3/d

Therefore, total amount of annual groundwater withdrawal by the population

= population × per capita demand of water per day × 365

= 579984 × 0.045 × 365

= 9,526,237 m^3

D. Total annual groundwater abstraction for various uses

Total annual volume of groundwater abstraction (D) = (A) + (B) + (C)

= (5,754,240 + 120,393,216 + 9,526,237) m^3

= 135,673,693 m^3

= 135.673 MCM or 135.7 MCM (approximately)

CHAPTER 5

Groundwater Environment in Delhi, India

Aditya Sarkar*, Shakir Ali*, Suman Kumar*, Shashank Shekhar*, and SVN Rao**
*Department of Geology, University of Delhi, India
**WAPCOS, Regional office, Hyderabad, India

5.1 INTRODUCTION

Delhi, the capital city of India, is the third largest city in the country by area and the second largest by population. It supports a population of over 16.7 million (Census of India, 2011). The city has a long history of political dominance by various dynasties. For a long time, its relevance as the political center of India led to the migration of different communities from all over the country, shaping its cultural heritage.

The Delhi region is part of the Indo–Gangetic alluvial plains with the Yamuna River flowing eastward through it (Figure 5.1). The plain is at an elevation of approximately 220 m above mean sea level. However, its maximum and minimum elevation levels can differ by as much as 60 m (Singh, 1999). The Yamuna River enters Delhi at Palla Village to the north and leaves at Okhla Barrage and is approximately 35 km long. It has 97 km^2 of active floodplain bounded by bunds (embankments), a zone with lots of potential from the groundwater point of view (CGWB, 2006a).

The National Capital Territory (NCT) of Delhi and its adjacent regions have traditionally faced water scarcity. Several water storage and groundwater recharge structures, some dating back to the eleventh century, can still be seen around Delhi (CSE, 2011). However, the last six decades have seen unprecedented population growth and extensive urbanization of the city, leading to a severe shortage of freshwater resources. The Delhi Jal Board (DJB), the agency responsible for water supplies in Delhi, provides approximately 650 million gallons a day (MGD) to the city. However, water demand is about 900 MGD and thus there is a shortage of 250 MGD (Shekhar and Prasad, 2009). This shortage has come about as a result of rapid migration of people into the city. The demand for water in the city is fulfilled by developing groundwater and diverting surface water from neighboring states.

Groundwater Environment in Asian Cities
http://dx.doi.org/10.1016/B978-0-12-803166-7.00005-2

Copyright © 2016 Elsevier Inc.
All rights reserved.

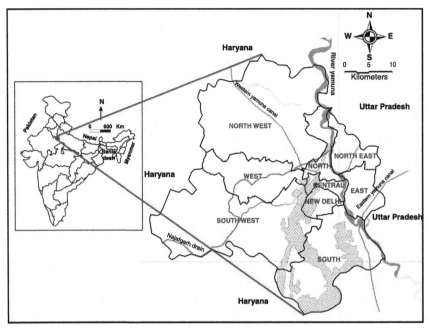

Figure 5.1 *Location of Delhi City and its Adjoining Areas.*

At present, Delhi City has limited surface water resources for drinking purposes and its groundwater resources are being widely abstracted to meet industrial, agricultural, and domestic needs. The large-scale abstraction of groundwater has not only disturbed the demand–supply balance of the resource, but has also raised serious questions about the quality and quantity of groundwater. Today, various critical issues are associated with water resources in Delhi including overexploitation and pollution of groundwater resources. Delhi has been categorized as overexploited in terms of groundwater development (Chatterjee et al., 2009). In light of many disturbing factors, it is important to get an understanding of the overall picture of the hydrogeological conditions as well as the driving forces, pressures, current state, impacts, and responses of the groundwater environment in the territory to prepare for sustained utilization and management of groundwater resources in the long term.

A number of earlier studies and government agencies have described the prevailing hydrogeological conditions in and around the city, and attempted to establish a comprehensive view on groundwater resource management to sustain the ever-growing need for fresh water. This chapter aims to present a comprehensive view on the groundwater environment of Delhi.

5.2 ABOUT THE CITY

5.2.1 Physiography and Climate

The NCT of Delhi is located between 28°24′17″ N to 28°53′00″ N and 77°50′24″ E to 77°20′37″ E. It encompasses nine revenue districts, and shares boundaries with two neighboring states: Haryana to the north, west, and south; and Uttar Pradesh to the east (Figure 5.1). Delhi City covers an area of 1483 km².

The climate of the city is humid subtropical to semiarid with generally dry winters extending from November to January (DES, 2014). The hot and humid summer period from April to July is followed by the monsoon season in July and August, characterized by heavy rainfall with winds blowing from the Arabian Sea. The average temperature of the city varies from 25°C to 45°C from April to July and drops from 22°C to 5°C in December and January (DES, 2014).

5.2.2 Morphology and Geology

Kaul and Pandit (2004) identified four major geomorphological units in this region (namely, the older floodplain, the active floodplain, the peneplains, and upland areas), while Chatterjee et al. (2009) discuss six major geomorphic units (Figure 5.2). The older floodplain in the northern and eastern parts of the city has geomorphic features such as paleochannels, marshes, and meander cutoffs, while southern upland areas are characterized by strike ridges, dissected hills, badlands with rills, and ravines (Kaul and Pandit, 2004). Quartzitic rocks are exposed in the central and southern parts of the Delhi region.

The Delhi Ridge, a prominent geological feature of the region, is the northernmost extension of the Aravali mountain range and extends for about 35 km along the southern border of Delhi, ending at the north on the west bank of the Yamuna River (Figure 5.2). It is a NNE–SSW-trending quartzitic ridge that cuts through the alluvial plains (Sett, 1964). Sediments occurring on the Delhi Ridge could be loess deposits of eolian origin (Tripathi and Rajamani, 1999).

The stratigraphic succession of the Delhi region (Table 5.1) shows that the major rock type occupying the region consists of quartzite, interbedded with minor schists that belong to the Delhi Supergroup.

These quartzites are gray to brownish-gray in color with thin to massively thick bedding, doubly plunging toward northeast and southwest as "a coaxially refolded regional anticline" (CGWB, 2006a). Post-Delhi intrusion of quartz veins has been reported in a few places in these rocks (Thussu, 2006). These Proterozoic quartzites act as the basement for the Delhi region (Chatterjee et al., 2009).

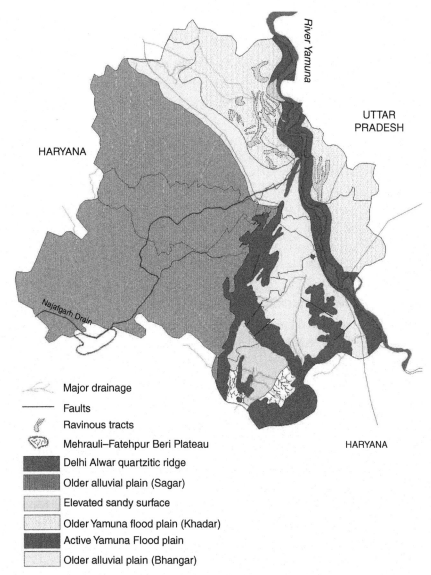

Figure 5.2 *Geological Map of Delhi. Adapted from Chatterjee et al. (2009).*

5.2.3 Hydrogeology

The Delhi Supergroup is unconformably overlain by unconsolidated sediments from the Quaternary period to recent times (Table 5.1). These unconsolidated sediments are broadly grouped as older and younger alluvium. The older alluvium mainly consists of clay admixed with silt and kankar (calcareous

Table 5.1 Stratigraphic succession of rocks in Delhi

Holocene	Yamuna channel alluvium	Gray, fine to medium sand, grit with coarse sand, silt, and clay	Point bars and channel deposits
	Yamuna older floodplain and terraces	Gray sand, coarse grit, pebble beds, and minor clay	Paleochannels, abandoned channels, meander scrolls, oxbow lakes
	Older alluvium	Sequences of sand–silt–clay with yellowish brown medium sand with silt, kankar with brown eolian sand	Abandoned channels, meander scrolls

No sedimentation

Neoproterozoic	Post-Delhi intrusive	Pegmatite, tourmaline–quartz veins, and quartz veins	
Mesoproterozoic	Delhi Supergroup	Quartzite with minor schists, tuff, and ash beds (Alwar group)	

Source: Thussu (2006).

concretions), while the younger alluvium consists of sand admixed with silt, clay, and gravel (Shekhar and Prasad, 2009).

The 97 km^2 active floodplains of the Yamuna River bounded by bunds comprise coarser sediments such as sand, gravel, and pebbles (CGWB, 2006a). They are therefore very important from the groundwater perspective. The younger alluvial floodplains have unconfined aquifers with fresh water zones overlying saline water (Shekhar, 2006; Rao et al., 2007). However, the older ones comprise relatively finer sediment (silt), and the presence of calcareous concretions distinguishes them from the younger alluvium (Shekhar and Prasad, 2009).

Geological cross-sections across South Delhi (Figure 5.3) expose quartzitic ridges that rise to about 300 m above mean sea level (masl) in some locations and drop to 100 m or so below ground level (mbgl) (CGWB, 2006a). They also reveal that the Chhattarpur Basin is filled with clay, silt, and kankar nodules and intercalated granular zones of sand (Figure 5.3).

Another geological cross-section of Delhi from the west to east (Figure 5.4) reveals a quartzitic ridge separating a predominantly sandy active floodplain from older alluvial plains of silt or clay. The thickness of unconsolidated

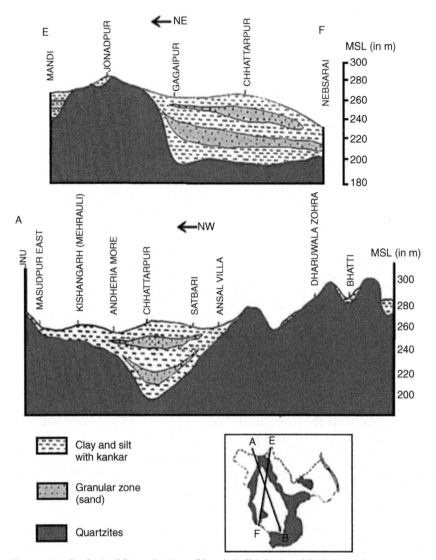

Figure 5.3 *Geological Cross-Section of South Delhi. Source: CGWB (2006a).*

sediments overlying the basement rock seems to gradually increase away from the ridges (Figure 5.4).

The cross-section also reveals that the bedrock seems to have a steeper slope toward the western flank from the central Delhi Ridge than the eastern flank in the trans-Yamuna area.

Another geological cross-section across the floodplain of the Yamuna River (Figure 5.5) reveals a quartzitic ridge separating a predominantly

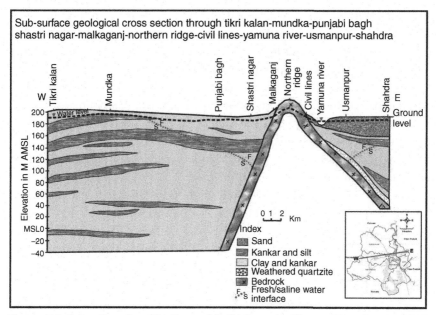

Figure 5.4 *Subsurface Cross-Sections of Delhi from the West to East.* Source: CGWB (2006a).

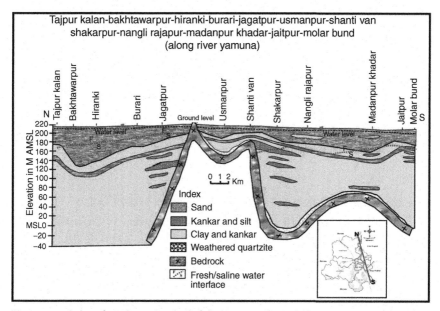

Figure 5.5 *Subsurface Cross-Section of the Yamuna River.* Source: CGWB (2006a).

sandy active floodplain downstream to Jagatpur and upstream to Usmanpur. The subsurface profile also shows the much deeper fresh–saline water interface south of the Delhi Ridge from Shantivan to the Molar Bund with respect to Burari and Jagatpur, north of the ridge (Figure 5.5).

The presence of groundwater has been reported in all the formations found in the Delhi region (Sett, 1964).This groundwater is controlled by hydrogeomorphic units in the region such as the rocky tracts of the ridge, pediments, alluvial uplands, valley fill, and floodplains (Bajpai, 2011). However, the alluvial floodplain of Yamuna is regarded as the most proficient area of subsurface fresh water resources in the city.

Shekhar et al. (2009) suggested high transmissivity and discharge for the younger alluvium (Table 5.2). In the floodplains of the Yamuna River, the freshwater aquifer ranges from 30 mbgl to 70 mbgl. The aquifer system in the alluvial plains can be classified into three groups based on the behavior of aquifer material and its depth: (i) Group I, which is unconfined and extends up to 60 mbgl from the surface; (ii) Group II, which is confined or semiconfined in nature, and comprises sandy horizons between 65 mbgl and 200 mbgl; and (iii) Group III, which is confined in nature and found in horizons between 200 mbgl and >300 mbgl (CGWB, 2012a).

An unpublished study on floodplain aquifers revealed that after 72 h of continuous pumping (Figure 5.6), 77% of water pumped was recovered in 48 h.The transmissivity value estimated using Neuman's approach was 1371 m^3/d and the specific yield was 0.24.

Lying above the bedrock beneath Delhi City are groundwater aquifers with alluvial deposits. The bedrock has an elevation range from about 190 to above 300 masl (Figure 5.7). The basement consists mainly of fractured and weathered quartzite hard rock (Bajpai, 2011). Wells sunk in the weathered and fractured aquifers of hard rock generally penetrate down to 80 mbgl but in some places they reach 150 mbgl (CGWB, 2012a).

Table 5.2 Groundwater features for geological formations in Delhi, India

Nature of formation	Depth of the well (mbgl)	Discharge (m^3/h)	Drawdown (m)	Transmissivitiy (m^2/d)
Younger alluvium	40–60	50–180	5.0–8	600–2000
Older alluvium	30–60	20–60	6.0–24	130–403
Quartzite	50–150	2–10	6.0–30	5–135

Source: Shekhar et al. (2009).

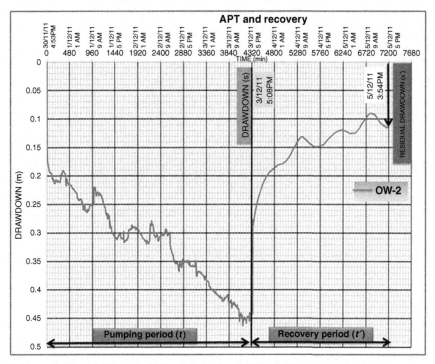

Figure 5.6 *Plot of 4320 min (t) Drawdown Data Followed by 2880 min (t′) of Recovery Data for a Well Pumped at the Rate of 1373 lpm (1.373 m³/min) in the Palla Wellfield.*

5.3 DRIVERS

The three key drivers contributing to changes in the groundwater environment for the NCT of Delhi are population growth, urbanization, and climate. The drivers and corresponding pressures for the groundwater environment of current state, impacts, and responses are quantified and tabulated in Table 5.3.

5.3.1 Population Growth

Population growth in Delhi has always been a critical factor for city planners and water supply organizations in the city. The earliest organized recorded census of the city was conducted in 1881 during the British era (Census of India, 2011). This practice was continued over the years as part of the decadal Census of India. The latest census report (Census of India, 2011) shows rapid growth in the population of the city over the years (Figure 5.8 and Table 5.3), mainly due to migration from

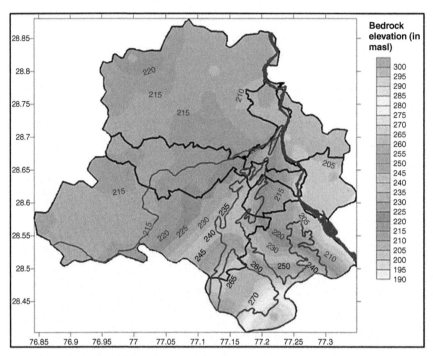

Figure 5.7 *The Bedrock Elevation Map of Delhi. Source: Shekhar and Sarkar (2013).*

other states. This is also reflected in the fact that the trend toward urban population growth in Delhi since 1901 has been much faster than in rural areas (Figures 5.8 and 5.9). Population growth is growing so fast that it has become a major concern for the city. It exerts pressure on the water supply, and thus a large amount of groundwater needs to be abstracted to fulfill various needs.

5.3.2 Urbanization

The city has seen a rapid rise in the rate of urbanization in the last 50 years. Until 1991 most of the city's population dwelt in rural areas (Figure 5.9). However, around 75% of the land in Delhi in the last few years has been urbanized to accommodate the increasing population (Figure 5.9). Urban areas in the city covered approximately 1113.65 km² in 2011, representing a significant increase over the years from 685.34 km² in 1991 (Table 5.3). This has resulted in the construction of many multistory buildings in the city. Furthermore, urbanization has led to changes in lifestyle, thus increasing water demand in the city.

Table 5.3 Status of DPSIR indicators in Delhi, India

Indicators		Year	1911	1961	2011
Drivers	Population ground: total population (million) (Census of India, 2011) (Figure 5.8)	Year	0.413	2.658	16.75
	Highly dense district (Census of India, 2011)				Northeast
	Highly populated district Census of India (2011)				Northwest
	Urbanization (Department of Planning, 2014) (km²) (Figure 5.9)	Year	1991	2001	2011
		Rural	797.66	558.32	369.35
		Urban	685.34	924.68	1113.65
	Climate change (DES, 2010, 2014)	Year	1990	2001	2011
		T_{max} (°C)	39.8 (June)	38.8 (May)	41 (May)
		T_{min} (°C)	8.3 (December)	6.7 (January)	6.4 (January)
		Total rainfall (mm) (Figure 5.10)	1119.2	693.8	708.1

(Continued)

Table 5.3 Status of DPSIR indicators in Delhi, India *(cont.)*

	Indicators		2005–2006	2011–2012
Pressures	Land use/land cover change (Figure 5.11)	Year	2005–2006	2011–2012
		Buildup	52.77%	54.32%
		Agricultural area	38.06%	36.76%
		Forest cover	2.05%	1.94%
		Barren lands	4.76%	4.62%
	Surface water and groundwater use (Department of Planning, 2012)(in BCM)	Delhi water requirement is fulfilled partially by the Bhakhra Beas Management Board (140 MGD), Yamuna (310 MGD) and Ganga (240 MGD). Also, 115 MGD of groundwater is abstracted to meet the water requirement		
State	Well statistics (Department of Planning, 2014)	2636 tubewells and 21 Ranney wells* for groundwater abstraction by the DJB		
	Groundwater abstraction (CGWB, 2006b, 2011) (in BCM)	Assessment year	2004	2009
		Irrigation	0.2	0.14
		Domestic and industrial	0.28	0.26
	Groundwater level	Water level ranges from 2 mbgl to over 70 mbgl (Figure 5.13)		
	Groundwater quality	Delhi is affected mainly by fluoride, nitrate, lead, chromium, cadmium, and high salinity (Table 5.4 and Figure 5.14)		
	Groundwater recharge (in BCM) (CGWB, 2006b, 2011)	Assessment year	2004	2009
		Rainfall (for both monsoon and nonmonsoon seasons)	0.15	0.13
		Other sources (canal, water bodies)	0.15	0.18
		Total groundwater recharge	0.30	0.31

	Depletion in groundwater level	The general decline in groundwater level in Delhi ranged between 4 m and 20 m from 1960 to 2000. (Figure 5.15)
Impacts	Degradation in groundwater quality	Fresh water resources are reducing due to groundwater pollution and high salinity in shallow groundwater (Figure 5.17)
Responses	Institutional arrangement	Several major government agencies such as the CGWB, DJB, and CPCB are responsible for the availability and portability of groundwater in the NCT of Delhi. The DJB is an institution that insures sustainable use of groundwater resources and control of groundwater abstraction within sustainable limits
	Groundwater monitoring (CGWB, 2012a)	162 groundwater-monitoring stations managed by the CGWB
	Groundwater regulation	The CGWA initially earmarked two districts of Delhi and Yamuna floodplains for regulated groundwater development. It further earmarked the entire NCT of Delhi for regulated groundwater development. Table 5.5 lists a few regulatory orders

BCM, billion cubic meters; CGWA, Central Groundwater Authority; CGWB, Central Ground Water Board; CPCB, Central Pollution Control Board; DJB, Delhi Jal Board; DPSIR, driver–pressure–state–impact–response framework; GW, groundwater; MGD, million gallons per day; NCT, National Capital Territory.
*Ranney wells are large-diameter structures with horizontal slotted pipes. From a few Ranney wells in Delhi as much as 4 MGD of groundwater are abstracted.

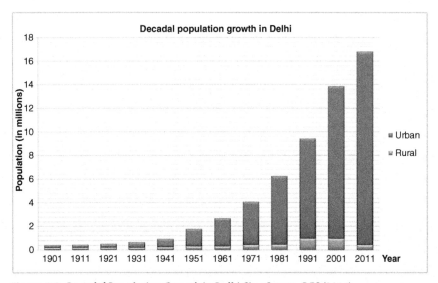

Figure 5.8 *Decadal Population Growth in Delhi City. Source: DES (2014).*

5.3.3 Climate Change

The climatic conditions in Delhi are extreme, with daily average tempera-tures ranging from lows of 5°C in winter to highs of 45°C in summer (DES, 2014). Annual rainfall in the city has fluctuated widely over the years

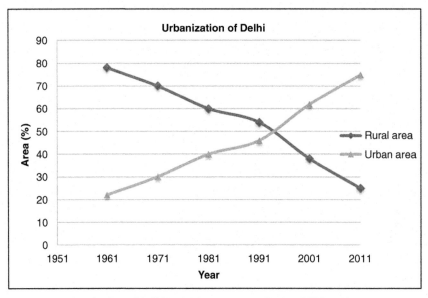

Figure 5.9 *Urbanization of Delhi in the Last 50 years. Source: DES (2014).*

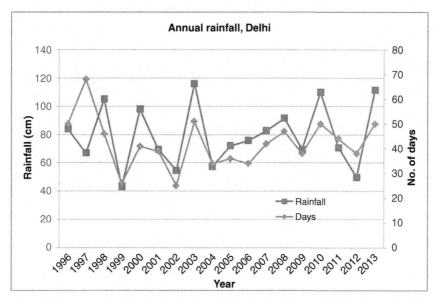

Figure 5.10 *Annual Rainfall and Number of Rainy Days. Source: DES (2014).*

(Figure 5.10). Rainfall has decreased considerably over the years. Rainy days have also shown a decreasing pattern in the past decade (Department of Planning, 2014; Table 5.3). Such a decrease in rainfall reduces groundwater recharge, and seriously impacts the groundwater environment of the city.

5.4 PRESSURES

5.4.1 Land Use/Land Cover (LULC)

Changes in LULC for the three seasons – kharif (monsoon), rabi (winter), and zaid (summer) – for 2005–2006 and 2011–2012 are shown in Figure 5.11. LISS-III data resolution of 23.5 m was used for LULC classification. Since migration to the city has grown extensively, built-up areas have increased from 52.77% to 54.32%, while forest cover has decreased from 2.05% to 1.94% (Table 5.3 and Figure 5.11) (NRSC, 2015). No significant changes can be seen in water bodies, rivers, streams, or canals.

LULC change puts both direct and indirect pressure on the groundwater system. On the one hand, conversion of forest and agriculture into urban areas reduces recharge and subsequently reduces the amount of groundwater recharge; on the other hand, water-intensive urban areas abstract more groundwater than their rural counterparts, and therefore deplete groundwater reserves. All these factors exert pressure on the groundwater environment.

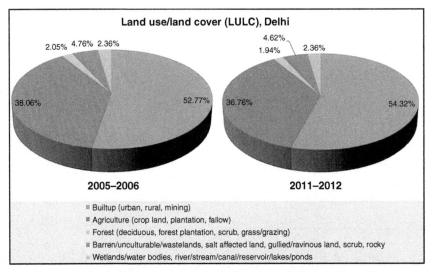

Figure 5.11 *LULC of Delhi, India. Source: NRSC (2015), available from: www.bhuvan.nrsc. gov.in.*

5.4.2 Surface Water and Groundwater Use

Major sources of groundwater recharge in the city include rainfall and seepage of water from canals, rivers, and irrigated water. DJB, the main water supply agency in Delhi, has reported that the water supply for the city has grown from 66 MGD in 1956 to 805 MGD in 2012 (Department of Planning, 2012). This includes 140 MGD of water from Bhakra Beas Management Board, 310 and 240 MGD from the Yamuna and Ganga rivers, respectively, and 115 MGD from groundwater abstraction. However, as per the norms of the Central Public Health and Environmental Engineering Organization, Ministry of Urban Development, Government of India, the estimated water requirement for Delhi is 1020 MGD (Department of Planning, 2014).

Population growth is the major concern of the city and requires not only huge abstraction of groundwater but also the use of available surface water. Increased water demand creates conflicts since the requirement is generally hard to achieve. Groundwater development, assessed on the basis of budgeting, also indicates the same.

The concept of groundwater budgeting involves quantifying all outflows (represented by annual groundwater draft for irrigation, domestic, and industrial use), all inflows (represented by net annual groundwater availability), and system storage changes over a specified period of time (Hutchison

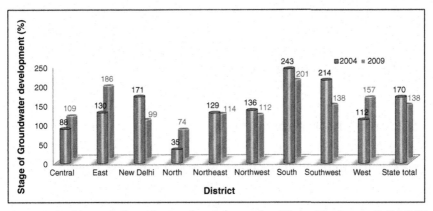

Figure 5.12 *Stage of Groundwater Development for All the Districts in Delhi NCT. Source: Chatterjee et al. (2009); CGWB (2011).*

and Hibbs, 2008; Chatterjee et al., 2009). Total annual groundwater draft for Delhi in 2004 was 479.45 million cubic metres (MCM), while annual replenishable groundwater resources were reported to be 281.56 MCM (Chatterjee et al., 2009). The groundwater resource estimation methodology defines groundwater development (percentage of total annual groundwater draft with respect to annual replenishable groundwater resource) as safe when it is less than or equal to 90% without any significant long-term decline in groundwater level (GEC, 1997). Groundwater development for 2004, however, ranged from 35% to 243% (Figure 5.12). All nine districts of the NCT of Delhi, except for the central and northern districts, fell into the overexploitation category (Chatterjee et al., 2009).

Groundwater development has increased over the years in central, eastern, northern, and western districts and decreased in the remaining five districts, ranging from 74% to 201% (Figure 5.12). In 2009 only the northern district was safe in terms of groundwater development (CGWB, 2011). Despite increasing demand for groundwater, groundwater development for the NCT of Delhi decreased from 170 to 138% from 2004 to 2009 (Chatterjee et al., 2009; CGWB, 2011).

5.5 STATE

5.5.1 Well Statistics

The Central Ground Water Board (CGWB) has established 162 hydrograph monitoring stations across the city, while the DJB has installed 2636 tubewells and 21 ranney wells (see Table 5.3 for its definition) for groundwater abstraction (Department of Planning, 2014). Most ranney wells are located in the Yamuna River vicinity. Around 120 tubewells and 5 ranney wells are operational in the Palla area of North Delhi. These Palla wells together supply approximately 30 MGD of drinking water.

5.5.2 Groundwater Abstraction

The DJB has estimated that about 115 MGD of groundwater is being exploited through their ranney wells and tubewells (Department of Planning, 2014). The remaining deficiency between water demand and supply is covered by groundwater abstraction through private tubewells and borewells. About 461,000 of the 3.34 million households in Delhi abstract groundwater, which helps bridge the gap between water demand and supply (Census of India, 2011). However, it has been observed that total annual groundwater draft has declined from 0.48 billion cubic meters (BCM) in 2004 to 0.4 BCM due to reduced groundwater abstraction for irrigation as well as domestic and industrial purposes (CGWB, 2006b; CGWB, 2011; and Table 5.3).

5.5.3 Groundwater Level

The groundwater level contours of the Delhi region are controlled by a natural aquifer. The groundwater level in the alluvial plains ranges from 192 masl to 216 masl (Figure 5.13) while the general groundwater level contour in the Delhi Ridge and its adjacent area is around 240 masl (Figure 5.13). A major water level depression can be observed in the southeastern part of the city (surrounded by the Delhi Ridge on three sides) with a general water level elevation of 170–180 masl. Groundwater level depressions can also be observed in the southwest district on both sides of the Najafgarh drain (Sahibi River) with a general water level elevation of 192 masl (Figure 5.13). The regional water trough near Kharkhari (west of the Najafgarh drain) and Papankalan (east of the Najafgarh drain) is the result of heavy groundwater abstraction in the southwest district (Sarkar and Shekhar, 2015). These depressions have a strong influence on groundwater flow, as can be seen by the convergence of groundwater toward the Najafgarh drain in the southwestern part of Delhi (Shekhar and Sarkar, 2013). Groundwater on the left side in this stretch of the drain seems to flow away from the drain, while groundwater on the right side flows toward the drain (Figure 5.13).

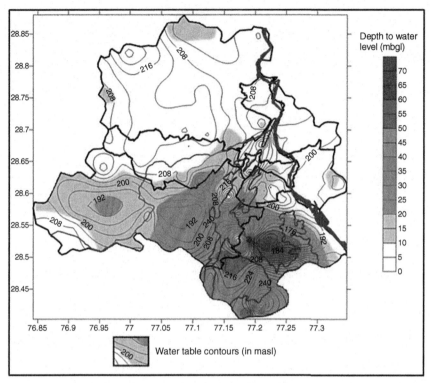

Figure 5.13 *The Groundwater Level Contour and Depth to Water Level Map of Delhi.*
Source: Based on CGWB dataset. Available from: http://gis2.nic.in/cgwb/Gemsdata.aspx.

5.5.4 Groundwater Quality

The groundwater chemistry in Delhi has shown variations both spatially and temporally. There are geogenic and anthropogenic sources of contamination influencing groundwater quality across the city. However, primary control on groundwater quality variation is geogenic. Geomorphology impacts the groundwater dynamics of the city, which in turn controls the groundwater chemistry. Younger alluvium found in the Yamuna floodplains and hard-rock ridge contains good volumes of fresh water. In the rest of the area, shallow groundwater is fresh and salinity increases with depth.

Lithology is another geogenic factor that plays a dominant role in the hydrochemical groundwater process in Delhi. The lithology of the region is known to control the major-ion chemistry of the groundwater system (Subramanian and Saxena, 1983). The weathering of carbonate concretions in parts of older alluvial plains, for example, is known to affect the evolution of hydrochemical facies of groundwater (Datta and Tyagi, 1996; Sarkar and

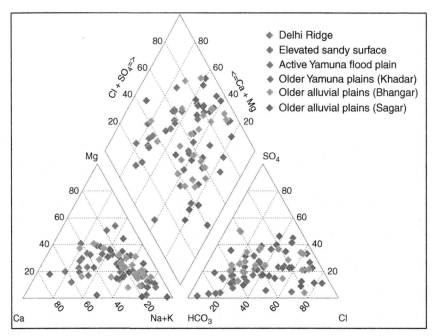

Figure 5.14 *Hydrochemical Facies in the Delhi Region for 2010. Source: CGWB (2012b).*

Shekhar, 2015). Hydrochemical facies of groundwater in the Delhi region (Figure 5.14) reveal them to be predominantly controlled by both changing geomorphology and anthropogenic influences. Facies variation in the alluvial plains clearly indicates that in moving laterally away from the Yamuna River, hydrochemical facies also evolve from bicarbonate-dominant facies in active floodplains into chloride-type facies in older alluvial plains (Figure 5.14).

Hydrochemical facies in active floodplains are known to be bicarbonate in shallower aquifers (Sarkar and Shekhar, 2013). It has also been observed that general hydrochemical facies in the Delhi Ridge area and its elevated sandy surface (Chhattarpur Basin) were of $Ca–Mg–HCO_3$ type, which could account for continuous recharge in parts of these areas (Sarkar and Shekhar, 2015). The presence of high concentrations of sulfate and chloride ions in the groundwater system could be attributed to bed seepage from drains and wastewater (Subramanian and Saxena, 1983). In areas of low salinity, a $Ca–Mg–HCO_3$-type hydrochemical facies has been reported that is affected by local recharge (Kumar et al., 2006; Sarkar and Shekhar, 2015).

General assessment of the groundwater quality for all districts of Delhi reveals the presence of groundwater contaminants such as fluoride, nitrate, and lead in some localities. (Table 5.4). The groundwater quality of the

Table 5.4 Maximum reported contaminants in different districts of Delhi

District	Fluoride (mg/L)	Nitrate (mg/L)	Lead (mg/L)*	Cadmium (mg/L)*	Chromium (mg/L)*
Northwest	7.24 (Khera Kalan)	710 (Rani Khera)	0.08 (Puthkalan)	0.017 (Bhalswa new landfill)	7.85 (Bhalswa old landfill)
North	3.08 (Majnu Ka Tila)	91 (Majnu Ka Tila)	2 (ST Nehru Vihar)	0.01 (Timarpur and ST Nehru Vihar)	0.051 (Nehru Vihar Mother Dairy)
Northeast	0.69 (Gokulpuri)	38 (Gokulpuri)	—	0.004 (Maujpur)	0.024 (Maujpur)
East	0.89 (Akshardham Temple)	306 (Krishna Kunj)	0.06 (Mayur Vihar and Chander Vihar)	0.012 (Ghazipur Dairy)	1.95 (Ghazipur Dairy)
New Delhi	0.75 (SS Bhawan)	533 (India Gate)	0.01 (Indraprastha)	0.003 (Lodhi Gardens)	0.015 (IP estate)
Central	0.94 (Parsi Dharmashala)	75 (Parsi Dharmashala)	0.04 (Jama Masjid)	0.008 (Jama Masjid)	0.068 (Karol Bagh)
West	4.32 (Tilangpur Kotla)	1500 (Tikri Kalan)	0.08 (Hiran Kudna, Nihal Vihar, Tilangpur Kotla)	0.006 (Hiran Kudna)	0.049 Hiran Kudna)
Southwest	8.04 (Dhansa)	575 (Raota)	0.076 (Naraina Phase 2)	0.09 (Shikarpur)	0.046 (Chhawla and Dwarka, Sector 19)
South	1.9 (Humayun Tomb)	1112 (Hauz Khas)	0.077 (Okhla, Phase 2)	0.003 (Jaitpur Extension)	0.071 (Jaitpur Extension)

*Compiled by the Central Ground Water Board (CGWB) from unpublished data gathered between 2003 and 2011.
Source: CGWB (2012b).

western district seems to be more problematic than that in other districts of Delhi.

5.5.5 Groundwater Recharge

A study conducted by the CGWB estimates that total surplus runoff of 457 MCM is available for groundwater recharge. This is total surface runoff generated in Delhi of 175 MCM combined with 282 MCM of surplus monsoon runoff available from the Yamuna River (CGWB, 2013). The topography of the city is a major factor controlling groundwater recharge. It has been observed that the general elevation of recharging sites varies from the northwest to southwest (Datta and Tyagi, 2009). Groundwater recharge in topographically low-lying areas of Delhi is sourced from surface runoff (Datta et al., 1996a). Seepage of surface water bodies is another source of groundwater recharge in Delhi (Datta et al., 1996a). Total groundwater recharge in Delhi, estimated as the sum of recharge from rainfall and other sources, was 300 and 310 MCM for 2004 and 2009, respectively (CGWB, 2006b; CGWB, 2011).

To store surplus runoff and aquifer recharge, the concerned authorities have constructed a number of rainwater-harvesting and artificial recharge structures in such locations as Rastrapati Bhavan, Lodhi Gardens, Shram Shakti Bhawan, and the Indian Institute of Technology Delhi (CGWB, 2008b). The CGWB estimates that the installation of 125,000 rooftop rainfall-harvesting systems in city buildings suitable for artificial recharge could recharge 14.69 MCM of rainwater to groundwater in the case of normal rainfall with 80% efficiency (CGWB, 2013). Additionally, the construction of four check dams in the Sanjay Van area, adjoining the Jawaharlal Nehru University (JNU) has resulted in a reservoir capacity of 49.05 MCM (CGWB, 2008a; Sarkar and Shekhar, 2015). This has led to an increase in groundwater levels for JNU of up to 2.55 m in the postmonsoon season (Chatterjee et al., 2009).

5.6 IMPACTS
5.6.1 Decline in Groundwater Level

The effect of heavy groundwater abstraction and changes in groundwater flow patterns in the city are reflected by changes in groundwater levels over the years (Figure 5.15).

Comparison of general groundwater levels between 2000 and 1960 (Figure 5.15) shows that declines in northern parts of the city range from 4 to 8 m, while in the south they are reported to be more than 20 m

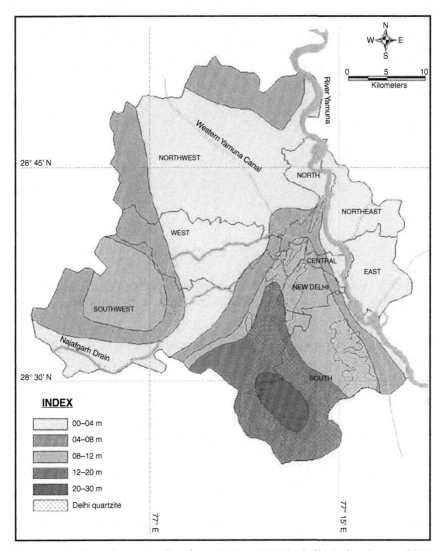

Figure 5.15 *Groundwater Decline from 1960 to 2000 in Delhi, India. Source: CGWB (2006c).*

(CGWB, 2006c). Localities close to surface water bodies in East Delhi (near the Yamuna River), parts of North Delhi (near the Western Yamuna Canal), and West Delhi (near the Najafgarh drain) show a minimum decline of less than 4 m.

Analysis of hydrographs at different locations across Delhi reflects decadal fluctuation in city groundwater levels (Figure 5.16). In the younger alluvial

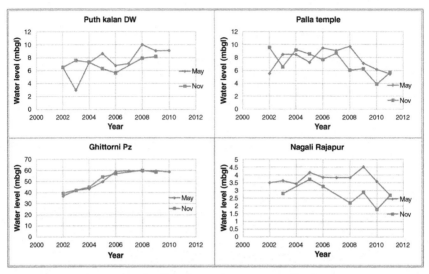

Figure 5.16 *Hydrographs of Selected Locations across Delhi. Source: Based on CGWB dataset (CGWB 2015a). Available from: http://gis2.nic.in/cgwb/Gemsdata.aspx.*

plains of the Yamuna, such as Palla and Nagali Rajapur, the groundwater levels for premonsoon (May) and postmonsoon (November) periods have been observed to fluctuate between 4–10 mbgl and 1.5–4.5 mbgl, respectively (Figure 5.16).

There has been a declining trend in groundwater level in Puth Kalan, a census town in the older alluvial plains. The rate of decline there is greater in the premonsoon period than in the postmonsoon period (Figure 5.16). The hydrograph for Ghitorni, a census town in pediments adjacent to the ridge, also shows a declining trend in groundwater level. However, this trend is gradual for both premonsoon and postmonsoon periods, with groundwater declining from 40 mbgl in 2002 to around 60 mbgl in 2010 (Figure 5.16).

5.6.2 Degradation in Groundwater Quality

Groundwater salinity is a common problem in Delhi. The depth to the interface between fresh and saline water (Figure 5.17), which also varies with contours in the groundwater level, has been fixed on the basis of varying electrical conductivity ranging from 1500 μS/cm to 2000 μS/cm for groundwater depth (Shekhar et al., 2005). The maximum depth of the fresh–saline interface is found in parts of the south and southwest districts of Delhi around the Chhattarpur Basin and in the area adjacent to Delhi

Ridge (Shekhar et al., 2009). The depth of the interface in the area around Delhi Ridge ranges from 50 mbgl to 100 mbgl (Figure 5.17). In other parts of Delhi, the depth of the fresh water interface is found to be less than 40 mbgl (Figure 5.17). Highly saline groundwater in the city is mainly linked to water logging, low-lying areas, and discharge zones (Shekhar et al., 2015). Changes in groundwater dynamics are also known to trigger salinity enrichment in groundwater, as observed in parts of Delhi and the adjoining state of Haryana. Irrigation and water logging in these areas also causes an increase in chloride and nitrate ion concentrations in groundwater (Lorenzen et al., 2012).

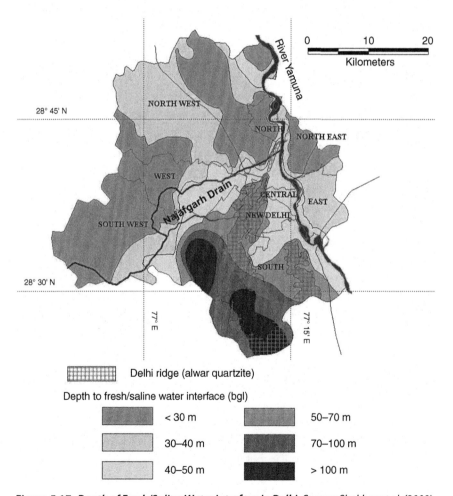

Figure 5.17 *Depth of Fresh/Saline Water Interface in Delhi. Source: Shekhar et al. (2009).*

Anthropogenic factors should not be neglected when considering changes in the groundwater quality of the city. Industrial areas located in the western, southern, and eastern parts of the city as well as the three coal-driven thermal power plants (Badarpur, Indraprastha, and Rajghat) in the city (CPCB, 2008) are among the major anthropogenic sources of groundwater contamination. Industrial and domestic effluents are known to increase nitrate concentrations in groundwater in some pockets of the west, southwest and northwest districts of Delhi (Table 5.4). These nitrate plumes in groundwater in the southern and northern parts of the city are indicative of the trend toward urbanization of the center (Datta et al., 1997).

High concentrations of arsenic in groundwater have been reported at a few locations across Delhi. They are attributed mainly to anthropogenic sources such as garbage from open landfill sites (Lalwani et al., 2004). Arsenic contamination in the Yamuna floodplains could be linked to anthropogenic contamination due to arsenic-rich coal ash from thermal power stations in Delhi (Dubey et al., 2012).

High concentrations of fluoride in the groundwater of Delhi are mainly found in the southwest (Shekhar et al., 2006) and along parts of the west and northwest districts (CGWB, 2006a). There is more fluoride concentration in the saline and brackish water zones than in the entire city (CGWB, 2012b). Fluoride enrichment in groundwater is usually associated with natural factors such as climate and the weathering of fluorine-bearing minerals (apatite, fluorite, fluorapatite, etc.) found in various rocks and soils (Handa, 1975; Pickering, 1985). The safe limit for fluoride in water given by the World Health Organization (WHO) as 1 mg/L, while the permissible limit of the Bureau of Indian Standards (BIS) is 1.5 mg/L (WHO, 1984; BIS, 2012). However, the sources of fluoride contamination in groundwater in Delhi are reported to be anthropogenic in nature. The contribution of fluoride in groundwater is believed to be from point sources such as brick industries in Delhi as well as nonpoint sources such as irrigation and surface water runoff (Datta et al., 1996b). Stable isotopic (^{18}O) studies investigating the origin of fluoride in the groundwater of Delhi indicate that fluoride leaching from the soil and lateral mixing of groundwater are responsible for such contamination (Datta et al., 1996b).

Heavy/trace metals in the groundwater of Delhi have also been reported in the past (Subramanian and Saxena, 1983). However, contamination of groundwater by trace metals in recent years has become a source of concern. Industrial waste in the Yamuna River catchment and Najafgarh drain system is a likely source of heavy metal contamination in groundwater

(Rawat et al., 2003; Shekhar and Sarkar, 2013). The use of water from the Yamuna River and its associated drainage systems for irrigation of agricultural lands leads to an accumulation of heavy metals such as chromium, copper, zinc, lead, nickel, and cadmium as a result of industrial waste and sewage in the subsurface (Singh, 2001; Kaur and Rani, 2006; Bhattacharya et al., 2015).

In addition to inorganic contaminants from natural and industrial sources, another major cause of groundwater contamination is the excessive usage of fertilizers and pesticides in the region.

An assessment of groundwater contamination by fertilizers in the Delhi area shows that nitrate and potassium concentrations in groundwater have exceeded general acceptability limits as a consequence of the regular application and indiscriminate discharge of waste into drains (Datta et al., 1997).

5.7 RESPONSES

Water resource exploitation in Delhi, particularly groundwater, has triggered serious socioeconomic and ecological problems in the city. The relevant authorities in Delhi have addressed these issues by taking appropriate steps to regulate and monitor water resources in the city.

5.7.1 Institutional Arrangements

Several major government agencies – such as the CGWB, DJB, and Central Pollution Control Board (CPCB) – are involved in insuring the availability and portability of groundwater in the NCT of Delhi. While the CPCB and CGWB monitor the quality of surface water bodies and groundwater on a regular basis, the DJB's primary responsibility is to regulate the supply of water to localities in Delhi. The DJB is therefore the appropriate institution to ensure sustainable use of groundwater resources and has the ability to control groundwater abstraction within sustainable limits.

5.7.2 Groundwater Monitoring

The CGWB has set up a large number of groundwater-monitoring stations across Delhi to observe both the quantitative and qualitative aspects of groundwater systems (Figure 5.18). It has 162 stations (Table 5.3) across Delhi to monitor groundwater levels. These include 25 dugwells and 137 piezometers (CGWB, 2012a). Groundwater quality in the first aquifer is generally monitored by the CGWB every year during the premonsoon season (Figure 5.18). Additionally, depending on the circumstances, sporadic

Figure 5.18 *Groundwater-Monitoring Stations Maintained by the CGWB. Source: CGWB (2008b).*

checking of thematic groundwater quality like arsenic fingerprinting is also carried out. In addition, CPCB has set up 15 monitoring stations for surface water bodies including 4 for the Yamuna River, 2 for canals, and 9 for drains (CPCB, 2010), which help in assessing vulnerable stretches.

5.7.3 Groundwater Regulation

Deteriorating groundwater conditions in many parts of India led to the establishment of a new regulatory body: the "Central Groundwater Authority (CGWA)." It was created under Section 3(3) of the Environment (Protection) Act (1986) with the primary objective of "regulating and controlling development and management of groundwater resources in the country"

Table 5.5 Notable groundwater regulations in Delhi

Notice number	Issuing authority	Subject
Public Notice No. 6 of 2000	CGWA	Regulation on construction of tubewells/borewells in South and Southwest districts of Delhi
Public Notice No. 9 of 2000	CGWA	Earmarking the Yamuna floodplain area within the NCT of Delhi as a "notified area"
CGWA order No. 4/4/ CGWA/2004-893	CGWA	Restriction on withdrawal of water for domestic, commercial, and industrial purposes in the entire NCT of Delhi

(CGWB, 2015b). Based on scientific assessment of groundwater reserves, the authority initially earmarked two districts of Delhi and the Yamuna floodplain as notified areas for regulated groundwater development (Table 5.5). Subsequently, the regulatory role was taken on by relevant state authorities of the government of the NCT of Delhi. As soon as the state government agency was set up, the CGWA further earmarked the entire NCT of Delhi as a notified area for regulated groundwater development (Table 5.5).

Regulatory measures have resulted in a decline in total annual groundwater draft from 480 MCM in 2004 to 400 MCM in 2009, as mentioned earlier.

5.8 SUMMARY

The groundwater environment in Delhi is largely affected by rapid urbanization and industrialization. This is further aggravated by regular fluctuations in rainfall and changes in land use patterns. The needs of Delhi's ever-increasing population have led to overexploitation of aquifers. The industrialization of Delhi has also led to contamination of aquifers by waste. The overall groundwater condition shows a decline in groundwater levels and deteriorating groundwater quality in most parts of the city. However, controlled groundwater abstraction imposed by regulators has reduced overall groundwater abstraction in the last few years. Rainwater harvesting and artificial recharge have failed to stop declining water levels. However, there are a few success stories regarding water quality improvement. The present situation warrants getting all concerned stakeholders — such as the DJB, CGWB, CPCB, and local municipal authorities — to work together to provide holistic remedial measures.

ACKNOWLEDGMENTS

A.S. acknowledges the Council of Scientific and Industrial Research (CSIR) for a financial grant toward a senior research fellowship. S.A. acknowledges the Department of Geology, University of Delhi for a non-NET scholarship. S.K. thanks the Department of Geology, University of Delhi for a CAS fellowship. The authors acknowledge Dr. Vishnu Prasad Pandey for his critical comments, which helped improve the chapter.

REFERENCES

Bajpai, V.N., 2011. Hydrogeological studies in the National Capital Territory of Delhi with reference to land use pattern and effective groundwater management. Proc. Indian Natl. Sci. Acad. 77, 31–49.

Bhattacharya, A., Dey, P., Gola, D., Mishra, A., Malik, A., Patel, N., 2015. Assessment of Yamuna and associated drains used for irrigation in rural and peri-urban settings of Delhi NCR. Environ. Monit. Assess. 187 (1), 1–13.

BIS, 2012. Indian Standard Specification for Drinking Water. BIS 10500.

Census of India, 2011. Available from: http://censusindia.gov.in/ (accessed 14.05.2015).

CGWB, 2006a. Hydrogeological Framework and Groundwater Management Plan of NCT Delhi, Report. Central Ground Water Board, Delhi, Ministry of Water Resources, Government of India.

CGWB, 2006b. Dynamic Groundwater Resources of India. Central Ground Water Board, Ministry of Water Resources, Government of India, Faridabad, Haryana.

CGWB, 2006c. Ground Water Year Book, 2005–2006. National Capital Territory, Central Ground Water Board, New Delhi.

CGWB, 2008a. Success Stories of Rainwater Harvesting and Artificial Recharge of Ground Water in NCT, Delhi, Report. Central Ground Water Board, Delhi, Ministry of Water Resources, Government of India.

CGWB, 2008b. Groundwater Year Book, 2006–2007. National Capital Territory, Central Ground Water Board, New Delhi.

CGWB, 2011. Dynamic Groundwater Resources of India. Central Ground Water Board, Ministry of Water Resources, Government of India, Faridabad, Haryana.

CGWB, 2012a. Ground Water Year Book, 2011–2012. National Capital Territory, Central Ground Water Board, Ministry of Water Resources, Government of India, New Delhi.

CGWB, 2012b. Groundwater Quality of NCT Delhi. Central Ground Water Board, Ministry of Water Resources, Government of India, New Delhi.

CGWB, 2013. Master Plan for Artificial Recharge to Ground Water in India. Central Ground Water Board, Ministry of Water Resources, Government of India, New Delhi.

CGWB, 2015a. Groundwater Information System. Available from: http://gis2.nic.in/cgwb/Gemsdata.aspx (accessed 23.04.2015).

CGWB, 2015b. Central Groundwater Authority. Available from: http://cgwb.gov.in/aboutcgwa.html (accessed 2.06.2015).

Chatterjee, R., Gupta, B.K., Mohiddin, S.K., Singh, P.N., Shekhar, S., Purohit, R., 2009. Dynamic groundwater resources of National Capital Territory, Delhi: assessment, development and management options. Environ. Earth Sci. 59, 669–686.

CPCB, 2008. Status of groundwater quality in India, Part II, Groundwater Quality Series: GWQS/10/2007–2008. Central Pollution Control Board, Ministry of Environment & Forests, Government of India.

CPCB, 2010. Status of water quality in India – 2009, Monitoring of Indian Aquatic Resources Series: MINARS/2009–10. Central Pollution Control Board, Ministry of Environment & Forests, Government of India.

CSE, 2011. Excreta Matters, vol. 2: 71 Cities: A Survey. Centre for Science and Environment, New Delhi.

Datta, P.S., Tyagi, S.K., 1996. Major ion chemistry of groundwater in Delhi area: chemical weathering processes and groundwater flow regime. J. Geol. Soc. India 47, 179–188.

Datta, P.S., Tyagi, S.K., 2009. Delineation of potential groundwater recharge and productive aquifer zones in Delhi area based on ^{18}O signatures and GPS. J. Agric. Phys. 9, 33–37.

Datta, P.S., Bhattacharya, S.K., Tyagi, S.K., 1996a. ^{18}O studies on recharge of phreatic aquifers and groundwater flow paths of mixing in the Delhi area. J. Hydrology 176, 25–36.

Datta, P.S., Deb, D.L., Tyagi, S.K., 1996b. Stable isotope (^{18}O) investigations on the processes controlling fluoride contamination of groundwater. J. Contam. Hydrol. 24, 85–96.

Datta, P.S., Deb, D.L., Tyagi, S.K., 1997. Assessment of groundwater contamination from fertilizers in the Delhi area based on ^{18}O, NO_3 and K^+ composition. J. Contam. Hydrol. 27, 249–262.

Department of Planning, 2012. Issues and challenges for Twelvth Five Year Plan – 2012–17. Government of NCT of Delhi, India. Available from: http://www.delhi.gov.in/ (accessed 14.05.2015).

Department of Planning, 2014. Economic survey of Delhi, 2012–13. Government of NCT of Delhi, India. Available from: http://delhi.gov.in/wps/wcm/connect/DoIT_Planning/planning/misc./economic+survey+of+delhi+2012–13 (accessed 14.05.2015).

DES, 2010. Statistical Abstract of Delhi, Report. Directorate of Economics & Statistics, Delhi. Available from: http://www.delhi.gov.in/ (accessed 14.05.2015).

DES, 2014. Statistical Abstract of Delhi, Report. Directorate of Economics & Statistics, Delhi. Available from: http://www.delhi.gov.in/ (accessed 14.05.2015).

Dubey, C.S., Mishra, B.K., Shukla, D.P., Singh, R.P., Tajbakhsh, M., Sakhare, P., 2012. Anthropogenic arsenic menace in Delhi Yamuna flood plains. Environ. Earth Sci. 65, 131–139.

GEC, 1997. Ground Water Resources Estimation Methodology, Report. Ground Water Resources Estimation Committee, Ministry of Water Resources, Government of India.

Handa, B.K., 1975. Geochemistry and genesis of fluoride containing groundwaters in India. Groundwater 13275–13281.

Hutchison, W.R., Hibbs, B.J., 2008. Groundwater budget analysis and cross-formational leakage in an arid basin. Groundwater 46 (3), 384–395.

Kaul, B.L., Pandit, M.K., 2004. Morphotectonic evaluation of the Delhi region in northern India, and its significance in environmental management. Env. Geol. 46, 1118–1122.

Kaur, R., Rani, R., 2006. Spatial characterization and prioritization of heavy metal contaminated soil-water resources in peri-urban areas of National Capital Territory (NCT) Delhi. Environ. Monit. Assess. 123, 233–247.

Kumar, M., Ramanathan, A.L., Rao, M.S., Kumar, B., 2006. Identification and evaluation of hydrochemical processes in the groundwater environment of Delhi, India. Environ. Geol. J. 50, 1025–1039.

Lalwani, S., Dogra, T.D., Bhardwaj, D.N., Sharma, R.K., Murty, O.P., Vij, A., 2004. Study on arsenic level in groundwater of Delhi using hydride generator accessory coupled with atomic absorption spectrometer. Ind. J. Clin. Biochem. 19, 135–140.

Lorenzen, G., Sprenger, C., Brandon, P., Gupta, D., Pekdeger, A., 2012. Origin and dynamics of groundwater salinity in the alluvial plains of western Delhi and adjacent territories of Haryana State. India. Hydrol. Process 26, 2333–2345.

NRSC, 2015. Bhuvan, National Remote Sensing Center. Available from: http://bhuvan.nrsc.gov.in/ (accessed 17.05.2015).

Pickering, W.F., 1985. The mobility of soluble fluoride in soils. Environ. Pollut. 9, 281–308.

Rao, S.V.N., Kumar, S., Shekhar, S., Sinha, S.K., Manju, S., 2007. Optimum pumping from skimming wells from the Yamuna river flood plain in north India. Hydrogeol. J. 15, 1157–1167.

Rawat, M., Moturi, M.C.Z., Subramanian, V., 2003. Inventory compilation and distribution of heavy metals in wastewater from small-scale industrial areas of Delhi, India. J. Environ. Monit. 5, 906–912.

Sarkar, A., Shekhar, S., 2013. An assessment of groundwater quality of lesser contaminated aquifers in North District of Delhi. Proc. Indian Natl. Sci. Acad. 79 (2), 235–243, 2013.

Sarkar, A., Shekhar, S., 2015. The controls on spatial and temporal variation of hydrochemical facies and major ion chemistry in groundwater of South West District, Delhi, India. Environ. Earth Sci.doi: 10.1007/s12665-015-4399-2.

Sett, D.N., 1964. Groundwater geology of the Delhi region. Bull. Geol. Surv. India Ser. B 16, 1–35.

Shekhar, S., 2006. An approach to interpretation of step drawdown tests. Hydrogeol. J. 14, 1018–1027.

Shekhar S, Purohit R, Kaushik YB 2009. Groundwater management in NCT Delhi, Technical paper included in the special session on Ground water in the 5th Asian Regional Conference of INCID, December 9-11, 2009 held at Vigyan Bhawan, New Delhi, available online at http://www.cgwb.gov.in/documents/papers/INCID.htmlShekhar. (accessed 27.08.2015).

Shekhar, S., Prasad, R.K., 2009. The groundwater in the Yamuna flood plain of Delhi (India) and the management options. Hydrogeol. J. 17, 1557–1560.

Shekhar, S., Sarkar, A., 2013. Hydrogeological characterization and assessment of groundwater quality in shallow aquifers in vicinity of Najafgarh drain of NCT Delhi. J. Earth Syst. Sci. 122 (1), 43–54, 2013.

Shekhar, S., Singh, S.B., Romani, S., 2005. The controls to the variation in depth to fresh/saline interface in the groundwater of South-West District, NCT Delhi – a case study. J. Geol. Soc. India 66, 17–20.

Shekhar, S., Mohiddin, S.K., Singh, P.N., 2006. Variation in concentration of fluoride in the groundwater of South-West district, NCT Delhi – A case study. In: Ramanathan, A.L., Bhattacharya, P., Keshari, A.K., Bundschuh, J., Chandrashekharan, D., Singh, S.K. (Eds.), Assessment of groundwater resources, management. I K International, New Delhi, pp. 370–376.

Shekhar, S., Mao, R.S.K., Imchen, E.B., 2015. Groundwater management options in North District of Delhi, India: a groundwater surplus region in over-exploited aquifers. J. Hydrol. Reg. Stud., http://dx.doi.org/10.1016/j.ejrh.2015.03.003.

Singh, R.B., 1999. Urban impacts on groundwater quality in the Delhi region. IAHS 259, 227–231.

Singh, M., 2001. Heavy metal pollution in freshly deposited sediments of the Yamuna River (the Ganges River tributary): a case study from Delhi and Agra urban centres, India. Environ. Geol. 40, 664–671.

Subramanian V., Saxena K.K., 1983. Hydrogeochemistry of groundwater in the Delhi region of India, Relation of Groundwater Quantity and Quality (Proceedings of the Hamburg Symposium, August 1983). IAHS Publ. no. 146: 307–316.

Thussu, J.L., 2006. Geology of Haryana and Delhi. Geological Society of India, Bangalore.

Tripathi, J.K., Rajamani, V., 1999. Geochemistry of the loessic sediments on Delhi ridge, eastern Thar Desert, Rajasthan: implications for exogenic processes. Chem. Geol. 155, 265–278.

WHO, 1984. Guidelines for Drinking Water Quality (Vol. II): Health Criteria and Supporting Information. World Health Organization, Geneva, Switzerland.

CHAPTER 6

Groundwater Environment in Hyderabad, India

Shashidhar Thatikonda
Department of Civil Engineering, Indian Institute of Technology, Hyderabad, India

6.1 INTRODUCTION

Hyderabad, a city with a history of more than 400 y, attracts extensive domestic and international tourists. The uncontrolled and haphazard growth of Hyderabad in the past four decades has drastically changed land use patterns, resulting in greatly reduced natural recharge to groundwater. In the recent past, Hyderabad City developed as an information technology (IT) hub. The establishment of many national and international IT companies and institutions has put tremendous pressure on the city's groundwater resources. In March 2015, groundwater levels in most parts of the city were at an all-time low even though the city received more seasonal rain than normal. Groundwater contributes about 25–30% of the city's total water requirement. Hyderabad has a hard-rock aquifer and reduced recharge. Extensive pumping has created water stress even at depths of the aquifer system ranging from 100 m to 300 m (CGWB, 2013). Hundreds of lakes in and around the city, which in the last few decades served as the main source of groundwater recharge, no longer exist. Over 200 water bodies within the jurisdiction of the Hyderabad Metro Development Authority have been encroached upon as water resources. The effects of urbanization and industrialization in Hyderabad have led to contamination of the aquifer. Due to an inadequate sewerage system and treatment capacity, domestic sewage and industrial effluents flow directly into streams, causing severe groundwater contamination. The Musi River, the main source of water in the area, receives estimated untreated sewage of 500 million liters a day (MLD), and more than 70% of the city's groundwater samples show nitrate concentrations above the maximum permissible limit (MPL) of safe drinking water standards (CGWB, 2011). At present, the city has a water deficit of about 400 MLD, which is going to increase to 585 MLD by 2031 (Anon., 2011). To compensate for current and future water requirements, the government has initiated state-scale megaprojects such as

Groundwater Environment in Asian Cities
http://dx.doi.org/10.1016/B978-0-12-803166-7.00006-4

Copyright © 2016 Elsevier Inc.
All rights reserved.

the Krishna Water Supply Project and Pranahita Chevella Project, which are currently in progress. Both central and state groundwater departments monitor groundwater levels and assess the groundwater resources of the city on a continuous basis. The government has initiated several environmental guidelines and standards to regulate domestic and industrial discharges, but proper enforcement is required to avoid illegal discharge. Steps toward rainwater harvesting, wastewater reuse, and monitoring of groundwater withdrawals have been taken at individual house/apartment level for effective water utilization to avoid further overexploitation of the resource. Several research organizations and groundwater departments have initiated studies to understand the hydrogeology of the city's aquifer on a regional scale since the 1970s; however, due to the high heterogeneity and anisotropic nature of the aquifer systems, more studies at a finer resolution are necessary. Considering the precarious situation, groundwater development and management has to be implemented by fully understanding the sociohydrology of the city.

Attempts are being made in this context to develop a systematic knowledge base of Hyderabad's groundwater environment by separating drivers (D), pressures (P), states (S), impacts (I), and responses (R)(DPSIR). The study results would be useful to all stakeholders, including policy makers, in understanding the extent of the situation and could serve as a basis in decision making for sustainable management of the city's groundwater resources.

6.2 ABOUT THE CITY

Hyderabad, a city in the southern part of India (Figure 6.1), became the capital of newly formed Telangana State on June 2, 2014. It is one of the fastest growing metropolitan areas and ranks as the sixth largest urban

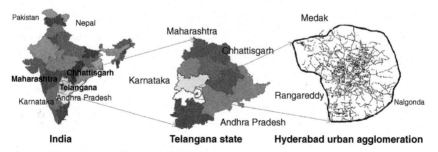

Figure 6.1 *Location of Hyderabad in India.*

agglomeration in the country. It is situated at 17°22′N and 78°27′E. Its undulating topography ranges between m 460 and 560 m above mean sea level (masl). In the recent past, Hyderabad emerged as one of the strong industrial, commercial, and technology hubs in the world. In December 2014 to January 2015 issue of the *National Geographic Traveler* magazine, Hyderabad was ranked second in a list of the best places to visit in the world. The city's population has grown rapidly through extensive migration from other parts of India as a result of IT sector development.

In a span of 90 y (1921–2011), the city has expanded haphazardly from 84 km² to the present area of 650 km². The present Hyderabad Urban Agglomeration (HUA) comprises the Municipal Corporation of Hyderabad (MCH), 10 peripheral municipalities, 8 panchayats in the Ranga Reddy District, and 2 municipalities in the Medak District, covering an area of 1864 km² under the jurisdiction of the Hyderabad Urban Development Authority (HUDA) (Prakash and Singh, 2013).

Hyderabad receives 75% of its 780 mm average annual rainfall during the southwest monsoon (January to September). It experiences a warm and temperate climate under semiarid tropical conditions. About 70% of the city's total water requirements are met by surface water and the remaining 30% by groundwater. The Hyderabad Metropolitan Water Supply and Sewerage Board (HMWSSB) supplies about 1545 MLD of surface water and 114 MLD of groundwater to the city. This water is mainly piped from the Manjeera and Krishna rivers some 100 km outside the metropolitan area.

The city is located on the Deccan Plateau. Some 97% of aquifer material is anisotropic and heterogeneous in nature consisting of mostly pink and gray granites with a weathered zone varying from 5 m to 25 m and discharges generally varying from negligible to 5 liters per second (lps) (Figure 6.2a,b). The remaining 3% of the area is underlain by alluvium. High-density fractures are observed in eastern, western, and northern parts of the area, while moderate to low-density fractures are observed in the central part of the main city area. In general, shallow fractures are more productive than deeper ones. However, the deeper fractures (127 and 172 m) in some of the western parts are more productive (with yields of 6 lps at Film Nagar and 10 lps at Borabanda) than the shallow fractures, indicating structural disturbances in some of the western parts.

The alluvial formations, though negligible both in areal and vertical extension, occur as isolated patches along the Musi River. This consists of medium to fine-grained sand, silt with thicknesses varying from a few meters

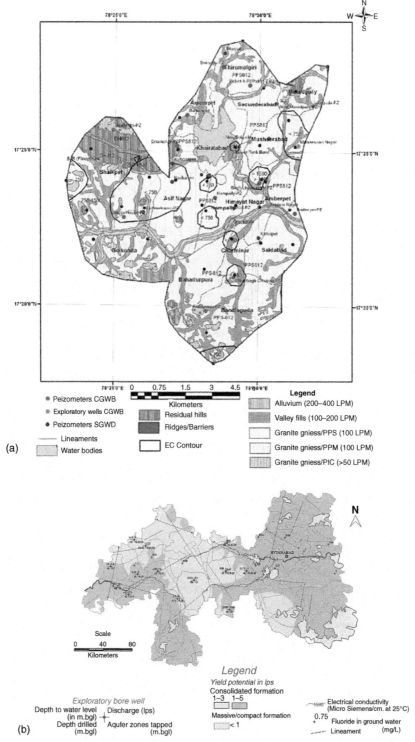

Figure 6.2 *Hydrogeology of Aquifer Systems in the GHMC.* (a) in the Hyderabad District (CGWB, 2013); (b) in the Ranga Reddy District (CGWB, 2007).

to about 5 m. Groundwater occurs under phreatic conditions. Groundwater was previously developed through large-diameter dugwells, but this is presently not the case due to contamination by pollutants from Musi River effluent (CGWB, 2011; http://www.waterandmegacities.org/hydrogeology-hyderabad-by-cgwb-india/).

Average well yields range from 60 to 90 liters per minute (lpm) and aquifer transmissivity ranges from 0.5 m^2/d to 200 m^2/d. Soil cover ranges from 0.25 m to 2.0 m with mostly red lateritic, yellow sandy–clay loam, and alluvial black soils (CGWB, 2011). An increase in the number of residential and commercial apartments has reduced the rechargeable area and completely overexploited groundwater resources in the city. Direct discharge of sewage and industrial effluent has polluted all previously useful surface water bodies in and around the city. Illegal discharge of various industrial effluents has also deteriorated the quality of precious groundwater resources in all industrial zones. From 1973 to 2013, most of the 934 lakes disappeared and 2.5% of the geographical area containing water bodies reduced to less than 1.5%.

6.3 DRIVERS

Population growth and urbanization are the main drivers behind increased structural, environmental, social, and economic challenges facing the city. Tourism and industrialization have also become drivers influencing the state of the groundwater environment. Recently, numbers of domestic and foreign tourists have increased dramatically as a result of the influence of IT sectors and improvements to recreational facilities such as Film City and Water World. All these factors have played parts in exerting more pressure on the city's groundwater environment (Table 6.1).

6.3.1 Population Growth

The very act of becoming the capital of Andhra Pradesh in 1956 triggered migration into the city. In six decades, the twin cities of Hyderabad and Secunderabad have spread from 172 km^2 to 1864 km^2, with the population exploding significantly (Table 6.1). According to 2011 statistics, the population density in the Greater Hyderabad Municipal Corporation was 18,489 persons/km^2. The highest density was observed in the Municipal Corporation of Hyderabad followed by the surrounding municipalities. Ward-wise density varied from above 76,000 persons/km^2 to less than 10,000 persons/km^2 (Figure 6.3).

Table 6.1 Status of DPSIR indicators in Hyderabad, India

	Indicators	1991	2001	2011
Drivers	*Population growth*			
	Population in millions★(Jawaharlal Nehru National Urban Renewal Mission, 2015; DCO, 2011) (within HUA area of 650 km²)	4.344	5.752	6.810
	Decadal growth rate (%) (Jawaharlal Nehru National Urban Renewal Mission, 2015; DCO, 2011)	70.66	32.41	18.39
	Population density (persons/km²)	5,978	7,391	18,489
	Urbanization			
	Urban population in millions (within HUDA area of 1864 km²)(Jawaharlal Nehru National Urban Renewal Mission, 2015; DCO, 2011)	4.670	6.380	7.749
	Urban area (km²)	55 km² in 1869; 84 km² in 1921; 172 km² in 1950; 650 km² in 2007(within GHMC), 1864 km² (within HUDA) (HMWSSB, 2015)		
	Tourism (India Tourism Statistics, 2010 and 2015)			
	Number of tourist arrivals (million): increased from 4.34 to 8.53 during 2010–2011			
	Number of hotel wells: 467 (in 2010) excluding company guest houses, guest houses run by trusts, dormitories, free dharamshalas, tourist bungalows, and free accommodation units			
	Industrialization			
	Most enterprises were established in the 1950s and 1960s. Since the 1990s various diversified service industries such as IT enterprises, biotech, insurance, and finance have grown. Of every 1000 people of working age, 770 males and 190 females are employed in industry as per 2005 statistics. Most of the water demands of those industries are met by groundwater abstracted from various aquifer depths.			

Pressures	*Inadequate surface water resources* As a result of various drivers, water demand is increasing but surface water is limited. To cater for this demand, the government has already initiated harvesting water from the KWSP and GDWP. Phase I and Phase II of the KWSP are delivering 820 MLD and Phase III is in progress to add a further 410 MLD water to city water resources.			
	Land cover change (%)(Jawaharlal Nehru National Urban Renewal Mission, 2015)			
	Built-up (residential/commercial) area (km²)	32.31	37.35	47.35 (in 2009)
	Agricultural areas (km²)	9.36	9.36	5.44 (in 2009)
	Nonagricultural areas (km²)	90.64	90.64	94.56 (in 2009)
	GW overexploitation (CGWB, 2013; CGWB, 2007)			
	Annual GW availability (ha–m)	1400 in 2004–2005, 906 in 2008–2009		
	GW withdrawal (ha–m)	12,099 in 2004–2005, 7370 in 2008–2009		
	GW balance (ha–m)	−10,699 in 2004–2005, −6,464 in 2008–2009		
	Annual GW recharge (ha–m)	1007 (in 2008–2009)		
	Provision for natural discharge (ha–m)	101 (in 2008–2009)		
State	*Well statistics* Number of wells (in 2010): dugwells = 18,416; shallow tubewells = 47,860; deep tubewells = 2,085 Well density: 5 km⁻² in 1982 and 20 km⁻² in 2012			
	GW level (CGWB, 2011) The average postmonsoon GW level ranges from 3.30 m (Manikeswarinagar) to 56.3 m (Film Nagar). The discharges in these wells vary from 0.21 lps to 6.9 lps with drawdowns of 6.0 m–20.6 m. The specific capacity of wells ranges from 10 lpm/m (New Boiguda) to 72 lpm/m (Borabanda) and transmissivity from 0.48 m²/d to 202 m²/d. In general, depths to water level during premonsoon and postmonsoon periods are 5–20 mbgl and 2–15 mbgl, respectively (Figure 6.5).			

(Continued)

Table 6.1 Status of DPSIR indicators in Hyderabad, India (cont.)

Indicators		1991	2001	2011
	GW abstraction 12,099 ha-m in 2004/2005 and 7370 ha-m in 2008/2009			
	GW quality (CGWB, 2011) EC (μs/cm): min = 200; max = 7500; general range: 700–3000; samples exceeding PML = 2% TH (mg/L): min = 70; max = 2780; general range: 150–1000; samples exceeding PML = 29% NO_3 (mg/L): min = 1.2; max = 760; general range: 10–300; samples exceeding PML = 72% SO_4 (mg/L): min = 4.8; max = 1392; general range: 25–460; samples exceeding PML = 2% Almost all the IDAs have a few industries, which are potential sources of heavy metal contamination of the environment. Concentration of Fe and Mn in some wells exceeds the MPL. Lead, nickel, zinc, and cadmium were detected in the IDAs of Balanagar, Sanatnagar, and Jeedimetla. Lead and cadmium were beyond the MPL.			
Impacts	*Depletion in GW level* (CGWB, 2011; CGWB, 2013) GW level is depleted over the years at many places like Boinpalli, Aghapura, Erragadda, Bashherbagh, Langar house, Jubilee hills, Begumpet, Koti, West Maredpalli, Gudimalkapur, Mushherabad, Sanath nagar, Picket, and Madhapur. GW level on March 28, 2012 was 25.35 mbgl, at Marredpally and 20.4 mbgl at Malkajgiri, a fall of 9.22 m since February 2011. In Nacharam, the level was 11.85 mbgl, higher than 7.74 mbgl observed in February 2011.			
	Land subsidence (CGWB, 2013) It is completely hard rock region and hence there is no possibility of land subsidence.			
	Public health (CGWB, 2011) GW caters 30% of the city needs. About 72% GW samples show a high NO_3, which may lead to methamanoglobinameia in childrens. All GW samples in IDAs contain heavy metals above MPL. Several carcinogenic cases were observed in these areas.			

Responses	*GW monitoring*
	A network of dugwells in both command (1404 wells) and noncommand areas (1582 wells) was established in 1974, and the GW level is being monitored on a bimonthly basis. The network has been expanded further in subsequent years.
	Environmental standards and guidelines
	The MoEF has set the MPL by developing several environmental standards and guidelines such as Water (Prevention and Control of Pollution) Act 1974 and Environment Protection Rules, 1986. However, implementation remains poor in practice.
	Megaprojects for water supply (HMWSSB, 2015)
	Megaprojects such as KWSP and GDWP are under progress to utilize surface source and reduce pressure on the groundwater environment. The KWSP alone will add 1230 MLD by 2021 and 1955 MLD by 2031 to the city's water resources.

GHMC, gram panchayats with the Municipal Corporation of Hyderabad; GDWP, Godvari Drinking Water Project; GW, groundwater; ha-m, hectare meters; HUA, Hyderabad Urban Agglomeration; HUDA, Hyderabad Urban Development Authority; IDA, Industrial Development Area; IT, information technology; KWSP, Krishna Water Supply Project; lps , liters per second; lpm , liters per minute; mbgl, meters below ground level; MLD, million liters a day; MoEF, Ministry of Environment and Forest; MPL, maximum permissible limit.

GHMC (2015)

*Decadal geometric growth rate $r = (P_{t+1} - P_t / P_t)$, where P_t = population at time t y from the base period, P_{t+1} = population at time $t + 1$ y from the base period.

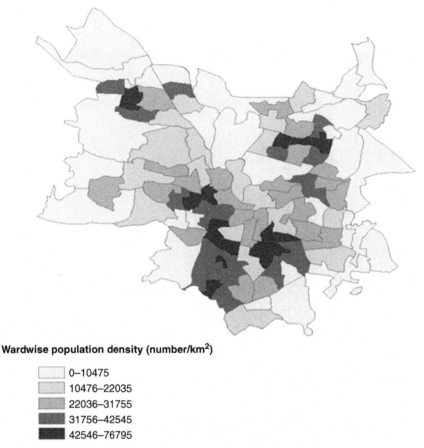

Wardwise population density (number/km²)

- 0–10475
- 10476–22035
- 22036–31755
- 31756–42545
- 42546–76795

Figure 6.3 *Spatial Distribution of Population Density in the Wards of HUA in 2011.*

6.3.2 Urbanization

Hyderabad, dubbed the City of Pearls, was founded in 1591 by Moham-
med Quli Qutub Shah. In 1869 Sir Salar Jung I, the then Prime Minister
of Hyderabad under the Nizam, constituted the Department of Municipal
and Road Maintenance and appointed a Municipal Commissioner. At that
time the city covered just 55 km² and had a population of 0.35 million. By
1921 the Hyderabad Municipality had increased to 84 km². In 1955 the
municipal corporations of Hyderabad and Secunderabad merged to form
the MCH. Hyderabad became the capital of Andhra Pradesh in 1956 and
on 2 June 2014 the state was split and Hyderabad became part of the present capital of the new state Telangana. The twin cities of Hyderabad and

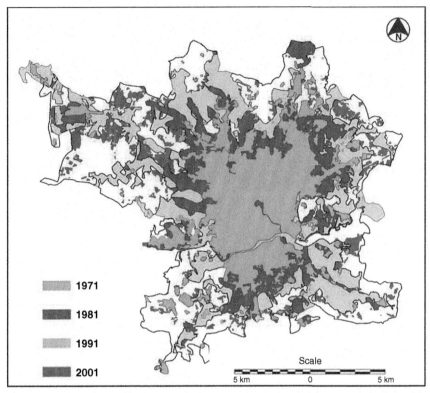

Figure 6.4 *Urban Sprawl in the HUA during 1971 to 2001. Source: Jawaharlal Nehru National Urban Renewal Mission (2015).*

Secunderabad in 1950 had a population of 4.5 million living in an area of 172 km². The new urban agglomeration sprawls across 650 km² and has a population of 6.7 million (Figure 6.4). The Greater Hyderabad Municipal Corporation (GHMC) was formed in 2007 by merging 12 municipalities and 8 gram panchayats.

The present HUA covers 1864 km². Most of the open areas around the widely scattered buildings have now been converted into apartments due to the heavy influx of people migrating from all corners of the state and country. During the last four decades, the residential area has increased from 22% to 50% and scrub and grassland have decreased to 2 from 56% (Figure 6.5). Built-up areas have continuously increased since 1971 (Table 6.2). This drastic change in land use patterns has greatly reduced the natural recharge of groundwater, and most of the 932 tanks listed in the 1973 survey records have disappeared in and around Hyderabad and their geographic area has reduced to less than 1.5 from 2.5%. From 1955 to 2009

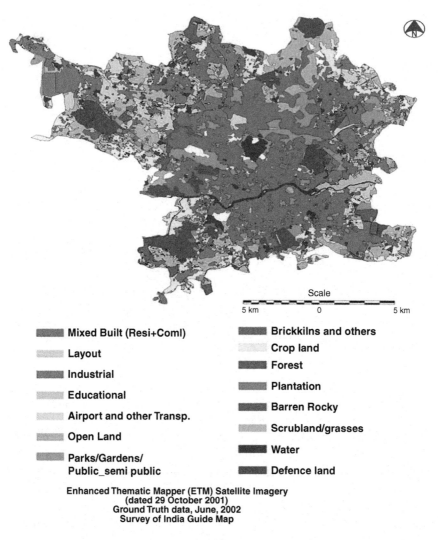

Figure 6.5 *Land Cover in the Hyderabad Urban Agglomeration 2001. Source: Jawaharlal Nehru National Urban Renewal Mission (2015).*

the net area irrigated by tanks around Hyderabad further reduced to 7.1% (i.e., from 21,564 ha to 1535 ha) (Sreedhar et al., 2009).

6.3.3 Tourism

Hyderabad is a historical city located in the southern part of India with excellent flight connectivity to the rest of the world. It has a richly mixed

Table 6.2 Land cover (%) in Hyderabad Urban Agglomeration (HUA) from 1971 to 2009

Land cover per year	1971	1981	1991	2001	2009
Mixed built (residential and commercial)	22.56	32.11	32.31	37.35	47.35
Layout	0.64	0.58	1.33	5.5	
Airport and other transportation	0.2	0.81	0.48	0.48	
Industrial	0	0	3.84	5.58	
Crop land	6.43	6.21	9.36	9.36	5.44
Plantation	0	0	1.19	1.19	
Forest	0.15	0.15	1.44	1.44	
Parks/gardens/public–semipublic	2.77	2.77	7.07	7.07	
Barren Rocky	3.12	3.12	1.21	1.21	
Scrub land/grasses	56.09	46.22	4.31	1.99	
Brick kilns and others	0	0	0.64	1.22	
Water	4.86	4.86	3.59	3.59	
Open land	3.16	3.16	26.05	16.85	
Defense land	0	0	7.19	7.19	

Source: Reddy and Reddy, 2009

culture and traditions spanning over 400 y. In 2012 the Indian Union Tourism Ministry declared Hyderabad the first ever "Best Heritage City of India." As well as places of historical interest, recreational facilities like Film City and IT hubs are major tourist attractions. Tourism has dramatically increased in the recent past. In 2010 about 4.34 million tourists visited the city, and this figure almost doubled in 2011 (Table 6.1). There are more than 470 hotels (excluding company guest houses), guest houses run by trusts, dormitories, free dharamshalas (religious rest house or sanctuary), tourist bungalows, and free accommodation units. Most of these hotels pump groundwater from deep aquifers to cater for water demand. However, the exact volume of abstraction from hotels is unknown.

6.3.4 Industrialization

Several industries existed prior to Hyderabad's independence. City-based industries started to grow in 1930 as a result of imported technology from the Western world. In the 1950s and 1960s, many public enterprises were established such as Bharat Heavy Electricals Limited (BHEL), National Mineral Development Corporation (NMDC), Hindustan Machine Tools

Limited (HMT), Bharat Electronics Limited (BEL), Indian Drugs and Pharmaceuticals Limited (IDPL), Electronics Corporation of India Limited (ECIL), Defence Research and Development Organization (DRDO), and Hindustan Aeronautical Limited (HAL). The electronic and pharmaceutical industries were established in the 1970s. Since the 1990s, various diversified service industries have grown in sectors such as IT, biotechnology, insurance, and finance. According to 2005 statistics, of every 1000 people of working age, 770 males and 190 females are employed within industry. The water demand for those industries is mainly met by groundwater abstracted from various aquifer depths.

6.4 PRESSURES

Inadequate surface water resources, overexploitation of groundwater resources, and land cover change (i.e., decrease in rechargeable areas) are pressures brought about by the drivers of Hyderabad's groundwater environment. In addition, improper disposal of domestic sewage and industrial effluent risk polluting groundwater aquifers.

6.4.1 Inadequate Surface Water Resources

Hyderabad needs to continuously expand its water supply system in order to meet the ever-changing gap between demand and supply. The HMWSSB is presently supplying piped water for two hours every other day (about 1550 MLD) to a GHMC service area of 680 km^2 (0.53 million connections), which reduces by half in summer (March to May). The Government of Andhra Pradesh has initiated several massive water supply projects on the Krishna River in a step-by-step process to overcome these deficits. Phases I and II of the Krishna water supply scheme have been completed – supplying about 820 MLD. Phase III is currently in progress to add a further 410 MLD to the city. The present deficit of 400 MLD may reach 585 MLD by 2031, which will put further pressure on already overexploited aquifers.

The quality of available water resources, both surface and subsurface, have been degraded due to the disposal of untreated effluent from industries and the domestic sector. About 70% of the area is covered by the sewerage network, collecting about 600 MLD of untreated sewage, which is less than 60% of total wastewater generation in the city. Treatment facilities, however, have the capacity to treat 113 MLD or 20% of total sewage collected by sewers. Most untreated sewage in the city is discharged into water bodies, resulting in poor water quality, high pollution, loss of habitat, and environmental degradation. The Musi River, which flows through the city, receives estimated untreated sewage of 500 MLD. The Hussain Sagar Lake, the biggest lake in the heart of the city, used to supply water for the city until 1930. It now receives untreated domestic and industrial sewage, causing high pollution levels. The fact that about 72% of samples contain excess nitrate confirms the vulnerability of shallow aquifers in coping with domestic pollution in the city. The HUA generates around 3379 t of solid waste every day, of which the MCH contributes 2040 t and the surrounding municipalities contribute 1139 t at a per capita generation rate of 600 g/capita/d. The inadequate solid waste disposal practices in the city are also responsible for the reduction in surface water resources by degrading their quality.

6.4.2 Land Cover Change

Land use patterns in the HUA have undergone tremendous changes over the past decades. Residential areas covered 22% of the city in 1971 but have risen to 50% in 2009, whereas "scrubland and grass" have decreased to 2% from 56% during the same period (Table 6.2). This clearly indicates that the recharge area has decreased in accord with change in land cover, adversely affecting groundwater levels. The expansion of residential areas has resulted in increased pumping in many parts of the city (Sreedhar et al., 2009).

6.4.3 Groundwater Overexploitation

Groundwater availability and withdrawal estimates from 2008 to 2009 were 906 and 7370 ha/y, respectively (Table 6.1). Hyderabad has been declared a severely overexploited area (CGWB, 2007) regarding groundwater development. Withdrawal with respect to availability is over 800%. In 2012, groundwater levels declined drastically to rock bottom when most of the deep borewells dried up and a drought situation prevailed in the city. Aquifer recharge in the area is estimated at 1007 ha/y, which is far less than annual groundwater withdrawal.

6.5 STATE

This section analyzes the state of the groundwater environment in Hyderabad from the perspective of well statistics, groundwater level, groundwater abstraction, and groundwater quality.

6.5.1 Well Statistics

The Groundwater Department, established in March 1971 in accordance with an agreement entered into by the Government of India and the World Bank, has been collecting bimonthly water level data since 1974. The use of groundwater for drinking and irrigation has been the case since ancient times. The history of wells shows that dugwells predominated up to the 1970s, the number of borewells increased after 1990, and the mean well yield subsequently decreased. The total number of dugwells, shallow tubewells, and deep tubewells in 2010 were 18,416, 47,860, and 2085, respectively. Well density dramatically increased from less than 5 km^{-2} in 1982 to over 20 km^{-2} in 2012.

6.5.2 Groundwater Level

The groundwater level has changed greatly over space and time. Central and state groundwater departments measure the groundwater levels in Hyderabad. Groundwater levels in the city range from 3 to 32 meters below ground level (mbgl) (Table 6.3), with an average fluctuation of about 10 m between the premonsoon and postmonsoon season (Figure 6.6). Most dugwells have dried up in the past decade. The majority of borewells dug recently have to drill down below 300 mbgl to abstract a reasonable volume of groundwater.

6.5.3 Groundwater Abstraction

Groundwater abstraction began prior to the establishment of the Groundwater Department in the 1970s. It took a few decades for groundwater levels to deplete as a result of the phenomenal increase of groundwater abstraction structures. Well density in the entire state went from 5 wells/km^2 in 1982 to 20 wells/km^2 in 2008. Table 6.4 shows groundwater potential and stage of development in 2004 and 2008 in and around Hyderabad. Abstraction, despite being less than 2004, is still much higher than recharge.

6.5.4 Groundwater Quality

The effects of urbanization and industrialization in Hyderabad have led to the contamination of groundwater resources. Due to the inadequate sewerage system and treatment capacity, domestic sewage and industrial effluents are discharged directly into streams, causing severe groundwater contamination. The concentration of nitrate significantly exceeds the MPL in many parts of the city. Excessive concentrations of sulfates and trace elements such

Table 6.3 Average groundwater level at Hyderabad in 2011 and 2012

Administrative unit (mandal)	May 2011	February 2012	May 2012
Balanagar	12.30	11.40	13.80
Bantwaram	19.07	19.87	31.57
Basheerabad	11.00	10.95	31.57
Chevella	19.14	11.22	31.54
Dharur	8.26	8.52	9.20
Doma	8.67	11.62	13.32
Gandeed	10.85	8.96	12.58
Ghatkesar	11.10	14.30	13.00
Hayatnagar	13.45	14.30	15.80
Ibrahimpatnam	13.45	14.30	15.80
Kandukur	13.50	13.70	13.40
Keesara	11.10	14.30	13.00
Kulkacherla	9.70	8.48	10.44
Maheshwaram	20.10	19.15	21.35
Malkajgiri	16.45	20.40	24.30
Manchal	13.20	15.20	16.05
Marpalle	19.50	20.56	22.74
Medchal	12.25	13.30	19.50
Moinabad	7.60	13.16	20.29
Mominpet	7.40	8.79	9.40
Nawabpet	10.60	12.40	13.98
Peddemul	12.74	13.02	14.56
Pargi	16.88	21.25	25.32
Pudur	10.98	12.16	14.42
Quthbullahpur	5.25	5.80	6.55
Rajendranagar	11.60	12.70	13.90
Ranga Reddy district average	12.03	13.13	16.17
Saroornagar	5.60	6.00	5.80
Serilingampally	4.95	4.71	7.30
Shahbad	16.36	21.40	23.26
Shamirpet	3.10	3.10	6.45
Shamshabad	10.86	13.39	14.18
Shankarpalle	12.83	13.47	16.57
Tandur	6.17	5.36	6.69
Uppal	10.50	11.86	12.95
Vikarabad	17.35	20.70	23.60
Yacharam	17.00	19.20	19.70
Yalal	14.62	15.63	16.97

Source: State Groundwater Department Data

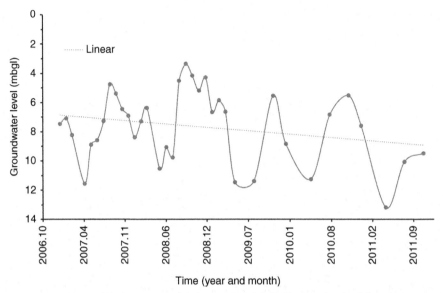

Figure 6.6 *Bimonthly Average Groundwater Level (mbgl) at Hyderabad from 2007 to 2011. Source: Groundwater Information System (2015).*

as copper, cadmium, zinc, manganese, and lead are found in and around industrial development areas (IDAs). The results of chemical analysis of groundwater samples (1998, 2010) from shallow borewells (40–60 m) indicate their alkaline nature. Total hardness ranges from 95 mg/L to 711 mg/L and groundwater type is moderately soft to very hard. Electrical conductivity (EC) ranges from 200 μS/cm to 7500 μS/cm, while the general range is 750–3000 μS/cm (Table 6.1). The areas with EC values beyond permissible limits (i.e., more than 3000 μS/cm) are located mostly around IDAs such as Bolarum, Suraram, Jeedimetla, Sanath Nagar, Kukatpally, and Saroornagar. Nitrate (NO_3) is observed to be higher than the MPL in drinking water standards in 72% of samples. Higher concentrations of sulfates (i.e., more than 400 mg/L) are noticed in areas such as Uppal Kalan, Sanath Nagar, Saroor Nagar, and near the Musi River, and are attributed to industrial pollution in these areas. The concentration of fluoride in deeper groundwater ranges from 0.50 mg/L to 2.57 mg/L, and at certain places the concentration exceeds the MPL. There are nine IDAs spread throughout the city intermingling with large residential colonies. Almost all the IDAs have industries, and hence are potential sources of heavy metal waste getting into the environment.

Table 6.4 Groundwater potential and stage of development in Hyderabad in 2004 and 2008

Mandal name	Area (ha)	2004					2008–09				
		GW availability (ha-m)	GW withdrawal (ha-m)	GW balance (ha-m)	Stage of development (%)	Category	GW availability (ha-m)	GW withdrawal (ha-m)	GW balance (ha-m)	Stage of development (%)	Category
Hyderabad (37 mandals)	9984	1400	12099	−10699	864	OE	906	7370	−6464	813	OE
BalaNagar	642	969	86	883	9	Safe	642	104	538	16	Safe
Ghatkesar	1299	1647	862	786	52	Safe	1299	496	803	38	Safe
Hayaat Nagar	1783	2283	1609	674	70	SC	1783	934	849	52	SC
Keesara	1193	1513	1649	−136	109	OE	1193	755	438	63	Safe
Malkajgiri	453	626	291	335	47	Safe	453	109	344	24	Safe
Medchal	1969	2275	3752	−1478	165	OE	1969	1994	−25	101	OE
Qutubullapur	1291	1782	1109	673	62	Safe	1291	627	664	49	Safe
Rajendra Nagar	1369	1896	808	1088	43	Safe	1369	446	923	33	Safe
Saroor Nagar	727	1101	856	245	78	SC	727	172	555	24	SC
Serilingampalli	1071	1463	301	1163	21	Safe	1071	193	878	18	Safe
Shamirpet	2272	3252	6093	−2841	187	OE	2272	1733	539	76	OE
Shamshabad	1881	2371	2724	−353	115	OE	1881	1757	124	93	OE
Uppal	313	437	130	307	30	Safe	313	86	227	27	Safe
RC Puram	6932	766	513	253	67	Safe	766	513	253	67	Safe
Patancheru	21433	2236	1665	571	74	SC	2236	1665	571	74	SC
Total	54612	26017	34547	−8529	133	OE	20171	18954	1217	94	OE

Groundwater availability m³/km² = 369,351; Population density = 18,489 persons/km²
Per capita groundwater resources = 20 (approx.) m³/y
GW, groundwater; OE, overexploited; SC, semicritical.
Source: CGWB, 2007; CGWB, 2011.

6.6 IMPACTS

Depletion in groundwater reserves (water level and well yield), land subsidence, and public health are considered indicators of impacts on the groundwater environment in Hyderabad.

6.6.1 Depletion in Groundwater Level

The Central Groundwater Board (CGWB) estimates Hyderabad's withdrawals at 178 MLD, which is lower than the unofficial estimate of 240 MLD abstracted by private borewells alone (CES, 2012). Due to uncontrolled groundwater abstraction in past decades, groundwater levels dropped from 5.8 mbgl to 31.57 mbgl in Greater Hyderabad in 2012, an all-time low, even though good rainfall was received in the previous year (Table 6.3). Groundwater in the city is being exploited to a great extent through borewells at various depths, some even deeper than 300 m, resulting in the premonsoon water level plummeting from 5 mbgl to 20 mbgl. Over the years the groundwater level has dropped at many places such as Boinpalli, Aghapura, Erragadda, Basheerbagh, Langar House, Jubilee Hills, Begumpet, Koti, West Maredpalli, Gudimalkapur, Musheerabad, Sanathnagar, Picket, and Madhapur. The groundwater level on March 28, 2012 was 25.35 mbgl at Marredpally and 20.4 mbgl at Malkajgiri – a fall of 9.22 m since February 2011. In Nacharam the level was observed as 11.85 mbgl, higher than the 7.74 mbgl observed in February 2011.

6.6.2 Land Subsidence

About 98% of Hyderabad stands on hard rock; therefore, land subsidence is less likely.

6.6.3 Public Health

Water pollution is a serious public health issue in Hyderabad. Most municipal solid waste, domestic sewage, and industrial effluent is being discharged directly into water bodies and/or onto open ground. Such practices are responsible for deteriorating surface and groundwater qualities. The Musi River in Hyderabad has become a polluted drain. The ground near the water is sticky and red, giving off a powerful ammonia smell and bacterial indicators capable of spreading waterborne diseases. The color of the water ranges from dull gray to black. The water carries various heavy metals and antibiotics, polythene bags, and animal carcasses. Men, women, and children suffer from back problems, pain in their legs and hands, poor eyesight, and skin lesions. Many women have had miscarriages, and medical expenses have increased. Despite its five water treatment plants, Hyderabad's water is contaminated. An analysis of data from the Ronald Ross Institute and Hospital for Tropical Diseases (a major referral hospital for poor people in the city) found that waterborne diseases were the major cause of mortality, with diarrhea cases increasing as well. In May 2009 the city reported 5 deaths and hospitalization of over 200 people due to the municipal water supply being polluted. Uncontrolled groundwater pumping has also increased fluoride levels and resulted in fluorosis problems for many who cannot afford bottled water.

6.7 RESPONSES

These issues are being addressed by the government, which has taken the initiative of monitoring groundwater levels as well as developing environmental standards and guidelines. It has also initiated megaprojects regarding water supply as well as lake and river cleaning.

6.7.1 Groundwater Monitoring

The Groundwater Department in Andhra Pradesh was established in 1970 to monitor groundwater levels and assess resources for their effective utilization. A network of dugwells in both command (1404 wells) and noncommand areas (1582 wells) throughout the entire state was established in 1974, and monitoring of groundwater levels on a bimonthly basis was started. Since then the Groundwater Department has expanded the network of monitoring wells year by year. The CGWB drilled about 10 borewells for groundwater exploration and established 17 wells to monitor groundwater levels. Furthermore, it is also establishing a digital water level recording system.

6.7.2 Environment Standards and Guidelines

To improve surface and groundwater quality, the Ministry of Environment and Forests (MoEF, the agency responsible for environmental standard guidelines) has set the MPL for effluent through the Water (Prevention and Control of Pollution) Act 1974 and Environment Protection Rules, 1986 (MoEF, http://www.envfor.nic.in/environmental_standards). The Water Act aims at preventing and controlling water pollution as well as maintaining or restoring the wholesomeness of water in the country. The act was amended in 1988. The Water Cess Act came into force in 1977, to provide for the levy and collection of a cess on water consumed by persons operating and carrying on certain types of industrial activities. This cess is collected with a view to augmenting the resources of the Central Board and State Boards for the prevention and control of water pollution constituted under the Water Act. The act was last amended in 2003. The Environment Protection Rules, 1986 have been amended several times to date. The second amendment in 1988 incorporated about 35 general standards for discharging effluent. These consist of industrial waste into surface water, wastewater from treatment plants into surface water, and industrial effluent into public sewers. In addition, a general standard has been determined as the MPL of wastewater effluent discharged into surface water and public sewers for various industries. The state government has also initiated the Water, Land, and Trees Act (WALTA, 2002). However, in practice the implementation of these standards remains poor when it comes to improving groundwater quality.

6.7.3 Megaprojects for Water Supply

The HMWSSB is currently supplying about 1550 MLD of piped water at a frequency of 2 hevery other day to the GHMC service area of 680 km^2 (0.53 million connections), which reduces by one half in summer (March to May). Hyderabad has to expand its water supply continuously to meet the ever-changing gap between demand and supply. Construction of the Krishna Water Supply Project (KWSP) was initiated in November 2002 and its first phase brought some 410 MLD into Hyderabad. The project will contribute another 1230 MLD by 2021, by which time the estimated demand of 1934 MLD would again outstrip supply. By 2031 the demand supply gap is expected to widen further as the city forecasts a water demand of over 2100 MLD against a supply of 1955 MLD. The HMWSSB initiated the Godavari Drinking Water Scheme (capacity of 681 MLD) in 2012, and in November 2013 acquired clearance from the MoEF for the construction of a water supply scheme. It was expected to be completed by May 2014 but was delayed due to the bifurcation of Telangana. These initiatives help conserve the groundwater environment by reducing groundwater abstraction. Furthermore, water levels in rivers are expected to increase, thus making them capable of carrying away their pollution loads and hence reducing pollution in groundwater resources.

The GHMC in collaboration with the Japan International Cooperation Agency (JICA) has made several attempts to protect Hussain Sagar, the biggest lake in Hyderabad. This has been done through various development programs, like lake remediation, in conjunction with organizational and research activities.

6.8 CONCLUSIONS AND RECOMMENDATIONS

From the DPSIR analysis, it is clear that population growth, urbanization, industrialization, and tourism are major driving forces exerting pressure on Hyderabad's groundwater environment. Hyderabad is currently under stress due to extensive groundwater mining, which has been carried out in the absence of a groundwater management policy, plans, management intervention, and public participation. Yearly groundwater abstraction is eight times greater than natural recharge. Overexploitation has already lowered yield and average groundwater levels to well below 32 m. Most of the city borewells, even those at a depth of 300 m, dry up in the summer. Institutions have been successful in developing various policies to control overexploitation. However, they have failed to implement those policies, subsequent laws, and regulations satisfactorily. Mandal Revenue Offices do not even maintain figures for the total number of borewells in their jurisdiction. However, efforts are being made to complete the KWSP and the Pranahita Chevella Project, which hopefully will fill the current water deficit and help in recharging the aquifers by 2021. Initiatives to develop microlevel aquifer maps through airborne surveys using electromagnetic equipment under the World Bank project are under way, which may help in understanding the heterogeneous anisotropic nature of aquifers in Hyderabad. To improve lake and stream water quality, the government has taken massive steps toward remediating it. Furthermore, initiatives have been taken toward establishing a national-level high-performance computational storage facility to support data dissemination and enhance research into climate modeling and impacts on sustainable water resource development.

REFERENCES

Anon., 2011. 71-City Water-Excreta Survey, 2005–06, Centre for Science and Environment, New Delhi.

Central Ground Water Board (CGWB), 2007. Groundwater Information, Ranga Reddy District, Andhra Pradesh. CGWB, Ministry of Water Resources, Government of India.

Central Ground Water Board (CGWB), 2011. Groundwater Scenario in Major Cities of India. CGWB, Ministry of Water Resources, Government of India. Available from: http://www.cgwb.gov.in.

Central Groundwater Board (CGWB), 2013. Groundwater Brochure, Hyderabad District, Andhra Pradesh, Southern Region, Hyderabad (AAP-2012-13). CGWB, Ministry of Water Resources, Government of India.

Directorate of Census Operations (DCO), 2011. Provisional Population Totals 2011. Directorate of Census Operations, Andhra Pradesh, Hyderabad, India.

Environment Protection Rules, 1986. Ministry of Environment and Forest, Government of India.

CES, 2012. Excreta Matters – Hyderabad: The Water–Waste Portrait, Centre for Science and Environment, New Delhi, pp. 331. Available from: http://cseindia.org/userfiles/hyderabad_portraits.pdf.

Greater Hyderabad Municipal Corporation (GHMC), 2015. Available from: www.ghmc.gov.in (accessed 28.03.2015.).

Groundwater Information System, 2015. Available from: http://gis2.nic.in/cgwb/ (accessed 10.03.2015.).

HMWSSB, 2015. 12th Annual Report. Available from: www.hyderabadwater.gov.in (accessed 4.04.2015.).

India Tourism Statistics, 2010. Tourism Survey for Andhra Pradesh. Available from: www.tourism.gov.in (accessed 8.03.2015.).

India Tourism Statistics, 2015. Available from: www.tourism.gov.in (accessed 17.03.2015.).

Jawaharlal Nehru National Urban Renewal Mission, 2015. Hyderabad Urban Agglomeration: Demography, Economy and Land Use Pattern, Chapter II. Available from: www.hyderabadwater.gov.in (accessed 13.03.2015.).

Prakash, A., Singh, S., 2013. Water security in peri-urban South Asia adapting to climate change and urbanization. SaciWATERs, South Asia Consortium for Interdisciplinary Water Resources Studies, Sainikpuri, Secunderabad, India. Available from: http://saciwaters.org/periurban/pdfs/popularpublication-march27-2014.pdf.

Reddy, A.G.S., Reddy, M.R.K., 2009. Evaluation of potential ground water zones in Hyderabad City, Andhra Pradesh, India. Int. J. Earth Sci. Eng. 2 (6), 561–568.

Sreedhar, G., Vijaya Kumar, G.T., Murali Krishna, I.V., Ercan Kahya, M., Demirel, C., 2009. Mapping of groundwater potential zones in the Musi basin using remote sensing data and GIS. Adv. Eng. Softw. 40, 506–518.

WALTA Act, Andhra Pradesh Water, Land And Trees Act-2002, http://www.apard.gov.in/walta-act.pdf.

CHAPTER 7

Groundwater Environment in Khulna, Bangladesh

Binaya Raj Shivakoti
Water Resources Management Team, Natural Resources and Ecosystem Area, Institute for Global Environmental Strategies (IGES), Japan

7.1 INTRODUCTION

Khulna (Khulna City Cooperation) is the administrative center of Khulna District under the broader administrative boundary of Khulna Division (Figure 7.1). It is also known as the "port city of the country" due to its proximity to the Bay of Bengal in the south and the Indian state of West Bengal in the west. However, most of its plain areas are low lying with an elevation range from 0.45 m to 5.40 m above mean sea level (ADB, 2010). It is the third largest city in Bangladesh covering an area of about 65 km^2 and is home to over 0.75 million people (as of 2011). Although the city area is less than 1.5% of Khulna District (4389 km^2), nearly 32% of people in the district live here (BBS, 2013).

The administrative boundary of Khulna consists of 31 wards. The Rupsha – Moyur river system flows through the city. The outskirts of the city are predominantly agriculture based. Agriculture and jute mills are the main contributor to Khulna District's economy. More than 46% of the district, mainly in the southern part, consists of thick forest such as mangroves (BBS, 2013). A network of rivers connected to streams and canals also passes through the district before draining into the Bay of Bengal. These waterways are important to the economy as they serve navigation and communication purposes. These rivers alone occupy nearly 14% of the district area.

Despite its extensive coverage of rivers and water bodies, groundwater is an important source of water for households, industries, and agriculture. All of the municipal water supply of the Khulna Water Supply Authority (KWASA) is sourced from groundwater. However, the municipal water supply is inadequate to meet the increasing demand for water in the city. The water supply gap is mainly covered through private abstraction by households, industries, and businesses. The city's increasing dependence on groundwater is also becoming a cause for concern as a result

Groundwater Environment in Asian Cities
http://dx.doi.org/10.1016/B978-0-12-803166-7.00007-6

Copyright © 2016 Elsevier Inc.
All rights reserved.

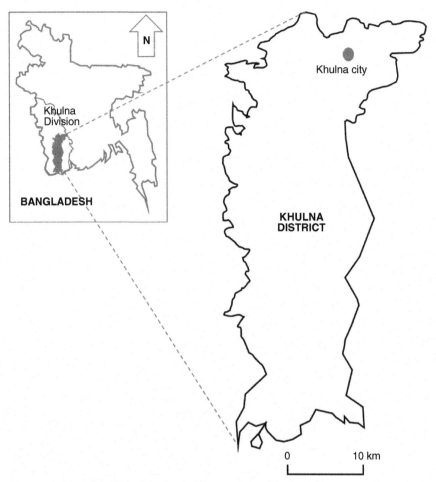

Figure 7.1 *Location Map of the Study Area.*

of drought, geogenic arsenic contamination, saline water intrusion, and pollution of aquifers by agrochemicals, industrial discharges, and poor sanitation and wastewater management.

7.2 GEOLOGY, GROUNDWATER AQUIFERS, AND CLIMATE

Khulna's geology, like most of the Bangladesh delta, was formed by the deposition of sediments from the extensive network of three major river systems: the Ganges, Brahmaputra, and Meghna. Geological processes in the Quaternary brought about different sequences of sediment depositions (ADB, 2010). The geology of the city area consists of late Holocene to Recent alluvium of the Ganges deltaic plain to the north and the tidal plain to the south (Roy et al., 2005). At the bottom of the alluvium there are layers of coarse sand and gravels. Above that layer, clay fine sand layers alternate with fine to coarse sand layers. The major soil types are calcareous alluvium, acid sulfate soil, peat, and floodplain soils (calcareous gray, calcareous and noncalcareous dark gray, and calcareous brown) (BBS, 2013). Soil near the coast is saline.

The deposition of sediments is favorable for the development of good-quality aquifers with high transmissivity and storage. There are three distinct aquifers separated by aquitards composed of clay layers (ADB, 2010; DPHE-JICA, 2010). These aquifer systems are not uniform in thickness in neither the east–west (EW) nor the north–south (NS) direction. Aquitards are thinner in the south and become thicker toward the north even exceeding 200 m. The first aquifer is shallow in type and is distributed up to about 50 m deep. The middle aquifer is distributed at around 100 m depth. The third aquifer is deep and is found between 250 m and more than 350 m below ground level. Figure 7.2 shows a typical lithological profile of a bore-well site in Khulna.

Khulna has a hot summer (April–June) and a mild winter (November–February). Maximum and minimum temperature in the summer of 2012 was 36.7 and 24.3°C, respectively, whereas in winter of the same year they were 29.3 and 13.7°C, respectively (Figure 7.3). Annual rainfall between 2008 and 2012 ranged between 1357 mm and 1948 mm. June–September is the rainy season when more than 60% of the rainfall occurs. The rainy season is often associated with flooding and inundation. Rainfall, flooding, and inundation are the major source of recharge to aquifers in the city.

7.3 DRIVERS

7.3.1 Population Growth

The urban population in Khulna has been rising gradually from nearly half a million in 1981 to more than 0.75 million in 2011 (Table 7.1; Figure 7.4) (BBS, 2012). Khulna District saw an almost one million increase in total population between 1981 and 1991 when its population passed the 2.5 million mark. However, the population decreased slightly by 0.25 million between 1991 and 2011.

Khulna is also home to slums, informal settlements consisting of a cluster of five to more than 100 households. In 2014 there were 1143 slums (made up of 20,536 households) in Khulna, which represents about 8% of slums in the whole country (BBS, 2015).

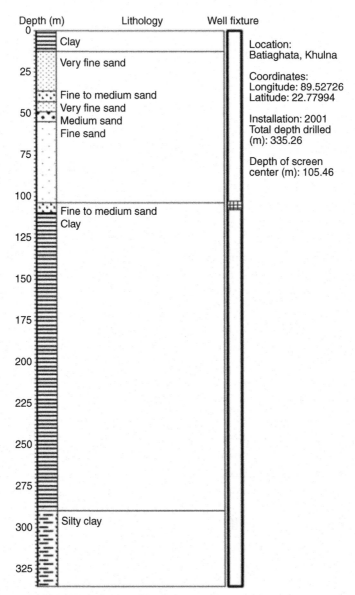

Figure 7.2 *Typical Lithological Profile of a Borewell Site in Khulna (DPHE-JICA, 2010).*

7.3.2 Agricultural Intensification

The alluvial flat plains of Bangladesh are highly fertile. Most of Khulna District, except the city, is devoted to agriculture consisting of staples (mainly rice), fruit, cash crops, and aquaculture. Nearly 78% of the land area

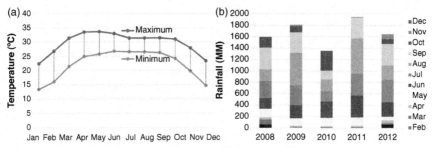

Figure 7.3 *Average Minimum and Maximum Temperature (a) and Monthly Rainfall Distribution (b) for 2008–2012 in Khulna (BBS, 2012).*

Table 7.1 Status of DPSIR indicators in Khulna, Bangladesh

	Indicators	Past	Recent
Drivers	Population growth	0.515 million (1981)	0.751 million (2011)
	Agricultural intensification	Area under agriculture: 1,786 km² (in 2008)	
	Climate change	Bangladesh is vulnerable to climate change impacts such as flood, water logging, droughts, cyclones, and sea level rise	
Pressures	Lack of alternative safe water sources	Increasing water demand, drought, pollution, and increased salinity of surface water sources and shallow aquifers; public water supplies are facing nonrevenue losses, inactive connections, and malfunctioning public hand pumps	
	Agriculture water use	More than 130,000 ha of land are irrigated by shallow and deep tubewells scattered across the district, which accounts for 71% of the irrigated area (BBS, 2013)	
	Sea level rise	Climate change-related melting of ice masses at the poles is going to raise the sea level permanently and inundate most coastal areas and cause increases in salinity	
State	Groundwater abstraction	25,000 m³/d in 1981 by KWASA (ADB, 2010)	209,704 m³/d in 2010 by KWASA and private abstractions (ADB, 2010)
	Groundwater level	3.04–6.84 m below ground level in the dry season of 2009 (ADB, 2010)	

(Continued)

Table 7.1 Status of DPSIR indicators in Khulna, Bangladesh *(cont.)*

	Indicators	Past	Recent
	Groundwater quality		Three water quality issues of importance in Khulna are natural contaminants (mainly, geogenic arsenic), salinity, and anthropogenic pollution. The arsenic concentration in 2001 was <6–130 ppb (BGS and DPHE, 2001).
	Borehole data		Data on 213 boreholes (whole district) and on about 100 boreholes within a 15 km radius of the city center (DPHE-JICA, 2010)
Impacts	Salinity intrusion		The increasing height of tides (2–3 m) is one of the causes of increasing salinity due to the low elevation of Khulna (<5.4 m). The salinity level in shallow aquifers is >10,000 mg/L. Farmers suffer from loss of crops as a result of salinity
	Health impacts		High arsenic (>50 µg/L) levels in shallow aquifers (<100 m). Contamination of groundwater from wastewater causes diarrhea, which is the most common health-related infections in the area
Responses	Legal and institutional reforms		The DPHE The WASA Act 1996 • National Policy for Arsenic Mitigation 2004 • National Policy for Safe Water Supply and Sanitation 1998
	Supply of arsenic-free groundwater		• Using water from deeper aquifers • Color marking for arsenic-free tubewells • Use of arsenic filters • Promoting use of nongroundwater sources (e.g., rainwater harvesting and pond water use)
	Supply augmentation		• KWASA is planning to develop an alternative water supply by constructing a dam at Phultala

(excluding forest areas) is used for agriculture (BBS, 2014). About 37% of cultivated areas are under irrigation mainly from shallow and deep tube-wells. Khulna Division is also famous for shrimp farming, which accounts for almost 78% of country-wide production.

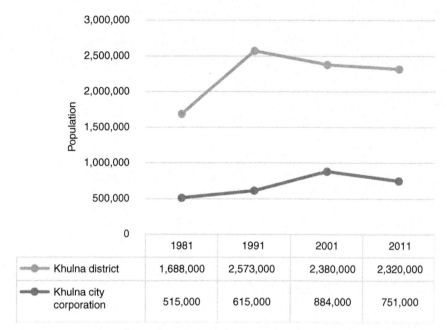

	1981	1991	2001	2011
Khulna district	1,688,000	2,573,000	2,380,000	2,320,000
Khulna city corporation	515,000	615,000	884,000	751,000

Figure 7.4 *Population Growth in Khulna District and City (BBS, 2012).*

7.3.3 Climate Change

Bangladesh has been identified as one of the countries most vulnerable to climate change impact mainly related to sea level rise, flooding and inundation, and tropical cyclones. The effect of climate change on such a low-lying country results in tidal waves penetrating deep inland and in this way increasing the salinity of rivers and land, let alone the devastating impact on people and property. Increased salinity, especially in the dry season, can significantly affect the city's water supply.

7.4 PRESSURES

7.4.1 Lack of Alternative Safe Water Sources

Khulna is facing acute water shortages due to increasing water demand, drought, pollution, and increased salinity of surface water sources and shallow aquifers. However, public water supplies also face the problem of nonrevenue losses, inactive connections, and malfunctioning public hand pumps.

7.4.2 Agriculture Water Use

Groundwater is used to irrigate crops throughout Khulna District. More than 130,000 ha of land are irrigated by shallow and deep tubewells scattered across the district (BBS, 2013). This accounts for 71% of the irrigated area.

7.4.3 Sea Level Rise

Sea level rise is becoming a reality in coastal areas of Bangladesh, including Khulna (Brammer, H. 2014). Climate change–related melting of ice masses in the poles is going to raise the sea level permanently, inundate most of the coastal areas, and cause increased salinity. In addition to the long-term impact, there will be a greater incidence of tropical cyclones and high tides and hence further incursions of sea water traveling inland.

7.5 STATE

7.5.1 Groundwater Abstraction

There are a number of sources of groundwater abstraction: domestic users, industries, agriculture, and KWASA. The daily abstraction by private users and KWASA is estimated to be 0.21 million cubic meters (MCM) (ADB, 2011). KWASA's production wells abstract about 16% of total abstraction while an additional 10 and 15% are pumped by its deep and shallow hand pumps, respectively. KWASA has 55 production wells that abstract water from the deep aquifer and supplies water to residents through 16,200 connections. However, the piped water supply from KWASA can only serve 18% of the population. In addition, KWASA has installed hand pumps (about 6000 deep tubewells and 4000 shallow tubewells) for public use at different places, which can additionally serve about 15% of the population. Nearly 59% of the water supply gap is fulfilled through groundwater abstraction from tubewells installed by individual households, slum communities, businesses, and industries.

Despite the water supply capacity of KWASA remaining basically static, private abstraction has been increasing by 2.2–2.9% annually, which is an approximate 27,000 m^3/d increase between 1995 and 2010 (Figure 7.5).

7.5.2 Groundwater Level

Satisfactory information on groundwater level, its temporal trend, and spatial variability is not available as a result of inadequate monitoring. However, the groundwater level in the dry season of 2009 ranged from 3.04 m to 6.84 m below ground level (ADB, 2010).

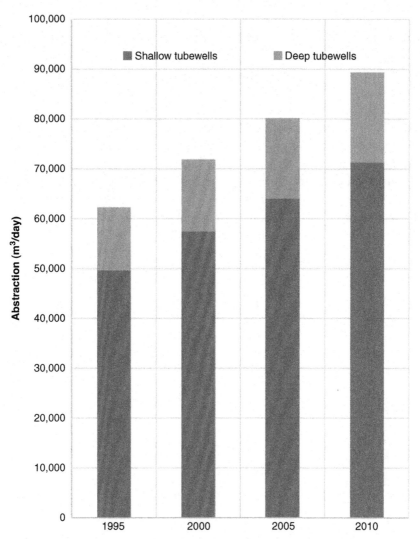

Figure 7.5 *Groundwater Abstraction by Private Wells in Khulna (ADB, 2010).*

7.5.3 Groundwater Quality

Three water quality issues are particularly important in the case of Khulna: natural contaminants, salinity, and anthropogenic pollution. Arsenic and iron are naturally occurring elements in groundwater aquifers in Khulna. Iron as such poses no health risk but its presence can affect the usability of groundwater as a result of its odor and development of a reddish yellow color. Geogenic arsenic contamination is a national problem in Bangladesh where millions of people are exposed to arsenic-related health risks through groundwater consumption. Arsenic has been detected in Khulna's shallow aquifers at a level exceeding 50 parts per billion (ppb), the World Health Organization's (WHO) standard, which has been adopted as Bangladesh's standard. As a part of a National Hydrochemical Survey, 11 wells in Khulna District were monitored in 2001 (BGS and DPHE, 2001). The survey found that 5 samples from shallow tubewells had arsenic above the WHO standard of 50 ppb (Figure 7.6).

Increased salinity in groundwater is becoming a major challenge facing water supply, industrial uses, and irrigation in Khulna. Although some parts of aquifers have a naturally high salt content, most of the salinity is brought by the ingress of sea water. High salinity has been observed in some shallow aquifers where chlorine concentrations exceeding 10000 parts per million (ppm) have been observed (Hasan et al., 2013).

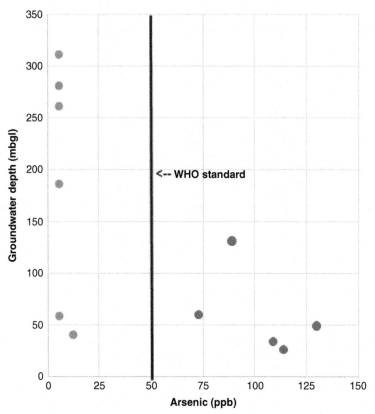

Figure 7.6 *Arsenic Contamination at Different Groundwater Depths (ppb is parts per billion) (BGS and DPHE, 2001).*

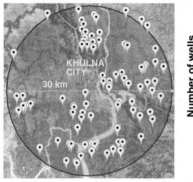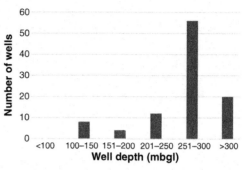

Figure 7.7 *Distribution of Borehole Database within 15 km Radius of Khulna City (DPHE-JICA, 2010).*

Anthropogenic pollution is a growing problem in Khulna where inappropriate disposal of waste, wastewater, agrochemicals (fertilizers, pesticides), pollution of water bodies, and unsafe sanitary practices combine together to cause groundwater contamination. The risk of contamination is especially high in the rainy season when poor drainage causes water logging and pollutants eventually infiltrate the aquifers.

7.5.4 Borehole Data

The Aquifer Database Inventory Program, set up by the Department of Public Health Engineering (DPHE), has so far maintained a database on 213 boreholes distributed in Khulna. These wells have depths ranging from 42 m to 352 m. In the case of the city, there is a database on nearly 100 boreholes, 76 of which have depths greater than 250 m (Figure 7.7). Each borehole database has records of location, date of installation, water quality data (such as arsenic, chloride, iron, manganese), depth and diameter of the well, and lithological profile.

7.6 IMPACTS

7.6.1 Salinity Intrusion

Salinity is becoming a serious problem in Khulna. It is mainly caused by the ingress of sea water. The increasing height of tides (2–3 m) is one of the causes of increasing salinity especially when coupled with Khulna's low elevation (<5.4 m). The problem becomes even more acute during the dry season when the river water level is normally low. In addition, there are frequent cyclones impacting the coastal areas of Bangladesh (BBS, 2014). These too are responsible for high tides invading inland areas. The impact salinity has on domestic uses, industries, and agriculture is a growing problem. Farmers in Khulna District suffer from the loss of crops due to salinity, while salinity in the shallow aquifers also prevents them from using groundwater irrigation. Industries too cannot use saline groundwater, hence their production is affected.

7.6.2 Health Impacts

Groundwater use in Bangladesh is associated with health risks mainly due to high levels of arsenic. Arsenic has been detected in the shallow aquifers of Khulna exceeding the safe recommended concentration of 50 ppb (Bangladesh standard). Apart from arsenic, shallow groundwater is prone to contamination from human waste and wastewater. This has affected safe access to groundwater from shallow depths, which has been a convenient and cheaper source of water for poor communities such as those living in slums. An investigation of five slum communities by the author revealed that hand pumps and toilets are often located close to each other (<10 m), further risk of contamination.

7.7 RESPONSES

7.7.1 Legal and Institutional Reforms

Bangladesh has introduced regulations and acts that deal with the safe access and use of groundwater. These include the National Policy for Safe Water Supply and Sanitation 1998 and the National Policy for Arsenic Mitigation 2004 (MLGRDC, 1998; DPHE, 2004). The former aims to increase that part of the population living within 150 m of the tubewells. However, to ensure adequate access it also aims to reduce the number of people per tubewell from 150 to 50 persons. The National Policy on Arsenic Mitigation 2004, which also complements the Safe Water Supply and Sanitation policy, aims to ensure access to safe water for drinking and cooking by implementing alternative water supply options in all arsenic-affected areas. The WASA Act 1996 was set up to oversee water supply and sanitation in cities. It authorizes the government to establish Water Supply and Sewerage Authorities (WASAs) in any area, with permission to perform any work relating to water supply, sewage systems, solid waste collection, and drainage (WSP, 2014).

The DPHE is the institution tasked with implementing both policies. However, water supply and sanitation are the responsibility of KWASA when it comes to Khulna City.

7.7.2 Supply of Arsenic-Free Groundwater

Government and other nongovernment organizations have been adopting different strategies to promote the safe use of groundwater. For that purpose a number of mitigation measures have been implemented. They include green marking of arsenic-free tubewells and arsenic removal filters. In addition to nongroundwater sources such as rainwater harvesting, pond water (after boiling) is also used.

7.7.3 Supply Augmentation

KWASA is also planning to construct a surface water supply system in Phultala area at about 20 km north of Khulna City, to reduce the water deficit by using upstream rivers and storing water in a dam yet to be constructed. Upstream rivers are less polluted and run a lower risk of salinity through direct mixing with seawater, while the storage dam will insure a stable water supply during dry periods when the river water level is low. Construction of this facility is expected to add about 110,000 m^3/d to existing supply capacity (ADB, 2010).

7.8 SUMMARY

Groundwater is the sole source of water supply in Khulna. It is abstracted both by municipality and privately. Furthermore, it is an important source of irrigation at the outskirts of the city and an important source of water supply during disasters that occur quite frequently. Despite the easy availability of groundwater, Khulna is facing acute water shortages as a result of drought, sea water intrusion through high tides, low water levels in local rivers, groundwater pollution, and geogenic occurrences of arsenic.

Several steps have been taken to improve access to groundwater and insure its safe use. No serious incidence of groundwater table depletion or land subsidence (except caused by natural tectonic movements) has yet occurred. Climate change is becoming a threat to long-term groundwater sustainability mainly from the potential threat from saline water intrusion and inundation of landmass. Judicious use of surface and groundwater should be promoted to reduce the high dependence on groundwater and enhance recharge in the dry season to insure stable discharge of groundwater.

ACKNOWLEDGMENTS

The author acknowledges the Kakenhi Project (No. 143010435206-0001) for supporting the field investigation; Dr Mohammad Abdullah, Managing Director of KWASA, for providing information about the water supply situation in Khulna; Mr Qazi Azad-uz-zaman,

Project Manager, Japan Association of Drainage and Environment (JADE), Khulna Office, for supplying additional information; and Prof Akira Sakai, University of Marketing and Distribution Sciences (UMDS), Kobe, Japan.

REFERENCES

ADB, 2010. Bangladesh: Strengthening the Resilience of the Water Sector in Khulna to Climate Change. Asian Development Bank (ADB), Manila.

ADB, 2011. Bangladesh Khulna Water Supply Project. The Asian Development Bank (ADB), Manila.

BBS, 2012. District Statistics 2011 Khulna. Bangladesh Bureau of Statistics (BBS), Dhaka.

BBS, 2013. The Statistical Yearbook of Bangladesh 2012. Bangladesh Bureau of Statistics (BBS), Dhaka.

BBS, 2014. The Yearbook of Agricultural Statistics of Bangladesh 2012. Bangladesh Bureau of Statistics (BBS), Dhaka.

BBS, 2015. Preliminary Report on the Census of Slum Areas and Floating Population 2014. Dhaka: Bangladesh Bureau of Statistics (BBS).

BGS and DPHE, 2001. Arsenic contamination of groundwater in Bangladesh. In: Kinniburgh, D. G., Smedley, P. L. (Ed.) Vol. 4: Data Compilation, *British Geological Survey Report WC/00/19*, British Geological Survey, Keyworth.

Brammer, H., 2014. Bangladesh's dynamic coastal regions and sea-level rise. Climate Risk Management. 1, 51-62.

DPHE, 2004. National Policy for Arsenic Mitigation 2004. Department of Public Health Engineering (DPHE).

DPHE-JICA, 2010. Final Report on Development of Deep Aquifer Database and Preliminary Deep Aquifer Map. Department of Public Health Engineering, Government of the People's Republic of Bangladesh.

Hasan, M.R., Shamsuddin, M., Hossain, A.A., 2013. Salinity status in groundwater: a study of selected upazilas of Southwestern Coastal Region in Bangladesh. Global Sci. Technol. J. 1 (1), 112–122.

MLGRDC, 1998. National Policy for Safe Water Supply & Sanitation 1998. Ministry of Local Government, Rural Development and Cooperatives (MLGRDC), Dhaka.

Roy, M., Datta, D.K., Adhikari, D., Chowdhury, B., Roy, P., 2005. Geology of the Khulna City Corporation. J. Life Earth Sci. 1 (1), 57–63.

WSP, 2014. Benchmarking to Improve Urban Water Supply Delivery in Bangladesh. Water and Sanitation Program (WSP), The World Bank.

CHAPTER 8

Groundwater Environment in Lahore, Pakistan

Muhammad Basharat
International Waterlogging and Salinity Research Institute (IWASRI), Pakistan Water and Power Development Authority (WAPDA), Lahore, Pakistan

8.1 INTRODUCTION

Lahore is the capital of the Pakistani Province of Punjab (Figure 8.1) and the second largest metropolitan city in Pakistan. With its proximity to the Kasur district in the south and Sheikhupura to the northwest, the area measures 1772 km² and its urban area largely comprises converted agricultural land. Lahore has expanded to almost double its area in the last 12–15 years. According to the 1998 population census, Lahore was home to nearly 6.32 million people resulting in a population density of 3566 persons/km², with the urban population representing 82.4% of the total. The literacy rate was 48.4% and 64.7% in 1981 and 1998, respectively. During 2007–2008, physical access to drinking water within homes was available to 98% of residents and improved sources for drinking water were consumed by 99% (Punjab Development Statistics, 2012). The estimated population of the Lahore district (including the rural population) was 8.83 m as at 31 December 2011 (Government of Punjab, 2011), while 84% of the population resided in the metropolitan city area (Government of Pakistan, 2011). An estimate provided in April 2014 put the population of the Lahore agglomeration at 9.6 million (Thomas Brinkhoff, 2014).

Lahore is the second largest financial hub in Pakistan and has important industrial areas such as Kot Lakhpat and the new Sundar Industrial Estate (near Raiwind). The major industrial trend in Lahore has shifted in recent decades from manufacturing to service industries. The local economy is also enhanced by Lahore's historic and cultural importance. Being the capital of the largest province in Pakistan provides the city with one of the largest development budgets in the country. It is also the most advanced in terms of infrastructure, having extensive and relatively well-developed road links to all major cities in Punjab and Khyber Pakhtunkhwa (KPK), a rail link with India and the province's biggest international airport. In the event that

Groundwater Environment in Asian Cities
http://dx.doi.org/10.1016/B978-0-12-803166-7.00008-8

Copyright © 2016 Elsevier Inc.
All rights reserved.

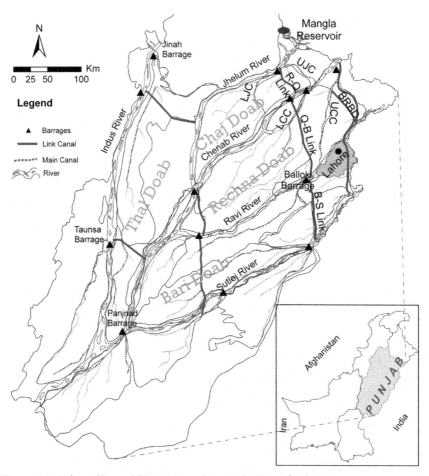

Figure 8.1 *Lahore City and District Boundary on the Map of Pakistan.*

relations between India and Pakistan improve and trade is accelerated, it will be the biggest trade route from Pakistan to India, through the Attari–Wagah border crossing. In a best-case scenario, Afghanistan and India could also use this as a trade corridor. Thus, the national and international importance of Lahore City is expected to rise in the future.

The city has almost doubled in area in the last 12–15 years. There is also a continuous reduction in recharge, mainly caused by desiccation of the Ravi River and land use change to urbanization. Abstractions from the aquifer are increasing, as all the existing and additional water demands are supposed to be met from the aquifer in the absence of any surface water allocation. Under the current state of affairs, recharge from rainfall in the Ravi River and surrounding irrigation system is less than 50% that of groundwater abstraction. Unregulated abstraction, without any recharging efforts and surface supplies, has led to an annual water table depletion rate of 0.57–1.35 m. The maximum depth of the water table reached 46 m in 2014. Thus, a lot of anthropogenic factors (regional to local in scale) are expected to have a continuous impact on the aquifer. Consequently, evaluation of these issues and possible consequences for the dynamics of the aquifer is urgently needed.

8.2 SURFACE AND GROUNDWATER HYDROLOGY OF THE AREA

8.2.1 Climate and Geography

The climate of this region is characterized by large seasonal variations in temperature and rainfall. Mean annual reference crop evapotranspiration (ET_O) at Lahore is 1649 mm. Mean annual temperature is about 24.3°C ranging from 33.9°C in June to 12.8°C in January (Pakistan Meteorological Department, 2006). The hottest month is June, where average highs routinely exceed 40°C, with the highest observed temperature of 48.3°C recorded on May 30, 1944, and the lowest recorded of −1°C observed on January 13, 1967. The wettest month is July, with heavy rainfall, evening thunderstorms, and the possibility of cloudbursts.

Annual normal rainfall in the 30 years from 1971 to 2000 for Lahore was 712 mm (Figure 8.2). About 75% of the annual total falls between June and September and in a typical year contributes approximately 40 mm to groundwater recharge.

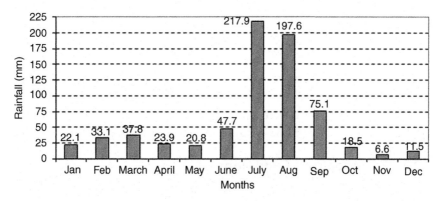

Figure 8.2 *Thirty-Year Normal Rainfall (1971–2000) for Lahore (Basharat and Rizvi, 2011).*

Areas upstream and downstream of Lahore are bounded by the Ravi River to the northwest and the Sukh–Beas drainage channel to the south. The Lahore District forms part of the Bari Doab – interfluvial lands between the Ravi River to the northwest and the Sutlej River to the south. The area is underlain by a significant thickness of alluvial deposits, up to 300 m in depth, proven by investigations carried out by the Water and Soils Investigation Division (WASID) between 1961 and 1962 (WAPDA, 1980). Recharge sources are the Ravi River (very occasional flows nowadays) to the northwest and the Bambawala–Ravi–Badian–Depalpur (BRBD) Canal to the east. Discharging boundaries are the Beas drainage channel to the south and again the Ravi River to the west. However, these discharging boundaries only function in the form of surface runoff during heavy rainfall; no groundwater discharge is possible due to excessive groundwater depletion in the area.

8.2.2 Surface Water Regime

The general altitude of the area is 208–213 m above mean sea level (masl). The area slopes toward the south and southwest at an average gradient of 1 in 3000. The uppermost part of the Bari Doab became Indian territory after partition, but being the integral part of the lower and central Bari Doab provides a natural route for regional groundwater flow. The irrigation system has been present in the area since 1879, with the construction of the first weir-controlled perennial irrigation channel: the Upper Bari Doab Canal (Pakistan National Committee of ICID, 1991) is an offtake of the Ravi River at Madhopur (now in India).

However, following the Indus Water Treaty (IWT) of 1960 between India and Pakistan, India stopped the flow of eastern rivers (Sutlej, Beas, and Ravi). In order to relieve this shortage, the BRBD link canal was constructed. With a water flow of 4000 cubic feet per second (cusec or cfs), it is fed by the Upper Chenab Canal offtaking from the Marala Barrage. The BRBD Canal provides canal water to the command area of the Upper Bari Doab Canal on the Pakistan side of the border. The area is also under the command of the Lahore Branch (403 cfs), Khaira Disty (141 cfs), Butcher Khana Disty (244 cfs), Main Branch Lower (1591 cfs), and other smaller channels (with a combined capacity of 81.68 cfs) directly offtaking from the BRBD Canal.

8.2.3 Hydrogeologic Conditions

In 1954, the US Geological Survey initiated hydrogeological investigations in the Punjab, in cooperation with the Pakistan Water and Power

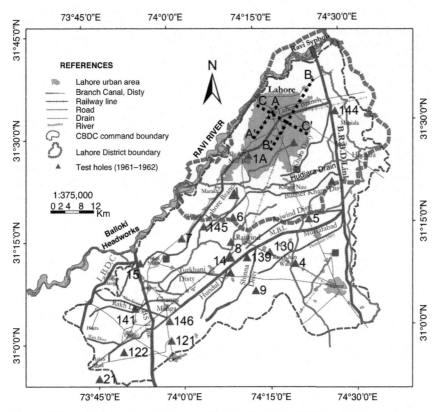

Figure 8.3 *Location of Test Holes and Test Wells (1961–1962), and Geologic Cross-Sections.*

Development Authority (WAPDA) (Greenman et al., 1967). The investigations included the drilling of test bores, construction of test tubewells, carrying out pumping tests, and data analysis. Many test holes were over 600 ft. deep and 33 were between 800 ft. and over 1000 ft. deep. Of the latter, 19 test holes were located in the northeastern part of the Doab, with only three test holes – BR 1A, BR 6, and BR 7 (all in the Lahore District, Figure 8.3) – reaching Precambrian basement rock at 1252, 1021, and 928 ft., respectively (WAPDA, 1980). The location of the test holes and test wells conducted in 1961–1962 is also shown in Figure 8.3. NESPAK and Binnie and Partners, 1987 collected borehole logs from 160 tubewells installed in Lahore by the Water and Sanitation Agency (WASA) to study the areal and vertical distribution of various subsurface lithological formations. Lithological cross-sections along the lines AA′, BB′, and CC′ (Figure 8.3) are shown in Figure 8.4.

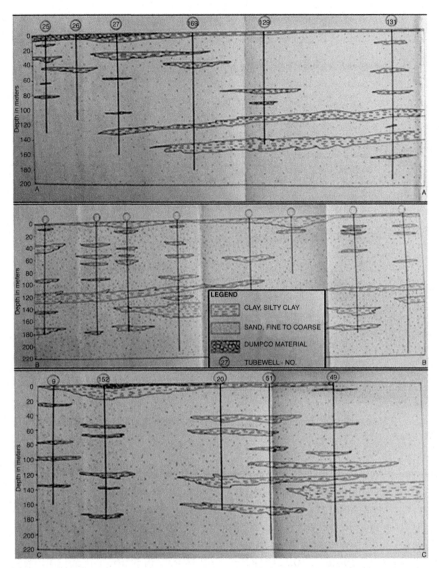

Figure 8.4 *Lithological Cross-Sections AA', BB', and CC' (top to bottom). Source: NESPAK and Binnie and Partners (1987).*

Analysis of the lithological logs of 149 test holes (600–1000 ft. deep) and 28 test tubewells (102–356 ft. deep) indicates that the Bari Doab consists of consolidated sand, silt, and silty clay, with variable amounts of kankars. The sands are fine to medium grained and subangular to subround. Very fine sand is common, finer grained deposits generally include sandy silt, silt, and silty clay, with appreciable amounts of kankars and other concretionary material. Except for a few local lenses a few feet thick, beds of hard-rock compact clay are rare in the area. Gravels of hard rock are not found within the alluvium, and there is little apparent lateral continuity in the sediments (Figure 8.4).

8.2.4 Aquifer Characteristics

The area is part of a flat alluvial plain known as the Indus Plain, which extends from the foothills of the outer Himalayas and the western ranges to the Arabian Sea. The alluvial sediments of the aquifer exhibit considerable heterogeneity both laterally and vertically. Despite this, it is broadly viewed that the aquifer behaves as a single contiguous, unconfined aquifer. The data of 21 pumping tests falling within the Bari Doab are reproduced in Table 8.1. Screened intervals of these test wells consisted predominantly of sand, silt, and thin beds or lenses of clay. Lateral permeability determined in these pumping tests is thus average for sediments at the pumping sites, exclusive of the thick deposits of clay. As such, these values are undoubtedly somewhat higher than the average lateral permeability of the section as a whole (Bennett et al., 1967). Din and Sheikh (1969) evaluated the variation of specific yield (S_y) with depth, from the lithological data of 24 tubewells (21 are in the Bari Doab) and 32 test holes (in the Chaj and Rechna Doabs). The analysis comprised 607 soil samples. According to the results, the shallow depth (0–25 ft.) contains about 34.1% of silt and clay components, whereas the deeper strata comprising medium sand ranging from 0.5 mm to 0.25 mm is 41.8% and medium sand ranging from 0.25 mm to 0.177 mm is 11.6 %. Thus, the medium sand in lower strata of the Bari Doab comprises about 50% of total soil particles. The report concluded that various pumping tests carried out by WASID (1953–1963) were performed in situations when the water table was shallower than 25 ft.; therefore, the specific yield values determined by these tests apply to the upper part (i.e., upper 25 ft. of strata). Hence, the specific yield values are likely to be higher than those determined from these pumping tests (i.e., 22–35%, average 26.3%) (Sheikh, 1971).

The lithology of the alluvium shows a heterogeneous character in vertical (depth wise) and transverse directions and a random distribution of clay zones (Greenman et al., 1967), which is also evident from geologic cross-sections (Figure 8.4). In spite of the heterogeneous composition, the aquifer is highly transmissive such that groundwater occurs under water table conditions. According to NESPAK and Binnie and Partners (1988),

Table 8.1 Results of aquifer tests as reported by Bennett et al. (1967)

Test No.	Site	K (m/d)	S_y	Screen length (m)	Total depth (m)	DTW (m)	Q (m³/d)	Duration (h)	Max. S_w (m)	Specific capacity (Q/S_w)
B-1	Kot Lakhpat	26.34		36.58	87.48	2.87	4893	456	5.83	839.21
B-2	Ljhenwali	15.80		30.48	62.48	4.18	4893	144	10.95	446.94
B-3	Bhai Pheru	34.24		36.58	68.58	10.45	4893	92	6.78	721.21
B-4	Kasur	34.24		36.58	68.58	2.04	6117	96	5.94	1029.10
B-5	Renala Khurd	39.50	0.06	36.58	68.58	3.08	4282	96	4.82	888.51
B-6	BS Link	65.84		36.58	62.48	3.38	4893	96	8.44	579.57
B-7	Gamber Rl Sn	255.45		38.10	79.25	6.46	7340	96	8.81	833.25
B-8	Pakpattan	31.60	0.24	46.33	71.63	6.00	7340	384	6.25	1174.69
B-9	Harrapa	36.87	0.04	36.58	96.01	10.67	6679	240	14.05	475.35
B-10	Arifwala	28.97	0.31	41.15	69.50	6.68	6117	244	6.75	906.39
B-11	Iqbal Nagar	63.20		36.58	62.48	8.23	6117	144	3.19	1920.34
B-12	Luddan	202.78		36.58	108.21	5.27	6117	87	7.80	783.89
B-13	Sheranwal R H	252.82		41.15	108.21	10.24	4893	96	6.55	746.70
B-14	Perowal Forest	65.84		36.58	76.20	9.27	7340	144	3.44	2131.07
B-15	Tiba Sultanpur	47.40		42.67	65.53	7.01	5970	187	3.68	1621.35
B-16	MS 621/4	39.50		36.58	69.50	7.13	6728	144	5.21	1290.89
B-17	Munirabad	50.04		36.58	70.10	4.88	6068	144	3.83	1586.22
B-18	Karor pacca	47.40		36.58	62.48	8.32	4893	96	7.45	656.61
B-19	Khajiwala	28.97		41.15	67.06	4.30	9322	68	6.87	1356.23
B-20	Chak pathan	34.24		41.15	65.53	7.74	6117	144	4.15	1475.55
B-21	Shujabad	31.60		42.67	67.06	3.05	6679	72	5.49	1217.43

DTW, depth to watertable; K, hydraulic conductivity; Q, discharge; S_y, specific yield; S_w, drawdown.

the sediments comprise fine to coarse sand with lenses of silty clay and clay. Borehole logs show that the lenses of less permeable material are neither thick nor continuous; so, the aquifer can be treated as single and unconfined.

8.2.5 Regional Groundwater Flow

Depth to groundwater is rapidly increasing in the area as a result of its use by the population and industry. In the city of Lahore the maximum observed depth to the water table in 2011 was 45 m. The aquifer under and around the city is very deep (more than 400 m) with high transmissivity of about 2100 m²/d (assuming 80 m thickness contributing to groundwater flow). The aquifer extends from Lahore to the piedmont area of the foothills in Jammu and Kashmir in a northeasterly direction (about 100 km). This is also the direction of the rivers and groundwater flow in general. In the downstream direction, it extends to the Arabian Sea for a distance of about 1100 km. The rivers and groundwater flow in a northeasterly to southwesterly direction in Punjab. Regional groundwater flow velocities are low (of the order of 1.5–2 km/100 y) due to mild surface and groundwater gradients (Basharat and Tariq, 2013).

The areas upstream and downstream of Lahore are bounded by the Ravi River to the northwest and the Sukh–Beas drainage channel to the south. So the area of interest with respect to Lahore Cty forms part of the Bari Doab – interfluvial land between the Ravi River to the northwest and the Sutlej River to the south. The area is underlain by a significant thickness of alluvial deposits, up to 300 m in depth, proven by the investigations carried out by WASID between 1961 and 1962 (WAPDA, 1980). Recharge sources are the Ravi River (very occasional flows nowadays) to the northwest and the BRBD Canal to the east. Discharging boundaries are the Beas drainage channel to the south and again the Ravi River to the west. However, these discharging boundaries only function in the form of surface runoff during heavy rainfall; no groundwater discharge is possible due to excessive groundwater depletion in the area.

Groundwater flow on the regional scale occupies the same flow direction as the canal irrigation network, mostly parallel to the river network (i.e., in a northeasterly–southwesterly direction). However, due to local variations in natural surface elevations as well as discharge and recharge patterns, depth to the water table varies considerably in the area. Consequently, groundwater elevations vary considerably at such locations. This is particularly true of the area under Lahore where, as mentioned earlier, the maximum depth to water table exceeds a depth of 45 m due to heavily concentrated pumping by tubewells installed by the WASA of the Lahore Development Authority (LDA) as well as private housing societies that pipe supplies to the local population, and individual pumping for dwellings and business. In addition, certain industries consume vast amounts of ground-water (e.g., the beverage industry).

8.2.6 Groundwater Quality

Depth-wise data on groundwater quality collected from the test holes (Figure 8.3) falling within the Central Bari Doab Canal (CBDC) command area is plotted in Figure 8.5. In general, groundwater lying at greater depths is of good quality when compared with shallow groundwater. Deep-groundwater quality data from 1961 to 1962 has revealed that a strip about 10 km wide between Pattoki and Chunian, starting from Raiwind and ending in approximately the middle of Okara and Sahiwal, has highly saline groundwater. Deep groundwater in the Raiwind area is highly saline (up to 9000 ppm) up to depths of 110 m. In the wake of an induced groundwater hydraulic gradient from Raiwind to Lahore (due to excessive pumping in Lahore), it is likely that saline groundwater may start intruding toward fresh groundwater, although this may still take about another 50 years to emerge as a serious saline intrusion issue. Greenman et al. (1967) concluded that the distribution of saline and fresh groundwater zones in the Bari Doab is a result of past and present hydrologic, climatic, and topographic factors. The present and former positions of stream channels representing sources of recharge, the high bluffs of bar uplands in the upper part of the Bari Doab,

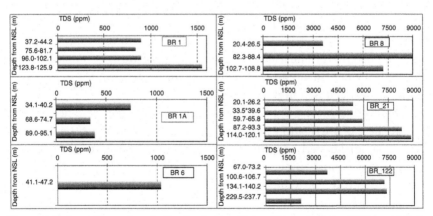

Figure 8.5 *Deep-Groundwater Quality in Test Holes (Figure 8.3) for 1961 to 1962 in Lahore.*

and differences in permeability within the alluvial aquifer are among the most important topographic factors.

8.3 DRIVERS

Population growth, conversion of agricultural land use for urbanization, and increased industry are the main drivers exerting pressure on the groundwater environment in Lahore City and its surroundings. Another important factor is the loss of canal water rights to areas converted for urbanization. All these factors point toward increased pressure on the groundwater environment, as discussed in subsequent sections.

8.3.1 Population Growth

Pakistan is an arid and water-scarce country; hence, the protection and sustainability of aquifers is critical, especially in urban cities. Population growth in Lahore continues to ascend. It increased from 3.54 million in 1981 to 6.32 and 8.83 million in 1998 and 2011, respectively (Table 8.2). The city's population more than doubled within a period of 30 y. Explosive growth of more than 2% per year is projected to continue for several decades, because the government has no plans to impede this urban growth rate. Instead, it is providing standard education and modern living facilities to small cities and rural areas. Population density within a 30-y period has increased from 1998 persons/km^2 to 4983 persons/km^2.

8.3.2 Urbanization

Urbanization can be defined as land use change from agricultural to residential (i.e., conversion of fertile agricultural land to populous urban areas with asphalt roads and concrete floors). The urban area was 7.8% in 1980, 54.3% in 2000, and 70.5% in 2010 (Table 8.2), demonstrating a 2% growth rate per year in land urbanization for the Lahore District. Due to continued expansion, much of the population shift involves movement from villages, small towns, and other cities in the province to the metropolitan city in search of quality education, jobs, and a lavish lifestyle.

8.3.3 Industrialization

Lahore is the second largest financial hub in Pakistan and has important industrial areas such as Kot Lakhpat and the new Sundar Industrial Estate (near Raiwind). The major industrial trend in Lahore has shifted in recent decades from manufacturing to service industries. The local economy is also enhanced

Table 8.2 Status of DPSIR indicators for 1981, 1998, and 2011 in Lahore City, Pakistan

	Indicators	1981	1998	2011
Drivers	Population growth (within the district, area: 1772 km²)	Population: 3.54 × 10⁶ Population density~1998 (persons/km²) Annual growth rate*	6.32 × 10⁶~3566 (1998–1981) 3.4%	8.83 × 10⁶ (estimated)~4983 (2011–1998) 2.57%
	Urbanization (within the district)	Urban area (Zaman and Baloch, 2011) (in 1980): 137.9 km² (7.8%) Urban population:** 2.95 × 10⁶	In 2000: 961.6 km² (54.3%) 5.14 × 10⁶	In 2010: 1249.7 km² (70.5%) 7.20 × 10⁶
	Industrialization(No. of registered industries)†		In 2000: 1399 (Bureau of Statistics, 2005)	In 2013: 3007 (Bureau of Statistics, 2012)
Pressures	Inadequate surface water resources	Ravi inflows (Basharat et al., 2014) 1922–1960: 7.0 MAF	1976–1999: 5.5 MAF	2000–2011: 1.2 MAF
		Surface water allocated to agricultural land in periurban areas since the inception of the irrigation system is no longer available due to land use change to urbanization; hence the consistent reduction in recharge to groundwater from Ravi flows and land use change		
	Land use change	Cultivated area (Zaman and Baloch, 2011) (in 1980): 1634 km²	(in 2000): 810.4 km²	(in 2010): 522.3 km²
	Groundwater overexploitation	Abstraction (Basharat and Rizvi, 2011): 1987: 1451 MLD (or 530 MCM/y)	2000: 1815 MLD (or 663 MCM/y)	2010: 3181 MLD (or 1161 MCM/y)
State	Well statistics‡ (only WASA wells)	Numbers: 1987: 200	2000: 316	2010:467
	Groundwater abstraction	Abstraction: (Basharat and Rizvi, 2011) 1987: 1451 MLD (or 530 MCM/y)	2000: 1815 MLD (or 663 MCM/y)	2010: 3181 MLD (or 1161 MCM/y)

	Thirty-year normal (1997–2000) rainfall in Lahore is 712 mm/y = 1262 MCM over an area of 1772 km²		
Groundwater level	1987: 7.6–19.8 m	2000: 10.7–32.0 m	2010: 12.2–44.2 m
Groundwater quality	Upper layer of groundwater is being rapidly polluted with disposal of untreated industrial effluents and leakage from the sewerage system. Comparison of TDS in groundwater samples with previous studies shows that the quality of groundwater is deteriorating with time. TDS and As contents are also less in deep wells. Therefore, in order to tap good-quality groundwater, the WASA and private groundwater users have increased bore depths from 150 m to 230 m in the past few years.		
Recharge	The Ravi River was once a major source of recharge, but nowadays flows only occasionally. Groundwater recharge is also decreasing due to canal irrigation of the fields ending (now converted to urban area), on the one hand, and rainfall recharge reduction particularly due to most of the area disappearing under buildings and roads, on the other hand. As a whole, there is drastic reduction in recharge. Currently, groundwater recharge is estimated as 1013 MCM (including flows from boundaries), compared with groundwater abstraction of 1161 MCM on annual basis. Thus, aquifer recharge is about 15% of total groundwater abstraction.		
Impacts	Depletion in groundwater level (Mahmood et al., 2013) (m/y)	1973–1980: 0.60	1980–2000: 0.65	2000–2011: 0.79
	Deterioration in groundwater quality	Deterioration of groundwater quality has to be constantly addressed; therefore, well depths are gradually increased in search of relatively good-quality groundwater at greater depths.		
	Decline in production capacity of wells (L/s)	Six of the 15 tubewells were reported as having reduced specific capacity of more than 50% (NESPAK and Binnie and Partners, 1987). Since well yields reduce depending on groundwater depletion, tubewell boring depths are successively increased to get the desired discharge.		
	Land subsidence	Land subsidence has never been monitored, but is now		

(Continued)

Table 8.2 Status of DPSIR indicators for 1981, 1998, and 2011 in Lahore City, Pakistan (cont.)

Indicators	1981	1998	2011
	Public health	Groundwater is the sole source of water supply in the city. However, its quality is under increasing threat due to domestic and industrial disposal of waste into surface water and drains, which ultimately enters the Ravi River at various points. Such effluent ultimately contributes to groundwater recharge, and these surface flows are used for irrigation farther downstream.	
Responses	GW monitoring	Proper groundwater monitoring was started in 1960, when a regional piezometric map was prepared. Monthly monitoring is carried out by the WASA in urban areas. Surrounding irrigated areas are monitored twice a year by the Irrigation Department. However, data is not easily available.	
	Environmental standards and guidelines	The provision of laws to protect the environment have never been implemented due to deficiencies in legalities, nonapplication of fines, public and official ignorance of environmental issues, and lack of political commitment. Sectoral legislation has totally failed to play even a limited role in environmental protection.	
	Arranging surface water from adjoining areas	There had been several proposals by professionals for improving the aquifer replenishment, like construction of a barrage on the Ravi River. A few years ago, the Government of Punjab announced construction of a lake on the Ravi River in order to raise and maintain the rapidly decreasing groundwater level of the city, in addition to constructing an international-standard recreational park for the city's citizens with the aim of making the atmosphere of the city pollution free. But, now the government has dismissed the plan in favor of a new city along the Ravi River at Shahdara. Such river course development pays little attention to depleting groundwater under the city, apart from creating ponding conditions in the river after channelization.	

* Exponential growth rate calculated using $r = (1/t) \times \ln(P_t/P_0)$, where P_t = population after t years from the base period, P_0 = base year population.
** Pakistan Bureau of Statistics, Statistics Division, Government of Pakistan, G-9/1, Islamabad.
† An establishment is said to be registered under Section 2 (J) of Factories Act 1934, (amended in 1973) if 10 or more workers are working on any day of the preceding 12 months and in any part of which manufacturing process is being carried on with or without the aid of power.
‡ Represents only about 55% of groundwater pumping.

by Lahore's historic and cultural importance. There were 1399 registered industries in Lahore in 2000, which increased to 3007 in 2013 (Table 8.2).

8.4 PRESSURES

Pressures on Lahore's groundwater environment are increasing with the passage of time due to inadequate surface water resources, overexploitation of aquifers, and land use changes resulting in a decrease in rechargeable areas (i.e., limiting supply to aquifers).

8.4.1 Inadequate Surface Water Resources

The Ravi River used to be associated with the identity of Lahore. The loss of water use from one of the three eastern rivers (Ravi) is a major setback for Lahore City, depriving the underlying aquifer a major source of recharge. Construction of the Thein Dam in 2000 on the Ravi River is having a profound impact by attenuating floods. This may explain why no substantial floods have been observed since 1997. After 2000, only a maximum discharge of 30,000–40,000 cusec was observed in the Ravi River. In future, except for such small flood events, no regular flows of appreciable amounts are expected in the reach, except that of the Marala Ravi (MR) link which releases from the Marala Barrage. Following the partition of India and Pakistan, annual flows into the Ravi decreased gradually from 7.0 million acre-feet (MAF) prior to the IWT (1922–1961) to 5.52 MAF, before construction of the Thein Dam, and finally to about 1.2 MAF, after the Thein Dam was built (Figure 8.6 and Table 8.3). The reality is that the Ravi

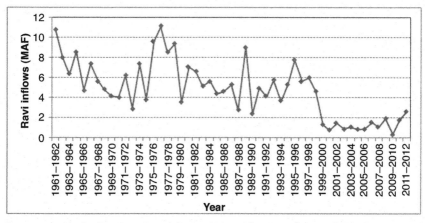

Figure 8.6 *Gradual Desiccation of the Ravi River after Completion of the Thein Dam.*

Table 8.3 Annual average flows (MAF) in the Ravi and Sutlej rivers (Basharat et al., 2014)

Ravi			Sutlej		
1922–1961	1976–1999	2000–2011	1922–1961	1976–1999	2000–2011
7.0	5.51	1.20	14.0	3.51	0.78

Figure 8.7 *Sewage and Solid Waste Flowing in the Ravi River at the Expense of Fresh Water.*

River no longer flows through Lahore; rather, it has become an effluent disposal drain (Figure 8.7). Thus, instead of contributing to groundwater recharge, it now pollutes the underground aquifer and has become a survival issue for the population of more than 8 million. According to flow changes in the Ravi and Sutlej rivers (Table 8.3), the impact of the IWT has struck home in the last decade, when it was realized that India had developed the capacity to divert almost all the flows of the Ravi and Sutlej rivers. Thus, the environmental flow requirements for the eastern rivers (Ravi and Sutlej) are now being essentially realized.

8.4.2 Land Use Change

Lahore has witnessed constantly increasing stress on the underlying aquifer as a result of increased groundwater use caused by the ever-increasing expansion of the city and by reduced aquifer recharge quantitatively and qualitatively. Reduced recharge is mainly caused by desiccation of the Ravi River and land use change toward urbanization. Since the inception of an irrigation system for agricultural land in periurban areas, surface water allocation is no longer available as a result of land use change toward urbanization. Thus, reduced groundwater recharge from the Ravi River and land use change contribute substantially to the negative water balance of the underlying aquifer. Due to a decrease in cultivated areas, land use with recharge potential has dropped from 1634 km² in 1980 to 810.4 km² in 2000 and 522.3 km² in 2010, which is a strong indicator of land use change and subsequent negative impacts on groundwater aquifer recharge.

8.4.3 Overexploitation of Groundwater Resources

With successive increases in groundwater pumping due to population growth and urbanization, the groundwater balance of the city first started to show negative trends in the 1970s. Uneven groundwater development and uncontrolled pumping through private tubewells have led to excessive drawdown and put stress on the aquifer. To visualize present groundwater flow directions, Basharat and Rizvi (2011) converted data on depth to the water table in and around Lahore City for 2009 to groundwater level elevations. The resulting groundwater elevation contours (Figure 8.8) show a major depression (about 36 m deep within a radius of about 25 km) under

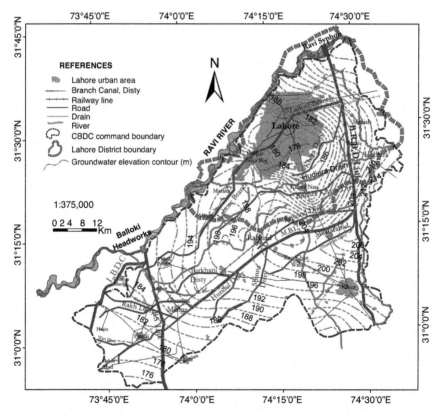

Figure 8.8 *Lahore District Boundary, Groundwater Elevation Contours (2009), and Irrigation Network in the CBDC (Basharat and Rizvi, 2011).*

the city, caused by intensive pumping from the aquifer and lack of recharge to it. Groundwater is flowing toward urban areas in Lahore. The area around Raiwind, with highly saline groundwater (about 9000 ppm), is lying at a higher elevation and currently behaving as a groundwater divide. Depth to groundwater is constantly increasing in urban areas. Thus, further groundwater depletion in Lahore City is likely to lead to a risk of saline intrusion from saline groundwater in the Raiwind area to fresh groundwater in Lahore City.

8.5 STATE

This section analyzes the state of the groundwater environment from the perspective of number of wells and their increasing depths, volume of groundwater abstraction, groundwater levels, groundwater quality, and annual recharge/inflow into aquifers under Lahore.

8.5.1 Well Statistics

Depth to the water table prior to the irrigation system was about 5–6 m in the present urban area. The installation of tubewells for the Lahore water supply was initiated by a Mr. Howell in 1930–1940, and turbine pumps were installed in 1945. Tubewells were pumping about 92 cfs of water in 1972, whereas total demand was 206 cfs. Total pumpage, including private wells, was 164 cfs. The Lahore urban area covered 450 km² in 1987 and had a population of 3.7 million, more than 80% of whom were connected to the WASA supply. Between 1960 and 1987, levels had fallen in parts of the city by up to 15 m (49 ft.). Typical falls over much of the city were of the order of 6 m (20 ft.). Historical data regarding abstraction, number of tubewells installed by the WASA, and depths to the water table for Lahore City are given in Table 8.4. In addition to these wells, provided by the government, there are more than 5000 privately owned agricultural tubewells. They are used to irrigate crops when there is a shortage of canal water.

Table 8.4 Groundwater development trends in Lahore City

Description/year	1987	2000	2010
Depth to water table	25–65 ft.	35–105 ft.	45–155 ft.
No. of tubewells	200	316	400
Total abstraction	320 MGD	400 MGD	696 MGD
Population	3.5 million	4.83 million	7.4 million

8.5.2 Groundwater Abstraction

Groundwater abstraction by the WASA in Lahore in 2007 was reported as 360 million gallons per day (MGD) (i.e., 668 cfs) with 406 tubewells pumping. This does not include private schemes, found mainly on the outskirts of the city, the majority of which have their own pumping system. In 2010 there were about 467 tubewells operated by the WASA. Including all other consumers – such as the Cantonment Board area, private housing societies, industries, and individual groundwater users – total groundwater withdrawal was about 1300 cfs (Basharat and Rizvi, 2011). At this rate of groundwater withdrawal, the annually pumped groundwater volume is 1161 million cubic meters (MCM) (0.94 MAF). The WASA supplies about 333 liters per capita per day (lpcd) of water (Iftikhar, 2007) to the area served. Using this water consumption rate and a population of 8.83 million in Lahore District, total annual water consumption comes out at 1093 MCM (0.885 MAF). Thus, including industrial consumption, the annual pumping volume of 1,161 MCM (0.94 MAF) seems to be a reasonable estimation of pumping from the aquifer.

8.5.3 Groundwater Level

The maximum depth to groundwater exceeds 45 m in the central area of the city (Shadman and Jail Road) as a result of excessive groundwater abstraction. Groundwater decline has left a major depression in its wake, which increases in size as a result of groundwater flows from surrounding areas. Depth to groundwater improves toward the Ravi River and surrounding agricultural areas. Table 8.5 gives spatial information regarding depth to the water table in Lahore City. Approximately 467 of the WASA's deep tubewells, whose capacities vary from 1.5 cusec to 4 cusec, along with others installed by private industries, housing societies, and individuals, are pumping groundwater entirely satisfactorily.

8.5.4 Groundwater Quality

In general, groundwater from greater depths is of good quality when compared with shallow groundwater in and around the Lahore City. Moreover, the quality of deep groundwater is good enough for human consumption in most areas of the district. Total dissolved solids (TDS) varies from 400 ppm to 1000 ppm, as proved by groundwater water samples collected during an investigation into deep test hole drilling carried out in 1961–1962 and thereafter in the region. Similarly, pH, Ca, Mg, SO_4, Cl, and Na were also

Table 8.5 Depth to static water level in WASA tubewells in 2009

#	Location of site	Capacity (cfs)	Static water table (m)	Pumping water table (m)	Drawdown (m)
1	A Block Gujjar Pura	4	19.81	27.74	7.92
2	A Block Muslim Town	2	40.59	43.21	2.62
3	Abbad Park Janaz Gah	2	32.06	35.64	3.58
4	Abbak Park	2	32.06	35.64	3.58
5	Adda Crown Bus Stop	4	34.75	39.32	4.57
6	Akbar Shaheed Road	4	38.71	43.79	5.08
7	Akbari Gate	2	33.12	36.13	3.01
8	Al Faisal Town Burji No. 9	2	31.09	35.66	4.57
9	Annand Road Upper Mall	2	38.20	42.98	4.78
10	APWA Lahore College	2	37.97	42.19	4.22
11	Bhogiwal Disposal	4	28.65	33.3	4.65
12	C Block Sabzazar Kunjpura	4	22.58	27.33	4.75
13	Chohan Road	4	23.88	29.14	5.26
14	D.O. Office North	4	26.21	31.09	4.88
15	F&V Market Multan Road	4	24.38	28.96	4.57
16	Faisal Park Shahdra	2	12.04	15.44	3.4
17	Farid Kot House	2	31.83	35.81	3.99
18	Fc College Kachi Abbadi	2	41.76	46.13	4.37
19	Gawala Colony	2	21.79	24.72	2.93
20	G block Johar Town	4	31.90	35.81	3.99
21	General Hospital	2	33.53	39.17	5.64
22	Ghazi Abbad Bus Stop	2	33.73	37.92	4.19
23	Gulshan-e-Hayat Shahdra	4	14.63	20.67	6.04
24	Gulshan-e-Riaz Munshi Hospital	2	26.21	30.79	4.57
25	H Block Sabzazar	2	28.50	32.31	3.81
26	Islam Pura	4	25.93	31.45	5.51
27	Jamsheed Park Singh Pura	4	35.36	40.97	5.62
28	Junior Model School DCO office	2	31.18	34.92	3.74
29	Kot Khawaja Saeed	4	30.18	35.18	5.05
30	Lady Sofia Park	2	26.21	33.43	7.21
31	Lytton Road Park	2	32.46	36.07	3.61
32	Napier Road	2	32.39	37.13	4.75
33	Nishtar Block	2	26.26	30.63	4.37
34	Paisa Akhbar	2	32.92	36.68	3.76
35	Patyala House G.O.R	2	38.51	41.76	3.25
36	Piran Wali	2	28.35	31.7	3.35
37	Shadman Park Mental Hospital	2	35.05	41.96	6.91
38	Sultan Pura	4	33.68	38.66	4.98
39	Takia Mahmood Shah	2	24.69	29.21	4.52
40	Tehsil Garden	2	35.46	41.86	6.4
41	Usman Park Singh Pura	2	35.66	39.01	3.35
42	Walton LDA Quarters	2	35.97	39.75	3.79
43	Waris Road	2	34.32	39.17	4.85
44	Zubair Park Raj Garh	4	26.16	31.17	5.67

found to be within acceptable limits. However, groundwater quality deteriorates toward the south, particularly in and around the Raiwind area (about 22 km from Thokar Niaz Beg), where pollutant measurements of 9000 ppm are found down to a depth of 110 m.

According to an interim report by the Pakistan Institute of Nuclear Science and Technology (PINSTECH) (PAEC and WASA, 2005), 43 samples were collected in the first sampling campaign from existing WASA tubewells (deep wells having a normal screen from 80 m to 200 m) and shallow pumps up to 40 m depth. In the second sampling campaign, about 100 samples were collected. Physicochemical parameters such as electrical conductivity (EC), pH, and temperature were measured in the field. The samples were analyzed for ^2H, ^3H, ^{18}O, ^{13}C, and other major chemical ions. Comparison of TDS in groundwater samples of the study with previous studies show that the quality of groundwater is deteriorating with the passage of time, as given in Table 8.6. The report also compares the quality of groundwater with the World Health Organization's (WHO) highest desirable level (HDL) and maximum permissible level (MPL), as reproduced in Table 8.7. Figure 8.9 further elaborates the distribution of arsenic in pumped groundwater.

According to PAEC and WASA (2005), nitrate concentrations in shallow and deep groundwater vary from 10 mg/L to 188 mg/L and 9 to 41 mg/L, respectively. Increased nitrate concentrations were also reported for both shallow and deep-groundwater sources, which exceed the WHO limit of 50 mg/L in almost all survey locations. However, nitrate concentrations in deep groundwater are less than those in shallow groundwater, making it safe to assume they are derived from currently active surface sources. According to the report, the shallow aquifer under Lahore is a major contributor to local sources – such as rain, rivers, irrigation canals, drains, and sewerage systems – but in different proportions. Accordingly, tritium data suggest that deep aquifer replenishment is very slow. The quality of shallow water

Table 8.6 Comparison of groundwater quality over time

TDS (mg/L)	(88 samples)	(> 200 samples)	(152 samples)
< 500	16%	12%	19%
501–1000	63%	40%	40%
1001–1500	17%	37%	21%
> 1500	4%	11%	20%

Source: PAEC and WASA (2005).

Table 8.7 Number of samples violating the WHO's HDL and MPL limits in wells from shallow and deep aquifers

| | | | | | No. of samples violating HDL and MPL | | | |
| | | | WHO limits | | > HDL | | > MPL | |
Parameter	No. of samples	Range (ppm)	(HDL)	(MPL)	Shallow	Deep	Shallow	Deep
As	38	0–129	0.01	0.05	16	19	6	1
Mn	42	7.5–1270	0.1	0.5	10	2	5	0
Fe	152	0.1–9	0.3	1	20	12	11	2
Ca	152	9.7–166	75	200	16	2	0	0
Na	152	2–1258	100	200	86	30	41	3
Mg	152	3.4–258	50	150	32	6	2	0
K	152	1.4–126	200	12			35	4
SO$_4$	152	4–565	200	400	10	3	4	1
Cl	152	8–2147		600	4	1	2	0
HCO$_3$	152	150–1199						

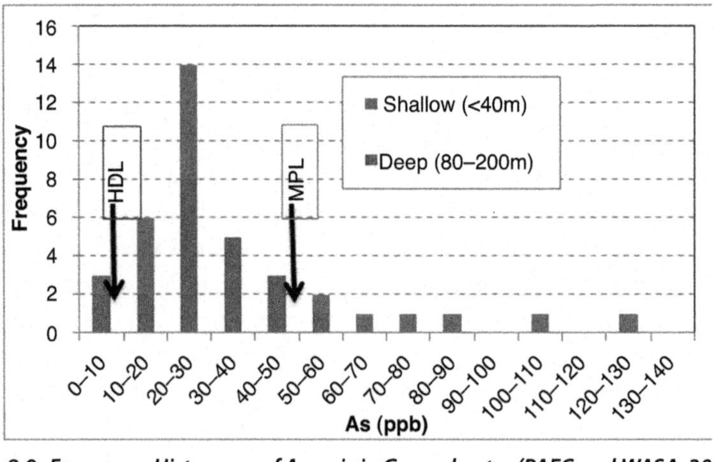

Figure 8.9 *Frequency Histogram of Arsenic in Groundwater (PAEC and WASA, 2005).*

has deteriorated over the years, and therefore the deep aquifer is under threat of pollution from top shallow groundwater. The study indicated an increasing trend of nitrate concentrations in groundwater. From Table 8.8 and Figure 8.9, it is evident that arsenic (As) contamination is present in groundwater at concentrations considerably above WHO standards. However, deeper wells are less contaminated than shallower ones.

Table 8.8 Pollution load entering the Ravi River (Hassan et al., 2013)

Site No.	Drain	EPD Discharge (cfs) (2008)	EPD TDS (ppm) (2008)	IRI TDS (ppm) (May 2011)	IRI TDS (ppm) (March 2012)
1	Mehmood Booti	20.8	312	775	1117
2	Shad Bagh	139.0	520	663	1067
3	Farrukh Abad	219.0	1000	1088	1627
4	Budha Ravi	42.0	690	1006	1100
5	Main outfall	193.0	560	627	1154
6	Gulshan-e-Ravi	246.5	660	897	1035
7	Babu Sabu	270.7	660	760	1135
8	Hudiara	535.7	1020	1197	1506

8.5.5 Recharge

Water supply and sewerage studies for Lahore began as early as 1960 (e.g., Mahboob Associates et al., 1962; Nihon Suido and ACE, 1969). Neither of these studies included development of a groundwater model. Camp Dresser and Mckee (1975) produced a master plan for the water supply, sewerage, and drainage of Lahore. This study included a first attempt at an analytical groundwater model, covering the city area of about 313 km². Against the background of rapidly increasing abstractions, falling groundwater levels, and future reduction in river recharge, NESPAK and Binnie and Partners (1988) developed a groundwater model of the aquifer under Lahore. The model grid covered an area measuring 30 × 37 km (1110 km²). According to the report, groundwater levels were in a state of quasiequilibrium in 1960. By 1976, however, groundwater levels were in steady decline. The model forecast a decline in water level of the order of 27.5 m by 2010 and proposed artificial recharge schemes for the city.

According to the water balance report by NESPAK and Binnie and Partners (1988), recharge was taking place from the Ravi River, rainfall, canal seepage, and drainage from field irrigation, whereas public and private tubewells for domestic and irrigation use as well as industrial wells were discharging. Major changes between 1960 and 1987 were abstractions and seepage from the aquifer (Table 8.9). In 1960 the Ravi River was receiving seepage from the aquifer along most of its length. By 1987 the river was a principal source of recharge and was feeding the aquifer over much of its reach (37 km) within the model area (30 × 37 km; 1110 km²). Various simulation runs were carried out for 2000 and 2010. A summary of aquifer

Table 8.9 Groundwater balance of the model area for 1960 and 1987

Year	Inflows (1000 m³/day)					Outflows (1000 m³/day)					
	From storage	Recharge*	Rivers/canals	Boundaries	Total in	Into storage	Wells	Drains	To river	Boundaries	Total out
1960**	65	107	175	119	466	133	108	40	123	62	466
1987	608	367	593	142	1710	33	1450	78	52	97	1710

* Recharge represents seepage from water supply, sewage, and minor irrigation canals. Average rainfall recharge was 38 mm/y in rural areas, 21 mm/y in suburban areas, and 16 mm/y in urban areas.
** 1960 was a dry year and there was negligible recharge from rainfall.
Source: NESPAK and Binnie and Partners (1988).

balances in the model area for December 2000 and 2010, in the case of the F5[a] scenario, is reproduced in Table 8.10.

A maximum drawdown of about 18 m in 2000 and 36 m in 2010 was simulated by the model. The study also anticipated future demand of 738 MCM/y for 2010. Accordingly, the water balance for 2010 indicated: areal recharge, 4.46 m^3/s; Ravi River and canals, 20.34 m^3/s; inflow from the boundaries, 5.69 m^3/s; and storage depletion, 5.99 m^3/s. As regards the impact of the Thein Dam on the Ravi River, the study assumed that floods with a peak discharge of less than 3250 m^3/s (115,000 cfs) would be reduced by 40%, and for floods greater than this the reduction would be less than 40%. However, the impact of the Thein Dam is presently much worse than predicted for Ravi River flows — so much so that the river remains dry for most of the year. The Ravi River flows for two months, July and August, in three out of every four years.

Alam (1994) developed a groundwater model for Lahore City covering an area of 1041 km^2. Model predictions were made for a period of 30 y (1991–2020). Abstractions estimated for 2000, 2010, and 2020 were 26.4, 36.4, and 38.3 cubic meters per second (cumecs) (932.3, 1285.4, and 1705.6 cusecs), respectively. According to the groundwater balance for the period 2011 to 2015, the Ravi River and canals in the area contribute 23.5 m^3/s, areal recharge is 5.3 m^3/s, well pumping is 39.7 m^3/s, and the flow from aquifer storage is 5.6 m^3/s. Various predicted scenarios were also simulated, one of which was the "do nothing scenario" in which "nothing would be done to arrest the decline in the water table. The population will increase in the future as in the past. The groundwater abstraction will increase in proportion to the increase in population. The natural recharge and discharge will remain the same as they were in the calibration period (1960–1990). Similarly, the stage values in the Ravi River and canals would be kept the same as in the calibration period."

According to the simulation, the depression cone would become progressively deeper and broader. Depth to the water table in the center of the depression cone would drop to 29.8, 35.8, 41.8, and 57 m in 2000, 2005, 2010, and 2020, respectively. Alam et al. (2012) conducted a postaudit study of the predictions of Alam's own 1994 model using actual monitoring data of groundwater levels by the WASA for an 11-year period (2000–2011). The prediction for 2010 regarding maximum groundwater depletion (44 m in 2009) matched well with observed data.

---

[a] Abstraction increased in the south/southeast to meet demand; demand was accordingly projected to increase from 16.8 m^3/s (593 cusec) in 1987 to 36.5 m^3/s (1289 cusec) in 2010.

Table 8.10 Summary of aquifer balances in December 2000 and 2010 for the F5 scenario

| | Inflows (m³/s) | | | | | Outflows (m³/s) | | | | |
	From storage	Recharge	Rivers / Canals	Boundaries	Total in	Into storage	Wells	Drains	To River	Boundaries	Total out
Year											
2000	1.62	3.50	15.21	2.66	26.03	0.0	25.91	0.0	0.08	0.04	26.03
2010	5.99	4.46	20.34	5.69	36.48	0.0	36.48	0.0	0.0	0.0	36.48

Source: NESPAK and Binnie and Partners (1988).

The model developed by NESPAK and Binnie and Partners (1988), which covered an area of 1110 km², indicated mean direct rainfall recharge to the aquifer on a long-term average basis from rural areas is 38 mm/y; suburban areas, 22 mm/y; and urban areas, 16 mm/y. Furthermore, irrigation and rainfall recharge were estimated as 40 mm/y and 38 mm/y. Hence, the total recharge rate in irrigated areas comes out at 78 mm/y. Thus, recharge from the area outside the model developed by NESPAK and Binnie and Partners (1988) is 1.64 m³/s (57.93 cusec). After adding recharge for the NESPAK model area (1110 km²) and the remaining irrigated area of Lahore District, total recharge (including inflow from boundaries) is 32.13 m³/s, which is equivalent to 1013.3 MCM/y. Thus, groundwater abstraction as a percentage of groundwater recharge is 114.6%.

8.6 IMPACTS

The impacts of excessive groundwater use in Lahore City are reflected by the following indicators: decline in groundwater levels, deterioration of groundwater quality, decline in the production capacity of wells, land subsidence, and effects on public health.

8.6.1 Decline in Groundwater Level

As a result of overexploitation and reduced recharge, the water table is lowering and aquifers are being depleted. Basharat and Rizvi (2011) analyzed dynamic water table (DTW) data from a few observation wells (Figure 8.10) in and around the urban area of Lahore. Depth to the water table near

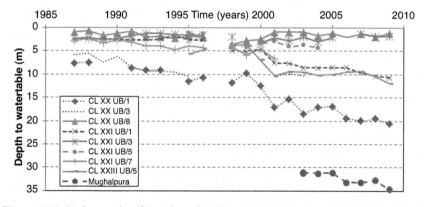

Figure 8.10 *Hydrographs of Depth to the Water Table in the Area.* Well locations are shown in Figure 8.12. *Source: SMO, WAPDA.*

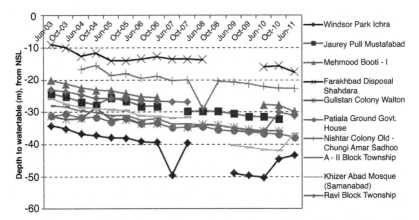

Figure 8.11 *Groundwater Trends in Lahore City (2003–2011).* Well locations are shown in Figure 8.12.

Bhobattian Chowk on Raiwind Road increased from 9.2 m in 1993 to 10.4 m in 2003. However, depth to the water table near the River Ravi, adjacent to Thokar Niaz Beg, increased by 10.8 m between 1999 and 2009 (observation well # CLXXUB/1, Figure 8.10). This was due to desiccation of the Ravi River, particularly after the construction of the Thein Dam in India. Similarly, in the city center of Mughalpura the water table dropped by 3.5 m over a period of 6 years. This means a depletion rate of 0.6 m/y in the city. On the other hand, no long-term aquifer depletion trend has been observed in the surrounding agricultural areas. Groundwater levels at selected representative tubewells in Lahore by the WASA for 2003–2011 are shown in Figure 8.11. According to the data, the decline of groundwater levels during the last 8 years in Lahore City ranged from 4.61 m (Gulistan Colony TW) to 10.85 m (Khizar Abad Mosque), equivalent to an annual groundwater depletion rate of 0.57–1.35 m. Maximum depth to the water table reached 45 meters below ground level (mbgl) in 2011 (Figure 8.12).

8.6.2 Deterioration of Groundwater Quality

The WASA of the LDA is responsible for the provision of water and sanitation facilities to almost 66% of the city area. There are 15 main drains in Lahore and Sheikhupura which discharge untreated municipal and industrial wastewater directly into the Ravi River. Their names and respective discharges (m³/s) are Farrukhabad Drain (5.13), Mehmood Booti (6.36), Shad Bagh (6.78), Khokhar Road (4.77), Bhati Gate (2.84), Main Outfall 1 (5.13), Main Outfall 2 (2.90), Main Outfall 3 (2.47), Gulshan-e-Ravi

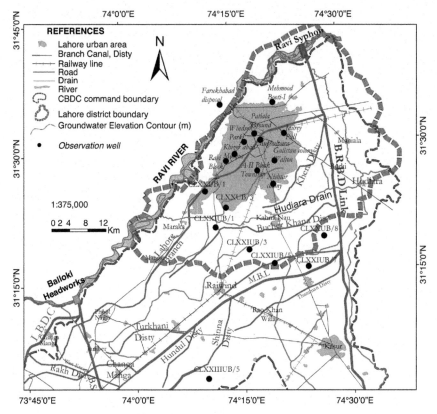

Figure 8.12 *Location Map of Observation Wells Shown in Figures 8.10 and 8.11.*

(15.86), Multan Road (6.78), LMP-Block Model Town Extension (3.55), and Nishtar Colony (3.15) (Hussain and Sultan, 2013).

According to Naqvi et al. (2012), many industries located near the Ravi River discharge their effluent into the river; in addition, the river brings some effluent from India. Previously, almost all the municipal waste of Lahore City was dumped into the Ravi River. However, nowadays the LDA has allocated three scientifically designed solid waste disposal sites. Nevertheless, the Ravi River is still being polluted with substantial quantities of industrial and municipal wastewater *via* untreated effluent disposal. Flow in the Ravi River is highly variable and reduces by as much as 1357.7 MCM (1.1 MAF) per annum, because of irrigation and hydropower diversions developed by India. Upstream of Balloki Head the river mostly discharges wastewater. These surface drains combined with the sewerage system in urban areas are polluting the upper shallow-groundwater layer. Despite insufficient data regarding the extent of groundwater quality deterioration of the upper shallow layer, it can be indirectly ascertained from data regarding surface drainage discharge and deteriorating quality, explained as follows:

1. *Ravi River pollution:* Groundwater pollution through solid waste, leakage, and disposal of sewage and industrial waste directly into surface drains is a major issue for the sustainability of good-quality groundwater from the aquifer. The dumping of solid waste directly into the drains at various locations is causing a reduction in their capacities; such a source of pollution is an environmental hazard (Aftab, 2005). Due to desiccation of the Ravi River and untreated discharge of municipal and industrial waste, it is now regarded more as a drain, particularly during the nonmonsoon season. Industrial areas adding wastewater to the Ravi River are Kalashah Kaku; the Gulberg industrial area; Kot Lakhpat; and the new Sundar Industrial Estate. It is only very occasionally that India spills excess flows downstream during floods, but the river washes away such year-round pollution. However, this extra floodwater also helps pollution to travel down into the aquifer by providing the extra solvent and head needed for solute transport. Hassan et al. (2013) pointed out that untreated drainage disposal into the Ravi River is deteriorating the quality of river water as well as the underlying groundwater; the situation is worst at Shahdara Bridge. The quality of drain water entering the river is deteriorating with the passage of time, as shown in Table 8.8.

2. *Hudiara Drain:* The Hudiara Drain has a total length of 98.6 km, of which 44.2 km lies in India and the remaining 54.4 km in Pakistan (Afzal et al., 2000); it joins the River Ravi about 5 km downstream of Thokar Niaz Beg. It enters Pakistan loaded with pollution from India. The drain carries mainly industrial and agricultural waste from both India and Pakistan. All along its route in India and Pakistan, wastewater, sewage, and industrial pollutants are discharged into the drain without any proper prior treatment. As a result, organic waste and toxic chemicals have badly affected aquatic life, both in the drain and the River Ravi. Farmers living near the drain frequently use its water for irrigation. Preliminary investigations have revealed that this water has high concentrations of metals. It flows all year round, carrying untreated sewage and chemical wastewater from 104 factories. There are around 100 factories located adjacent to the Hudiara Drain on the Indian side, so it is already quite toxic when it enters Pakistan. Until about 30 years ago, the Hudiara Drain used to be a stormwater drain used for irrigation and domestic purposes, draining

into the Ravi River and adding to the river's aquatic health but now it is the major source of pollution for the Ravi River.

Hudiara Drain carries about 5 cumec (177 cfs) of discharge (estimated value using the float method since no measured data are available, Basharat and Rizvi, 2011). "The people living along the drain in Pakistan, especially in areas bordering India, are afflicted by the hazardous effects of the untreated water, and we have been facing this problem for the last 20 years. The pollution fluctuates according to the volume of the water," says Muhammad Jamil, a farmer. Mian Mahboob, a local politician, even suspects that contaminants are being transmitted through cattle milk because domesticated water buffaloes and cows drink from the drain too. Experts say that many of these fears are not unfounded. "Cadmium, chromium, and copper in chemical waste from factories located alongside the drain are making vegetables oversized. The heavy chemicals, all carcinogenic, eventually end up in the food chain" (WWF Pakistan, http://www.adb.org/water/actions/REG/hudiara-drain.asp). Heavy metals such as Pb, Cu, Ni, and Zn have been detected in the Hudiara Drain (DLR, 2007). Their presence coupled with other salts pose a major threat to the quality of underlying groundwater.

8.6.3 Decline in the Production Capacity of Wells

Lahore has a total of 15 tubewells, 6 of which were reported as having reduced specific capacity of more than 50% (NESPAK and Binnie and Partners, 1987). Many tubewells have had to be abandoned as a result of reduced well yield and the excessive and gradual decline in groundwater levels (Figure 8.13). Tubewell boring depths are successively increased by the public and private sector to get the desired discharge and quality of pumped water in an effort to insure long-term supply of potable groundwater from the wells.

8.6.4 Land Subsidence

No information whatsoever is available on land subsidence because no monitoring is being undertaken. No publication or study is available regarding this in Lahore City, despite groundwater depletion progressing at an alarming rate and maximum depth to groundwater of 45 m being reached.

8.6.5 Public Health

Groundwater is the sole source of water supply in the city. However, its quality is under increasing threat due to domestic and industrial disposal

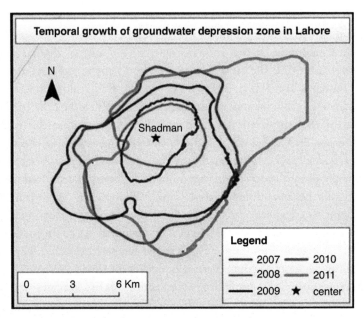

Figure 8.13 *Expansion of the Depression Zone Contour by 38 m since 2007. Source: Kanwal et al. (2015).*

into surface drains, which ultimately enters the Ravi River at various points. This effluent contributes to groundwater recharge. Farther downstream these surface flows are used for irrigation. Therefore, the degradation of surface and groundwater quality has badly impacted human health since this water is directly used or abstracted from the aquifer (using shallow pumps).

8.7 RESPONSES

Responses from different levels of society – such as groups of individuals, governments, or nongovernmental sectors in Lahore City – are discussed in terms of the following indicators: groundwater monitoring, development of environmental standards and guidelines, and arranging surface water from adjoining areas.

8.7.1 Groundwater Monitoring

Proper groundwater monitoring was started in 1960 to prepare a regional piezometric map. Continuous monthly monitoring is carried out by the WASA in urban areas. Monitoring is carried out twice a year in surrounding irrigated areas by the Irrigation Department and SCARP Monitoring

Organization (SMO) of the WAPDA. Thus, monitoring is already well established. However, a means of reporting the groundwater level situation to relevant concerned agencies and raising public awareness is strangely lacking. This could easily be achieved by allowing academia and research institutions to have easy access to the data. In this way, analysis of the situation could be carried out and reported in the literature; hence, the importance of building a data repository in a format that is easily accessible by stakeholders.

8.7.2 Environmental Standards and Guidelines

Various laws and regulations have been framed aimed at resolving the various adverse environmental impacts. The Pakistan Environmental Protection Agency was established under section (5) of the Pakistan Environmental Protection Act 1997 (Pak–EPA). The basic functions of Pak–EPA are to prepare, revise, and establish National Environmental Quality Standards (NEQS); take measures to promote the research and development of science and technology which may contribute to the prevention of pollution, protection of the environment, and sustainable development; identify the needs for, and initiate legislation in various sectors of the environment; provide information and guidance to the public on environmental matters; specify safeguards for the prevention of accidents and disasters which may cause pollution; and encourage the formation and work of nongovernmental organizations, community organizations, and village organizations to prevent and control pollution and promote sustainable development. The agency may undertake inquiries or investigate environmental issues, either of its own accord or upon complaint from any person or organization (http://environment.gov.pk/aboutus/Brief-Pak-EPA.pdf).

Improving the quality of water for drinking purposes, domestic consumption, personal hygiene, and certain medical situations has always been a top priority of the Government of Pakistan. *"The Guidelines and Criteria for Quality Drinking Water"* published by the WHO (1996) have made it possible to review, evaluate, and further improve the quality of water in Pakistan by measuring it against these standards. Through a combination of lectures, discussions, intense work sessions, and provision of relevant reading literature by the WHO and Ministry of Health, quality standards for drinking water in Pakistan have been finalized (Government of Pakistan, 2008).

However, the provisions of these laws to protect the environment have never truly been implemented as a result of deficiencies in legalities, nonapplication of fines, lack of public awareness, official ignorance of environmental issues, and political commitments. Sectoral legislation has totally failed to play even a limited role in environmental protection. Therefore, it is recommended that proper implementation of these laws be carried out to insure long-term sustainability of the aquifer.

8.7.3 Allocating Surface Water from the Irrigation Sector

Since groundwater is the sole source of water supply, it is being pumped by public and private sectors in an unrestricted and unregulated way. The only restriction is the electricity burden, which is borne by the WASA and private housing societies, whereas individual households normally pay at a flat rate. So, there is no economic impact on anyone regarding overuse of this precious resource. The Lahore aquifer is clearly under stress. Strict regulation and monitoring of quality and potential are the orders of the day to assure sustainability of this precious resource., The government does not presently seem to be prepared to adopt economic instruments as regulation tools to insure sustainability of the groundwater situation. A few years ago, the Government of Punjab announced construction of a lake on the River Ravi in an attempt to raise and maintain the rapidly decreasing groundwater level of the city. There were plans to construct an international-standard recreational park for its citizens at the same time. However, the government dismissed the plan in favor of a new project called the "Ravi River Development Front," the purpose of which is to build a new city along the Ravi River at Shahdara. This new city will be built on both banks of the river and will be 33 km in length. The project entails building a 36 km long concrete channel for the Ravi River and developing a new city on its banks with green belt areas, parks, and recreation facilities. A 1 km wide and 15 ft. deep river would be channeled from the siphon on the Bambawali–Ravi–Bedian (BRB) Canal to the Maraca area.

This project entails development of a number of high-quality residential schemes, colonies, educational institutes including colleges and universities, markets, trade centers, and high-rise plazas on land acquired in Lahore and Sheikhupura districts on both sides of the River Ravi. The project is likely to be completed in four to five years as per the wishes of the Punjab Chief Minister. The Ravi River will also be cleaned as part of the project. There is provision for wastewater treatment according to NEQS. In the master plan (yet to be developed), creation of lined channels and lakes, provision of extra surface supply is also being considered. Despite 100 cusec of surface water from the BRBD Canal being requested from the Irrigation Department for allocation to Lahore, it was agreed to provide only 25 cfs of canal water.

This river course development idea has paid little attention to issues of concern to the city, other than the groundwater depletion issue and keeping the water within the bounds of the river channel. Without major allocation of river water (at least 0.5 MAF) from the Indus Basin Irrigation System for water supply and additional recharge to the aquifer, this kind of business activity is not expected to improve the aquifer situation. This is because the additional amount of groundwater to be pumped for residents of the new river city would be more than additional recharge from the river channel. However, the LDA has initiated a rainwater-harvesting system in an effort to save depletion of underground water resources in the provincial metropolis. This system entails establishing 39 points for collecting rainwater at various appropriate locations, which will be used to recharge the underground aquifer. Two recharge wells have been announced for construction as pilots.

8.8 CONCLUDING SUMMARY AND RECOMMENDATIONS

Data analysis and use of the driver–pressure–state–impact–response (DPSIR) framework show that the city infrastructure is expanding and the population is increasing. Hence, drivers are constantly imposing increased demands on the aquifer. To make matters worse, recharge to the aquifer (the sole source of water supply) is decreasing, both in quantity and quality, as a result of urban housing development since the 1980s. Thus, per capita water availability from the safe yield of the aquifer system is decreasing with the passage of time. Reduced recharge is mainly caused by desiccation of the Ravi River and land use change to urbanization. Unregulated abstraction in the absence of recharging efforts or surface supplies has led to lowering of the water table. Thus, anthropogenic activities are successively putting pressure on the aquifer and the resulting impacts are very serious. However, society as a whole has failed to recognize the seriousness of the situation. The main conclusions regarding the groundwater water environment are as follows:

- Urban areas are growing at a rate of about 2% per year, which is expected to continue in the future. Population shift toward the metropolis from villages, small towns, and other cities in the province in search of quality education, jobs, and lavish lifestyles will intensify further.
- Recharge from surface and groundwater inflows from their surroundings is less than 15% of water abstracted from the aquifer.
- On average, groundwater depletion varies from 0.57 m/y to 1.35 m/y. Bearing in mind groundwater level decline in the city, the WASA regularly monitors groundwater levels in its wells. However, the non-WASA

part of the urban area is not being monitored by the respective agencies or the WASA, the biggest public stakeholder in the city.

- The quality of shallow water has deteriorated over the years. Hence, heavy exploitation of the deep aquifer makes it vulnerable to pollution from top shallow groundwater, which is being polluted by top surface recharge.
- The Hudiara Drain and Ravi River are major sources of pollution for the Lahore aquifer. Another concern is municipal waste disposal, which is mostly untreated and likely to be a major health hazard for users of underlying fresh groundwater and downstream surface water.
- The standard of monitoring the quality of deep groundwater being pumped is good enough to pick up toxic elements. However, there is a lack of prompt action when abnormal values are found (e.g., arsenic levels exceeding safe standards for drinking water quality).
- Water supply to users is mostly based on flat rates. Hence, economic instruments for groundwater abstraction, regulation, and quality protection are not being given due consideration as a result of there being no central regulatory authority imposing such economic instruments as binding.

Recommendations for making the water supply of the city sustainable are as follows:

- Each and every user of existing groundwater consuming more than 0.20 cusec should be registered, get a permit before the installation of pumping equipment, and incur a groundwater development surcharge.
- Building byelaws should be amended to achieve maximum possible rainwater harvesting and maximum aquifer recharge by imposing mandatory regulations for the dimensions of buildings coupled with certain incentives.
- Wastewater disposal by all entrepreneurs should be monitored. There should be the option of treating it at source or, otherwise, paying extra charges for untreated loads or for its collection.
- Sewage that has to be released into the Ravi River should be conveyed in a lined channel along both banks to avoid polluting the aquifer.
- Two small weirs (at the very least) should be constructed, one at Shahdara and the other about 20 km farther downstream, to enhance groundwater recharge with good-quality river water, especially during floods.
- Surface water should be allocated (at least 0.5 MAF) from the Indus Basin Irrigation System for water supply and additional recharge to the aquifer to improve the groundwater situation.
- Groundwater modeling should be undertaken to study and suggest appropriate management interventions under current and expected future conditions.

- Environmental laws already exist, but they are not being implemented because political commitment is lacking. Therefore, it is recommended that proper implementation of these laws be carried out to ensure the long-term sustainability of the aquifer.

REFERENCES

Aftab, M.P., 2005. Environmental profile of Lahore. J. Pakistan Eng. Congr. 42 (5), 9–15.

Afzal, S., Ahmad, I., Younas, M., Zahid, M.D., Khan, M.H.A., Ijaz, A., Ali, K., 2000. Study of water quality of Hudiara drain, India–Pakistan. Environ. Int. 26, 87–96.

Alam, K., 1994. Groundwater flow modeling of Lahore city and surroundings. MPhil thesis, Centre of Excellence in Water Resources Engineering, University of Engineering and Technology, Lahore, pp. 127.

Alam, K., Ahmad, N., Khan, M.R., 2012. Predictive accuracy of a groundwater flow model developed for Lahore city and surroundings – a post audit study. In: Proceedings of International Conference: Water, Energy, Environment and Food Nexus: Solutions and Adaptation under Changing Climate.

Basharat, M., Rizvi, S.A., 2011. Groundwater extraction and waste water disposal regulation – Is Lahore aquifer at stake with as usual approach? Presented on World Water Day "Water for Cities – Urban Challenges" organized by Pakistan Engineering Congress on April 16, 2011, Lahore, Pakistan.

Basharat, M., Tariq, A.R., 2013. Long term groundwater quality and saline intrusion assessment in an irrigated environment: a case study of the aquifer under LBDC irrigation system. Irrig. Drain. 62, 510–523.

Basharat, M., Ali, S.U., Azhar, A.H., 2014. Spatial variation in irrigation demand and supply across canal commands in Punjab: a real integrated water resources management challenge. Water Policy 16 (2), 397–421.

Bennett, G.D., Ata-ur-Rehman, Sheikh, I.J., Ali, S., 1967. Analysis of Aquifer Tests in the Punjab Region of West Pakistan. Geological Survey Water-Supply Paper 1608-G.

Bureau of Statistics, 2005. Punjab Development Statistics, 2005. Bureau of Statistics, Government of Punjab, Lahore, Pakistan.

Bureau of Statistics, 2012. Punjab Development Statistics, 2012. Bureau of Statistics, Government of Punjab, Lahore, Pakistan. Available from: http://www.documents.pk/docs/detail/list-of-industries---lahore_289

Camp, Dresser and Mckee Ltd., 1975. Lahore water supply and sewerage and drainage project, Final Report.

Din, Z.U., Sheikh, I.A., 1969. Study of lithological variations of the materials from the shallow depths to deeper zones in the ex-Punjab area and its impact on the specific yield. Technical Paper No. 36, WASID Publication No. 76.

DLR, 2007. Surface water quality monitoring in Punjab. Annual Report 2007, Directorate of Land Reclamation, Irrigation and Power Department, Punjab, Canal Bank Moghalpura, Lahore.

Government of Pakistan, 2008. National Standards for Drinking Water Quality (NSDWQ). Pakistan Environmental Protection Agency, Ministry of Environment, Government of Pakistan, Islamabad.

Government of Pakistan, 2011. Agricultural Statistics of Pakistan 2010–2011. Statistics Division, Pakistan Statistics Bureau, Government of Pakistan, Islamabad.

Government of Punjab, 2011. Statistical Pocket Book of the Punjab, 2011. Bureau of Statistics, Government of Punjab, Lahore. Available from: http://www.bos.gop.pk/?q=system/files/Statistical_poket_book_2011.pdf. 2011.

Greenman, D.W., Swarzenski, W.V., Bennett, G.D., 1967. The groundwater hydrology of the Punjab, West Pakistan. WAPDA, Water and Soils Investigation Division, Bulletin No. 6.

Hassan, G.Z., Shabir, G., Hassam, F.H., Akhtar, S., 2013. Impact of pollution in Ravi River on groundwater underlying the Lahore city. Proceedings of Seventy-Two Annual Session, Pakistan Engineering Congress, Lahore, Pakistan.

Hussain, F., Sultan, A., 2013. Existing situation of sewerage in Lahore city and its impact on River Ravi. The Urban Gazette, The Urban Unit, 503 Shaheen Complex, Egerton Road, Lahore, Pakistan.

Iftikhar, P., 2007. Presentation on the initiative taken in WASA Lahore regarding benchmarking of UWSS Water.

Kanwal, S., Gabriel, H.F., Mahmood, K., Ali, R., Haidar, A., Tehseen, T., 2015. Lahore's groundwater development – a review of the aquifer susceptibility to degradation and its consequences. Tech. J. 20 (1–2015), 26–38.

Mahboob Associates and Hydrotechnic Corporation, 1962. Lahore Water Supply, Sewerage and Drainage Project, Feasibility Report.

Mahmood, K., Rana, A.D., Tariq, S., kanwal, S., Ali, R., Ali, A.H., Tahseen, T., 2013. Groundwater levels susceptibility to degradation in Lahore metropolitan. Sci. Int. (Lahore) 25 (1), 123–126.

Naqvi, S.A., Asghar, S.N., Mehmood, M.A., Abbas, A., Kashmiri, S.A., 2012. Report of River Ravi Commission Sub-committee: Findings of visits to bioremediation sites, Islamabad and Waste Stabilization Plant, Faisalabad. A report submitted to River Ravi Commission, pp. 1–38.

NESPAK and Binnie and Partners, 1987. Groundwater resources evaluation and study of aquifer under Lahore. Interim Report, National Engineering Services of Pakistan in joint venture with Binnie and Partners Consulting Engineers, London, UK.

NESPAK and Binnie and Partners, 1988. Groundwater resources evaluation and study of aquifer under Lahore. National Engineering Services of Pakistan in joint venture with Binnie and Partners Consulting Engineers, London, UK.

Nihon Suido and Associated Consulting Engineers, 1969. Master Plan for Water Supply, Sewerage and Drainage in Lahore.

PAEC and WASA, 2005. Interim Report on Hydrological Investigations of Lahore Aquifer. Head Radiation and Isotope Application Division, Pakistan Institute of Nuclear Science and Technology, Pakistan Atomic Energy Commission (PAEC), P.O. Nilore, Islamabad and Water and Sanitation Agency (WASA), Lahore.

Pakistan Meteorological Department, 2006. Agrometeorological Bulletins of Pakistan for 2006. National Agromet Centre of Pakistan Meteorological Department, Islamabad.

Pakistan National Committee of ICID, 1991. Irrigation and drainage development in Pakistan. Country Report, Asia.

Punjab Development Statistics, 2012. Bureau of Statistics, Government of Punjab, Lahore.

Sheikh, I.A., 1971. Long term values of specific yield in Bari Doab. WASID Publication No. 95, WAPDA.

Thomas Brinkhoff, 2014. Major Agglomerations of the World – Population Statistics and Maps. Citypopulation.de, July 1, 2014. Available from: http://www.citypopulation.de/world/Agglomerations.html (accessed 12.08.2014).

WAPDA, 1980. Hydrogeological Data of Bari Doab, vol. 1. Basic Data Release No. 1, Directorate General of Hydrogeology, WAPDA, Lahore.

WHO, 1996. Guidelines for Drinking-Water Quality. Recommendation 1, 16–17.

Zaman, K., Baloch, A.A., 2011. Urbanization of arable land in Lahore city in Pakistan: a case study. J. Agric. Biotech. Sustain. Dev., Vol. 3(7), pp. 126–135.

SECTION III

Groundwater Environment in Southeast Asia

CHAPTER 9

Water Environment in Southeast Asia: An Introduction

Vishnu Prasad Pandey*, and Sangam Shrestha****
*Department of Civil Engineering, Asian Institute of Technology and Management (AITM), Nepal
**Water Engineering and Management, Asian Institute of Technology (AIT), Thailand

9.1 PHYSIOGRAPHY AND CLIMATE

The 12 countries that make up Southeast Asia cover an area of approximately 4.83 million km^2 (WB, 2015) within the geographical boundary between 12°S to 29°N and 92°E to 160°E (Figure 9.1). The subregion consists of continental margins and offshore archipelagos of Asia that lie geographically south of China and east of India. Elevation ranges from sea level to 5881 m above mean sea level (masl) at Hkakabo Razi in Myanmar. In terms of geographical area, climate, population size and density, and per capita income, there are large variations among the subregional countries (Table 9.1).

There are, of course, many differences in the physical environment of mainland and island Southeast Asia. These include the long rivers that begin in the highlands separating the subregion from China and northwest India and the extensive lowland plains separated by forested hills and mountain ranges. These fertile plains are highly suited to rice growing. They are home to several ethnic groups such as the Thais, the Burmese, and the Vietnamese, who developed settled cultures that eventually provided the basis for modern states.

Virtually, the entire subregion lies between the tropics, hence the similarities in climate as well as plant and animal life throughout. Except for the humid subtropical climate in the north, the climate is generally tropical, exhibiting hot and humid conditions throughout the year, and greatly influenced by monsoons originating in the South China Sea (UNEP, 2004). The economic wellbeing of the entire subregion is dependent on monsoon winds that blow regularly from the northwest and then reverse to blow from the southeast. These wind systems bring fairly predictable rainy seasons. The subregion has wet and dry seasons, caused by seasonal shifts in the monsoons. The tropical rain belt results in heavy precipitation during the

Groundwater Environment in Asian Cities
http://dx.doi.org/10.1016/B978-0-12-803166-7.00009-X

Copyright © 2016 Elsevier Inc.
All rights reserved.

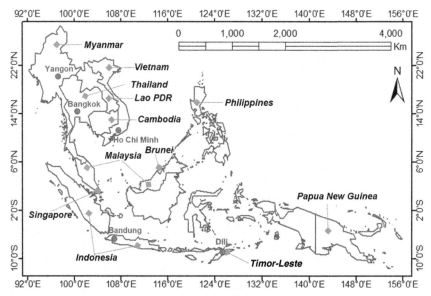

Figure 9.1 *Location of Countries and Case Study Cities in Southeast Asia.*

monsoon season. The coastlines of this subregion border the Andaman Sea, Gulf of Thailand, and South China Sea. The mountainous areas in the north, where higher altitudes lead to milder temperatures and drier landscapes, are the exception to this climate and vegetation (Babel and Wahid, 2009). The region receives ample amounts of rainfall, which varies from 1.8 billion cubic meters (BCM) per year in Singapore to 5163 BCM/y in Indonesia (Table 9.2). Temperatures are generally warm, although it is cooler in highland areas.

9.2 SOCIOECONOMICS AND ENVIRONMENTAL ISSUES

Total population living in mainland Southeast Asia (as of 2012) is estimated at 618.7 million (FAO, 2015), which varies from 0.41 million in Brunei to 246.84 million in Indonesia. Most people live along the major rivers and lakes, such as the Mekong River and Tonle Sap Lake. Population density varies from 15 persons/km^2 in Papua New Guinea to 7406 persons/km^2 in Singapore. This subregion comprises countries varying widely in area such as the small city-state of Singapore and Indonesia, the largest in the subregion. Land resources in the subregion are limited: despite covering only 3.35% of the world's total land area it is home to 8.8% of the world's population (as of 2012).

Table 9.1 Annual water withdrawal by sector and source in Southeast Asia

	Annual water withdrawal				With-drawal as % of TRWR	Source of fresh water	
	Agriculture sector		Total withdrawal			Groundwater	
Country	Volume (MCM)	% of total	Volume (BCM)	m³/ person		Volume (MCM/y)	% of total
Brunei	5.3	5.8	0.09	296.8	1.1	0.5	0.5
Cambodia	2053.0	94.0	2.18	158.9	0.5		
Timor-Leste	1071.0	91.4	1.17	1131.0	14.3		
Indonesia	92,760.0	81.9	113.30	526.9	5.6	17,610.0	15.5
Lao PDR	3193.0	91.4	3.49	580.9	1.0		
Malaysia	2505.0	22.4	11.20	417.7	1.9	165.0	1.5
Myanmar	29,570.0	89.0	33.23	674.6	2.8	2991.0	9.0
Papua New Guinea	1.0	0.3	0.39	61.3	NA	NA	NA
Philippines	67,070.0	82.2	81.56	843.4	17.0	3206.0	3.9
Singapore	7.6	4.0	0.19	82.0	31.7		
Thailand	51,790.0	90.4	57.31	867.3	13.1	9827.0	17.1
Vietnam	77,750.0	94.8	82.03	947.7	9.3	1402.0	1.7
Southeast Asia	327,775.9	84.9	386.2	624.1	5.4		

BCM, billion cubic meters; MCM, million cubic meters; NA, not applicable.
Source: FAO (2015).

Table 9.2 Water resource availability in Southeast Asia in 2012

		TRWR		
Country	P (BCM/y)	BCM/y	m³/person/y	TRGWR (BCM/y)
Brunei	15.7	8.5	20,631	0.10
Cambodia	344.7	476.1	32,028	17.60
Timor-Leste	22.3	8.2	7374	0.89
Indonesia	5163.0	2019.0	8179	457.40
Lao PDR	434.3	333.5	50,181	37.90
Malaysia	951.0	580.0	19,836	64.00
Myanmar	1415.0	1168.0	22,122	453.70
Papua New Guinea	1454.0	801.0	111,762	211.60
Philippines	704.4	479.0	4953	180.00
Singapore	1.8	0.6	113	—
Thailand	832.3	438.6	6567	41.90
Vietnam	602.7	884.1	9737	71.42
Southeast Asia	11,941.2	7197	11,632	1537

BCM, billion cubic meters; P, precipitation; TRGWR, total renewable groundwater resources; TRWR, total renewable water resources.
Source: FAO (2015).

Economic development measured in terms of per capita GDP in 2012 illustrates a distinct variation among the subregion's countries: Cambodia (946 USD/person) and Timor-Leste (1216 USD/person) have low GDPs, whereas Singapore (54,120 USD/person), Brunei (41,150 USD/person), and Malaysia (10,431 USD/person) have high GDPs (FAO, 2015). Agriculture contributes only 12.5% to the GDP of the entire region; however, the figure in some countries is as high as 28% (in Lao PDR) and 36% (in Myanmar).

The growing population and urbanization trend is expected to create a range of environmental problems that will put pressure on the subregion's water resources.

9.3 WATER AVAILABILITY AND WITHDRAWAL

The subregion is drained by a number of great rivers such as the Mekong, Salween, Irrawaddy, Sepik, Chao Phraya, Red, and Ma, which are endowed with abundant fresh water resources. It receives 9.5% of total global precipitation every year and is blessed with 16.2% of the world's total renewable water resources (Babel and Wahid, 2009). Average annual precipitation and renewable water resources vary widely in the subregion (Table 9.2). Despite increasing population pressures, per capita water resource availability is well above 5000 m^3/y in all the subregion's countries except Singapore (Table 9.2). Singapore, with a population density of 7400 persons/km^2, is severely water stressed. Total renewable water resources in the entire subregion are about 7197 BCM/y, about 21.4% of which is contributed by groundwater (Table 9.2). Total groundwater resources in the subregion vary from 0.1 BCM/y in Brunei to 457.4 BCM/y in Indonesia. However, water resources are stressed as a result of rapid urbanization and industrialization. The situation is made worse by water-related disasters, climate change, and poor governance. This has led to many rivers and lakes being polluted and, in some places, seriously contaminated at an alarming speed.

Subregional total freshwater withdrawal is 5.4% of total available renewable water resources, which alarmingly is well below the withdrawal threshold of 40% proposed by the WMO (1997). Total withdrawal in the subregion varies from 0.5% in Cambodia to 31.7% in Singapore (Table 9.1). Per capita withdrawal in the entire region is 624.1 m^3/y, which varies from 61.3 m^3/y in Papua New Guinea to 1131 m^3/y in Timor-Leste. Agricultural withdrawal accounts for 84.9% of total withdrawal, which varies from less than 5% in Papua New Guinea, Singapore, and Brunei to above 90%

in Vietnam, Cambodia, Lao PDR, Timor-Leste, and Thailand (Table 9.1). Although information on groundwater withdrawal of countries like Cambodia, Timor-Leste, Lao PDR, Papua New Guinea, and Singapore is not available in the FAO (2015) database, groundwater is the primary source of drinking and irrigation in the subregion. It accounts for 17.1% of total annual water withdrawal in Thailand, 15.5% in Indonesia, and 9% in Myanmar.

9.4 CASE STUDY CITIES

Groundwater is the primary resource supporting overall development in the subregion. The subregion has to address a number of groundwater environmental issues – such as groundwater reserve depletion, aquifer contamination, land subsidence, and salinity intrusion in coastal aquifers – because of uncontrolled abstraction in many aquifers. This part of the book analyzes the status of the groundwater environment in the aquifers of five cities, which are selected randomly. Selection was made as diverse as possible subject to the availability of contributors. The cities in alphabetical order are Bandung (Indonesia), Bangkok (Thailand), Dili (Timor-Leste), Ho Chi Minh City (Vietnam), and Yangon (Myanmar). They are detailed in Chapters 10–14. General characteristics of the cities are provided in Table 1.1.

ACKNOWLEDGMENT

The authors would like to acknowledge SEA-EU-NET II Project (http://www.sea-eu.net/about/aims_results) for supporting the development of this chapter.

REFERENCES

Babel, M.S., Wahid, S.M., 2009. Freshwater Under Threat: Southeast Asia, Vulnerability Assessment of Freshwater Resources to Environmental Change. United Nations Environment Programme and Asian Institute of Technology, Bangkok.

FAO, 2015. AQUASTAT Database. Food and Agriculture Organization of the United Nations (FAO). Available from: http://www.fao.org/nr/water/aquastat/data/query/index.html?lang=en (accessed 13.06.2015.).

UNEP, 2004. Environmental Indicators: South East Asia. United Nations Environmental Programme (UNEP), Regional Resource Center for Asia and the Pacific, AIT, Thailand.

WB, 2015. World Development Indicators. The World Bank (WB). Available from: http://data.worldbank.org/indicator (accessed 14.04.2015.).

WMO, 1997. Comprehensive Assessment of the Freshwater Resources of the World. Overview document, World Meteorological Organization (WMO), Geneva.

CHAPTER 10

Groundwater Environment in Bandung, Indonesia

Haryadi Tirtomihardjo
Center of Groundwater Resources and Environmental Geology (CGREG), Ministry of Energy and Mineral Resources (MEMR), Indonesia

10.1 INTRODUCTION

Groundwater is used as the water supply for various purposes in major Indonesian cities. Its use has been continuously increasing in recent decades. In several large cities in Java, such as Bandung (Figure 10.1), deterioration in both the quantity and quality of groundwater has been observed, while degradation of the environment such as land subsidence due to groundwater overexploitation is readily evident (Schmidt et al., 2005). Bandung, located in the Bandung–Soreang Groundwater Basin (BSGB), is the capital of West Java Province. The city is one of the major economic and industrial development centers in the province. The Bandung Metropolitan City witnessed an increase in population from 2.24 million to 2.46 million between 2004 and 2012. With a population growth of 1.2% per annum, population density has increased from 13,270 person/km^2 to 15,640 persons/km^2.

The urban water supply of Bandung until 1979 was dependent on 10 captured springs, 11 drilled wells, and a surface water source. These systems provided an annual total supply of 33 million cubic meters (MCM), up to 40% of which was lost as nonrevenue water (NRW) and the remainder served only 23% of the population. Since 1978 an additional 18.5 MCM/y of supply has flowed into the central urban water supply system by drilling 22 deep wells. With increased supply and a reduction of pipe losses, NRW dropped to 35% and the system served 42% of the population. By 1996 the central water supply system served 75% of the population of Bandung (Priowirjanto, 1985; Wagner et al., 1991).

In the early 1990s some 28 MCM/y of groundwater was abstracted by 25 deep wells for the central urban water supply of Bandung, Cimahi, Majalaya, and Cicalengka including piped discharge of five springs captured on the slopes of Tangkuban Perahu Volcano. Up to 50 MCM/y of

Copyright © 2016 Elsevier Inc.
All rights reserved.

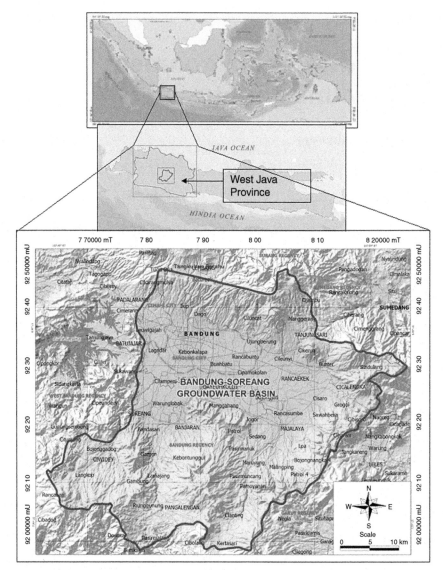

Figure 10.1 *Location Map of the BSGB.*

groundwater was abstracted to meet industrial water demand. However, a high proportion of the population was still supplied by a large number of shallow dugwells (Schmidt et al., 2005).

Total annual domestic and industrial water demand in the Bandung Basin was projected to increase from under 50 MCM in 1990 to around 300 MCM in 2006. However, implementation plans to meet incremental demand by large-scale expansion of well yields was restricted by the high cost of investment and operations. Instead, it was preferred to increase the utilization of treated surface water even for the supply of drinking water.

Deep-groundwater exploitation is problematic as a consequence of rather low aquifer hydraulic conductivity and heavy groundwater abstraction, particularly in industrial areas. Yields of individual wells are relatively low and drawdowns in pumped wells are high as a result of the hydraulic conditions. Intensive abstraction of deep groundwater has already resulted in a rapid decline in groundwater levels in many areas, affecting exploitation of the deeper aquifer as well as the overlying shallow aquifer.

10.2 GEOLOGY, AQUIFERS, AND GROUNDWATER POTENTIALS

The BSGB is an intermontane basin bounded by mountain ranges to the north, east, south, and west. The basin covers Bandung City (100%), Bandung Regency (72.4%), West Bandung Regency (3.7%), Cimahi City (80.8%), Garut Regency (0.6%), and Sumedang Regency (8.8%) in the West Java Province (Table 10.1). Lateral extent of the basin is about 1.7 km^2 (Figure 10.1).

The three morphological units of the basin are alluvial plain, stratovolcanic, and old volcanic (Annex 10.1). The alluvial plain covers the southern part of Bandung City and the northern part of Bandung Regency with an average elevation of 600 m above sea level. A range of stratovolcanoes occupies the northeast and southern part of the basin. Among these are Mt. Bukittunggul (2209 m), Mt. Mandalawangi (1575 m), Mt. Kendang (2608 m), and Mt. Kencana (2182 m).

Table 10.1 Area of regency/city within the BSGB

Site No.	Name of regency/city	Total area of regency/city (km^2)	Area of regency/ city within the basin (km^2)	Area percentage (%)
1	Bandung City	168.80	168.80	100.00
2	Bandung Regency	1762.00	1276.01	72.4
3	West Bandung Regency	1249.00	46.45	3.7
4	Cimahi City	41.32	33.37	80.8
5	Garut Regency	3103.00	17.48	0.6
6	Sumedang Regency	1564.00	137.60	8.8

10.2.1 Geology

The geology of the Bandung area (Annex 10.2) as outlined by Alzwar (1989), Silitonga (1973), and Sudjatmiko (1972) is followed in this chapter. During the Lower Pleistocene, volcanic activity and uplifting was continuous and a sequence of volcanic conglomerates, breccias, tuffs, and andesitic interlayers (Cikapundung Formation) up to 350 m thick was deposited. This formation or unit widely outcrops to the north and east of Dago as well as along the southeastern and northwestern slopes of Tangkuban Perahu, where there is very coarse-grained tuff. At the beginning of the Upper Pleistocene, the first tectonic movements took place along the Lembang fault. During the Upper Pleistocene to Holocene, the volcanic products of Tangkuban Perahu were extruded and filled the Lembang and Bandung depressions with an alternating sequence of tuffs and breccias as well as pumice deposits up to 180 m thick (Cibeureum Formation). Within this unit there are several vertically graded volcanosedimentary sequences, which are interpreted as lahars. The bottom of the Cibeureum Formation is in places built of a volcanic conglomerate underlain by volcanic clay. In places coarse-grained tuffs or basaltic lava flows mark the bottom of the unit. It covers a sizable area stretching between Cicaheum, the center of Bandung City, and the south of Cimahi City. The Cibeureum Formation is poorly consolidated and cemented, resulting in relatively good hydraulic conductivity. To the south, the rocks of the Cibeureum Formation interfinger with those of the Kosambi Formation, indicated by a decrease in grain size from volcanic sandstones to volcanic tuffs. To the north of Bandung City, the rocks of the Cibeureum Formation are overlain by Holocene basalt lava flows and tuff deposits erupted from Tangkuban Perahu. Basalt lava flows typically occur within the river valleys of Cikapundung, Cibeureum, and Cimahi and range from 8 m to 16 m in thickness. Outside of the valleys, this unit called the Cikidang Formation, is represented by coarse-grained tuff layers up to 65 m thick.

Evidence of lahar flows can be seen in the Cibeureum Formation. These flows dammed the Citarum River during the Late Pleistocene filling the Bandung area with a large lake that was 720 masl. Limnic deposits up to 80 m thick (Kosambi Formation) consist of tuffaceous clays, volcanic sandstone, and silt as well as organic matter. This formation covers the central part of the basin east of Bandung City (Dayeuhkolot–Majalaya–Cicalengka–Ujungberung). As a result of the low hydraulic conductivity of lake sediments, this area is frequently flooded during the rainy season. Late Holocene terrace sediments up to 5 m thick join the Citarum River valley.

10.2.2 Climate and Rainfall

The BSGB, like other areas of Western Indonesia, has a tropical monsoon climate characterized by a rainy season extending from October to May and by average monthly rainfall of more than 100 mm, and a short season of low precipitation between June and September with average rainfall around 50 mm. Mean annual precipitation in the Bandung Basin, depending on altitude, is between 1500 mm/y and 3500 mm/y (Annex 10.3). Mean annual precipitation in the western part of the basin is about 2000 mm/y and higher up in the volcanoes about 3500 mm/y. Monthly mean relative humidity, which is greatly affected by the monsoon wind, is 70–80% in the plains and 85–95% in the mountains.

10.2.3 Aquifer Systems and Hydrogeology

The areal distribution of the lithological units of the Bandung Basin and their hydrogeological significance regarding hydraulic conductivity and groundwater productivity (Annex 10.4) are shown on the Hydrogeological Map of Bandung Quadrangle at a scale of 1:250,000 (Soetrisno, 1983). The extent of the zone under artesian pressure and heavy groundwater exploitation in early 1983 are also shown.

The groundwater system in the BSGB can generally be divided into an upper part "the shallow groundwater system" and a lower part "the deep groundwater system" (Bender et al., 1981; Soetrisno, 1983). Normally, the shallow system occupies the upper 10–20 m below ground level (mbgl).

The BSGB has three main aquifer systems: a shallow-aquifer system up to 35 m deep, an intermediate-aquifer system between 45 m and 90 m deep, and a deep-aquifer system slightly less than 100 m deep (Figure 10.2). The aquifers are heterogeneous in composition and show vertical and horizontal

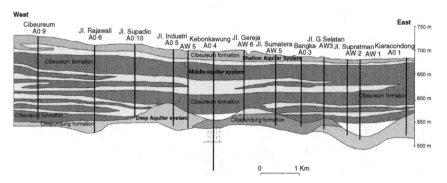

Figure 10.2 *Hydrogeologic Cross-Section of the BSGB.*

variations. It is generally safe to assume that no direct critical connections exist between the shallow and deeper aquifer systems (Soetrisno, 1983).

In addition to rainfall, direct infiltration from small creeks, canals, and paddy fields contribute to recharge of the shallow-groundwater system. Most probably comes from paddy fields.

The intermediate and deep aquifer systems are confined: water exploited from existing wells in 1983 was mostly artesian and piezometric heads were generally near or above the land surface. Groundwater flows in the deep-groundwater system in the northern part of the basin were from north to south.

The distribution of transmissivity values in the deep-aquifer system indicates that transmissivity of the central part of the basin is relatively high, ranging from 500 to more than 1500 m²/d. The aquifers in this central part are highly productive. The area is bound in the east and west by zones with low transmissivity values of less than 250 m²/d and classified as productive to moderately productive (Figure 10.3). Bender et al. (1981) identified typical hydraulic discharge and recharge situations (Figure 10.4) along a section from the Tangkuban Prahu Volcano in the north to the Citarum River (Schmidt et al., 2005).

10.2.4 Groundwater Potential

Based on the quantity and quality of groundwater used for drinking water, the BSGB can generally be subdivided into two zones as shown in Annex 10.5 (Ruchijat and Sukrisna, 2000). The map shows that moderate groundwater potency in both shallow and deep aquifers is distributed in the center of the basin. Optimum well yields of the deep-aquifer system with a well distance of 300–450 m might be of the order of 2.0–9.5 L/s.

10.3 DRIVERS

Population growth, urbanization, increase in tourism, and the development of industrial areas are the main drivers exerting pressure on the groundwater environment of the BSGB. The development of industrial areas is a less significant driver than the others.

10.3.1 Population Growth

Based on data available in 2004 from the Central Statistics Agency of West Java Province (2013), total population in the basin was 4.723 million, 47.42% of which were residents of Bandung City, 40.70% of Bandung Regency, 1.04% of West Bandung Regency, 8.63% of Cimahi City, 0.26%

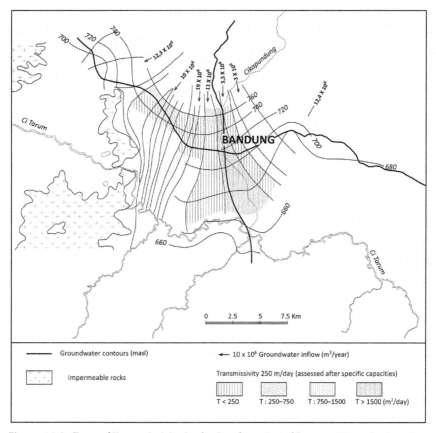

Figure 10.3 *Zone of Transmissivity in the Bandung Area (German Water Engineering, 1980) after Soetrisno, 1973 (written at explanatory note of thale hydrogeological map of Bandung Sheet); Soetrisno, 1983).*

of Garut Regency, 1.91% of Sumedang Regency, and 0.03% of Subang Regency. Between 2004 and 2012 the population grew by 1.9% per annum, resulting in an increase in population density from 2808 persons/km² to 3259 persons/km². During the same period, Bandung Metropolitan City witnessed an increase in population from 2.24 million to 2.46 million. Population density increased from 13,270 persons/km² to 15,640 persons/km² as a result of population growth of 1.2% per annum (Table 10.2).

10.3.2 Urbanization

Based on official data in 2012, the average number of immigrants arriving in Bandung was 50,000/y. They came in search of a better lifestyle as a result of the city's economic development. Unfortunately, detailed information

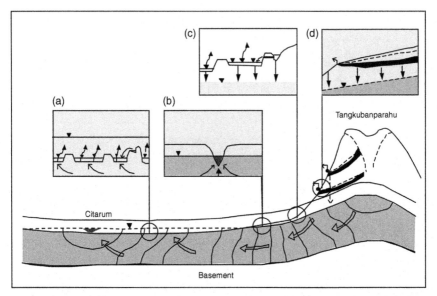

Figure 10.4 *Hydraulic Situation of the Bandung Basin. (a) Discharge Groundwater to Paddy Fields; (b) Discharge Groundwater to Rivers; (c) Recharge from Paddy Fields; (d) Leakage from Perched Aquifer (Bender et al., 1981; Schmidt et al., 2005).*

about urbanization development in Bandung Metropolitan City is not currently available.

10.3.3 Increase in Tourism

Based on data from the Central Statistics Agency of Bandung City (2011), tourist numbers increased from 4,320,634 to 4,951,439 between 2008 and 2010. During that period the number of tourists staying overnight in hotels increased from 2,638,535 to 3,385,872. In addition, data from the Central Statistics Agency of West Java Province (2013) show that tourists visiting Bandung increased from 90,278 to 146,736 people between 2010 and 2012 (Table 10.3).

10.3.4 Development of Industrial Area

Based on data from the Central Statistics Agency of West Java Province (2013), the number of businesses within the BSGB between 2007 and 2013 decreased significantly from 1870 to 1676. These were located in Bandung City, Bandung Regency, West Bandung Regency, and Cimahi City (Table 10.4).

Contributions to government revenue between 2008 and 2012 from the industrial sector in Bandung City decreased from 28.14% to 22.72%. This situation was brought about by increased activity in such areas as construction, development, commerce, hotels, restaurants, transportation, and

Table 10.2 Population growth in the BSGB between 2004 and 2012 (persons/km²)

City/Regency	2004	2005	2006	2007	2008	2009	May 2010	June 2011	June 2012
Bandung City	13,269	13,649	13,853	14,054	14,220	14,382	14,188	14,442	14,585
Bandung Regency	11,389	11,537	11,731	11,922	12,085	12,245	13,637	13,882	14,190
West Bandung Regency	291	299	304	309	313	317	333	339	345
Cimahi City	2,415	2,624	2,739	2,857	2,973	3,093	2,589	2,636	2,682
Garut Regency	73	73	74	75	76	76	80	81	82
Sumedang Regency	535	531	536	541	545	549	570	580	586
Subang	8	8	8	8	8	8	9	9	9
Total (BSGB)	2,808	2,882	2,935	2,987	3,033	3,078	3,152	3,208	3,259

BSGB, Bandung–Soreang Groundwater Basin.
Source: See Table 10.1 for areas of the Cities/Regencies.

Table 10.3 Numbers of tourists visiting Bandung through Husein Sastranagara Airport, Bandung

Site No.	State	2010	2011	2012
1	Singapore	5,948	10,930	26,016
2	Malaysia	79,589	96,978	109,205
4	Japan	233	406	477
5	South Korea	198	214	691
6	Taiwan	33	57	869
7	China	166	216	511
8	India	194	502	127
9	Philippines	189	542	575
10	Hongkong	61	107	886
11	Thailand	339	746	156
12	Australia	478	682	1024
13	United States	406	512	1048
14	United Kingdom	324	445	641
15	Netherlands	339	404	585
16	Germany	202	291	419
17	France	131	186	433
18	Russia	24	19	34
19	Saudi Arabia	48	84	151
20	Egypt	11	6	17
21	United Arab Emirates	6	12	11
22	Bahrain	0	2	13
23	Others	1359	1945	2847
	Total	90,278	115,286	146,736

Table 10.4 Number of businesses in the BSGB

Site No.	City/Regency	2007	2008	2009	2010	2011	2012	2013
1	Bandung City	503	503	429	597	670	560	570
2	Bandung Regency	1018	1018	1018	977	865	793	822
3	West Bandung Regency	179	179	179	158	160	154	154
4	Cimahi City	170	170	140	130	130	130	130
5	Garut Regency	–	–	–	–	–	–	–
6	Sumedang Regency	–	–	–	–	–	–	–
7	Subang	–	–	–	–	–	–	–
	Total (BSGB)	1870	1870	1766	1862	1825	1637	1676

BSGB, Bandung–Soreang Groundwater Basin.

communication. Moreover, increased competition at national and global levels influenced the development of local industry in Bandung City (RPJMD Kota Bandung 2014–2018).

10.4 PRESSURES

Pressures can be defined as direct stresses brought about by expansion in the anthropogenic system and associated interventions in the natural environment. Pressure on the BSGB is the result of inadequate surface water, overexploitation of aquifers in industrial areas, and a decrease in rechargeable areas limiting water supply to aquifers.

10.4.1 Inadequate Surface Water

Increases in the clean water requirements of Bandung City are closely linked to population growth, increasing social prosperity, and economic activity (be it industrial or otherwise). Flowing through Bandung City, the Cikapundung River has the potential to provide raw water for the Bandung drinking water supply. The irrigation area of the city in 2010 had been converted from agricultural to urban as a result of human settlement and economic activity. Hence, water allocated for irrigation purposes can be transferred for raw water provision to the Bandung drinking water supply. However, increased demand has not only put pressure on Bandung's drinking water supply, it has also led to water shortages between May and December. Between 2015 and 2020, 90% of dependable flow will be insufficient for the Bandung drinking water supply due to increased water demand.

The deficit between water demand and supply for the local water supply utility Perusahaan Daerah Air Minum (PDAM) can only be fulfilled from alternative sources (mainly groundwater). Although there will be water shortages for the Bandung area between 2015 and 2020, it is predicted there will be enough for industrial requirements and maintenance flow. Water from the Cikapundung-Gandok River was sufficient for maintenance flow, industry, irrigation, and Bandung drinking water in 2005. However, this is not expected to be the case between 2015 and 2020 because of the large amount of water required.

10.4.2 Overexploitation of Aquifers in Industrial Areas

Groundwater levels for industrial areas in 2009 ranged between 40% and 60% of maximum drawdown, calculated from the initial groundwater level

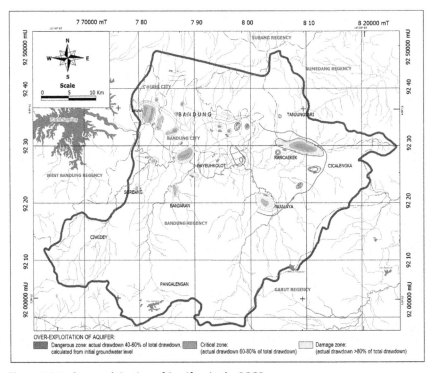

Figure 10.5 *Overexploitation of Aquifers in the BSGB.*

(dangerous zone), between 60% and 80% (critical zone), and more than 80% (damage zone). These three zones occur when groundwater abstraction is greater than potential groundwater flow in the area. Overexploitation of aquifers mainly occurred in the industrial areas (Figure 10.5) of Leuwigajah (Cimahi City), Bandung City, Dayeuhkolot, Banjaran, Majalaya (Bandung Regency), and Rancaekek (Sumedang Regency).

10.4.3 Decrease in Rechargeable Areas

Theoretically, the volume of water that can be recharged to the ground depends on such factors as the hydraulic permeability of the soil/rock, land coverage, and depth to groundwater level. Change in land coverage (e.g., vegetated land to residential) will decrease groundwater recharge as a result of fewer rechargeable areas. Land use and change in land coverage in the Bandung area are shown in Figure 10.6. There has been rapid development of housing estates in the area since 1985.

Figure 10.6 *Land Cover in (a) Groundwater Recharge Area and (b) Groundwater Discharge Area of the BSGB.*

10.5 STATE

State describes the condition and trends of the groundwater environment as a result of the pressure brought by drivers. This section analyzes the state of the groundwater environment from the perspective of number of wells in shallow and deep aquifers, volume of groundwater abstraction, groundwater levels, groundwater quality, and annual recharge into aquifers.

10.5.1 Groundwater Abstraction

The urban water supply of Bandung up until 1979 was dependent on a system of 10 captured springs, 11 drilled wells, and one surface source. With total production of 33 MCM/y, this system could only supply 23% of the population. More groundwater has been added to the urban water supply system since then. By the mid-1980s, 18.5 MCM/y were added to the system by drilling 22 deep wells to increase water supply coverage to 42%. Coverage increased further in the early 1990s by adding some 28 MCM/y of groundwater abstracted through 25 deep wells in the urban water supply system. The many shallow dugwells that supply water to most of the population were not counted in the above statistics (Aust et al., 1994; Wagner et al., 1991; Schmidt et al., 2005). In addition to increased abstraction for domestic use, up to 50 MCM/y of groundwater was abstracted to meet industrial water demand.

As shown in Figure 10.7, intensive groundwater abstraction in the BSGB began in 1970 when about 10.8 MCM of groundwater were abstracted from productive aquifers by 96 registered production wells. Maximum abstraction of about 76.8 MCM occurred in 1996 with the help of 2628-registered production wells. A significant drop in groundwater abstraction occurred in

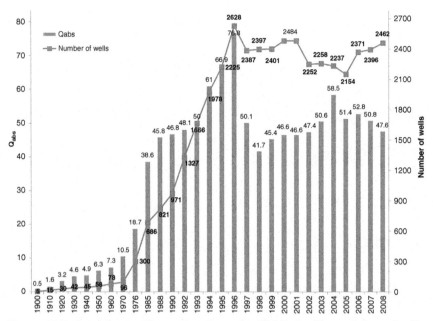

Figure 10.7 *Groundwater Abstraction from the Middle-Aquifer and Deep-Aquifer Systems.*

1996 (76.8 MCM) and 1998 (41.74 MCM) due to the economic crisis. In the period that followed, groundwater abstraction tended to increase yearly, reaching 58.5 MCM in 2004. Groundwater abstraction began to decrease between 2006 and 2008 with the lowest abstracted volume being about 47.6 MCM in 2008.

10.5.2 Groundwater Level

Data on the groundwater level of the shallow-aquifer system (water table) and confined aquifer system (piezometric heads) was based on measurements from dugwells in 2009 and drilled wells in 2010. The lateral distribution of phreatic levels and piezometric heads is shown in Figures 10.8 and 10.9. Groundwater generally flows from north to south in the basin. Groundwater depression cones have been found in industrial areas (i.e., Leuwigajah, Dayeuhkolot, Rancaekek, and Banjaran).

10.5.3 Groundwater Quality

Shallow groundwater in the Bandung region is characterized by having generally low salinity. Expressed in electrical conductivity (EC), groundwater salinity is mostly below 500 μS/cm on slopes at the foot of volcanoes and in the Batujajar Plain. In lower parts of the Bandung Plain, salinity

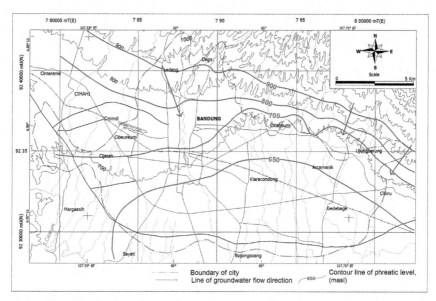

Figure 10.8 *Distribution of Phreatic Levels in the Bandung Area for 2009. Source: Modified from Salahudin and Arismunandar (2009).*

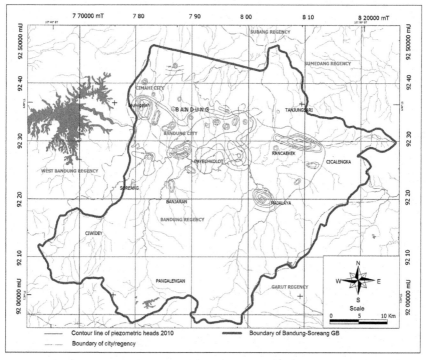

Figure 10.9 (a, b) *Distribution of Piezometric Heads in the Bandung Area for 2010. Source: Modified from Salahudin and Arismunandar (2009).*

exceeds 750 μS/cm in places. Evapotranspiration might play a major role in increasing salinity, especially where intensive irrigation occurs (Bender et al., 1981; Schmidt et al., 2005).

The shallow groundwater of the Bandung Basin is characterized by a predominance of HCO_3 with major cations such as Na, Ca, and Mg scattered in a range from 15 meq% to 30 meq%, the proportion of sulfate being generally low (<20 meq%). However, relatively high SO_4 concentrations occur on volcanic slopes, particularly in the Lembang area and south of Majalaya. In industrial areas around Majalaya and Rancaekek, as well as in the urban areas of Bandung–Cimahi, a general shift from a mixed bicarbonate water type via Na–HCO_3 to Na–Cl can be recognized together with an increase in salinity (>1000 μS/cm) (Bender et al., 1981; Schmidt et al., 2005).

Perennially high biological activity and CO_2 production in the soil zone, low availability of carbonates in soils and rocks, and high proportion of organic matter at shallow depths in places affect groundwater characteristics as follows:

- relatively high concentrations of free aggressive carbon dioxide;
- elevated iron and manganese concentrations;
- reducing environment due to poor oxygen.

Unsewered domestic and industrial wastewater, especially from the many textile factories, causes widespread diffuse contamination of shallow groundwater in the urban and industrial zones of the Bandung Basin. Contaminated groundwater shows elevated concentrations of chloride, sulfate, and nitrogen compounds. There is generally less nitrate as a result of the low oxygen content; however, nitrite and ammonia concentrations range from 0.05 mg/L to >1 mg/L, often exceeding maximum admissible levels for drinking water. Elevated concentrations of halogenated hydrocarbons and trace elements locally occur downstream of waste and industrial compound disposal.

The deep artesian aquifers tapped by wells for the central urban water supply were unaffected by contaminated shallow groundwater in the early 1990s. However, the high water pressure of artesian aquifers is likely to be reduced with a heavy increase in the abstraction of groundwater for industrial water supplies (Aust et al., 1994; Schmidt et al., 2005).

The hydrochemical pattern of deep artesian groundwater tapped at depths ranging from 30 to 160 mbgl within the urban area can be characterized as follows:

- low salinity and oxidizing conditions in the west;
- slightly increased concentrations of bicarbonate and chloride toward the east;

- reducing conditions in the east combined with a reduction in sulfate and nitrogen compounds.

This hydrochemical pattern reflects differences in the deep-groundwater flow system: relatively high groundwater flow velocity under semiconfined aquifer conditions in the west and low groundwater flow velocity associated with highly confined aquifer conditions in the east. This gives rise to longer residence times as indicated by higher salinities and ^{14}C water ages (Wagner et al., 1991; Schmidt et al., 2005).

The lateral distribution of EC values in the shallow-aquifer system during 2009 is shown in Figure 10.10 (Salahudin and Arismunandar, 2009). The highest EC values of groundwater can be seen to occur southeast of Bandung City.

10.5.4 Groundwater Recharge

Total groundwater recharge of the BSGB upstream of the inflow of the Citarum River into the Saguling Reservoir (i.e., excluding the Batujajar Basin) is estimated to be around 750 MCM/y. Groundwater flow through

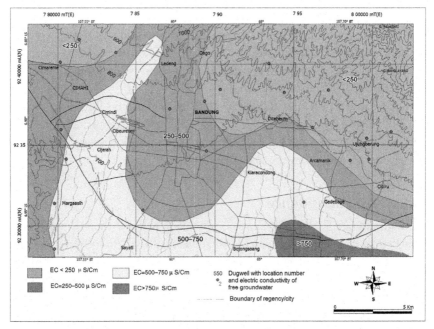

Figure 10.10 *EC of Free Groundwater in the Bandung Area.* *Source: Modified from Salahudin and Arismunandar (2009).*

more central parts of the basin – an area covering the Bandung Plain and lower mountain slopes – is of the order of 470 MCM/y. Groundwater abstraction by deep wells in the same area amounted to 48 MCM/y in 1990, approximately 10% of total groundwater flow (Schmidt et al., 2005).

Ruchijat and Sukrisna (2000) calculated total groundwater recharge for an unconfined aquifer system by considering rainfall and recharge percentage. The results show total groundwater recharge to be of the order of 795 MCM/y.

10.6 IMPACTS

Impacts deal with effects on the human body and the environment as a result of changes in states, thus contributing to the vulnerability of both natural and social systems. The adverse effects of overexploitation include a decline in groundwater levels, land subsidence, and poor-quality drinking water.

10.6.1 Decline in Groundwater Levels

Around 60 monitoring wells were built in the BSGB to assess changes in groundwater levels as a direct impact of groundwater abstraction. Monitoring well depths ranged from 60 m to 180 m. Groundwater hydrographs at some monitoring wells were constructed based on recorded data between 1990 and 2004 (Annexes 10.6–10.8).

Groundwater levels in industrial areas are on the decline: Leuwigajah (3.11–15.12 m/y), Dayeuhkolot (3.00–12.26 m/y), Rancaekek (0.52–3.85 m/y), and Banjaran (0.89–4.57 m/y). Table 10.5 shows the location and rate of drawdown in the Bandung Basin.

10.6.2 Land Subsidence

Land subsidence is clear for all to see in some parts of the Bandung area. Subsided land generally coincides with zones that have groundwater depression cones. The geological settings beneath Bandung, combined with major groundwater abstraction, constitute ideal conditions for land subsidence.

Land elevation as a result of subsidence dropped in some industrial areas between 2000 and 2002: Leuwigajah (52 cm), Dayeuhkolot (46 cm), and Rancaekek (42 cm). Table 10.6 shows the locations and magnitude of land subsidence.

Table 10.5 Drop in groundwater level in some areas of the Bandung Basin (Hasyim, 2006)

Site No.	Location	Groundwater level (mbgl)	Rate of drawdown (m/y)
1	Cimahi	12.0–25.0	7.19
2	Leuwigajah	41.0–71.4	3.1–15.1
3	Cijerah	29.9–51.5	1.3–4.3
4	Buahbatu	22.7–49.5	1.6–3.1
5	Dayeuhkolot	25.7–66.6	3.0–12.3
6	Mohammad Toha	–	1.47
7	Cicaheum	16.0–49.5	1.6–2.1
8	PT. Grandtex		1.63
9	PT. Bintang Agung		2.12
10	Rancaekek	6.8–23.6	0.5–3.9
11	PT. Kewalram		2.01
12	Bojongsalam		0.44
13	Majalaya	31.7–50.2	0.3–3.9
14	Ciparay	7.7–29.4	0.9–4.6
15	Soreang	1.5–30.9	0.4–1.6
16	Bojongkunci		0.77
17	Cipedung		0.38

Table 10.6 Land subsidence in the Bandung Basin from 1996–2000 (Hasyim, 2006)

Site No.	Location	Total land subsidence (cm)	Rate of land subsidence (cm/y)
1	Cimahi–Leuwigajah	84.5	21.1
2	Bojongsoang	83.9	20.9
3	Kopo	18.9	4.7
4	Banjaran	63.9	15.9
5	Dayeuhkolot	20.8	5.2
6	Gedebage	24.3	6.1
7	Ujungberung	20.6	5.2
8	Majalaya	8.4	2.1
9	Rancaekek	11.8	2.9
10	Cicalengka	44.5	11.1

10.7 RESPONSES

Several attempts have been made since 1978 to improve the groundwater environment by getting to know the groundwater system better as well as by designing and implementing management interventions. Between

1978 and 1996, the German Federal Ministry for Economic Cooperation and Development (BMZ) commissioned three technical cooperation projects in Indonesia in relation to hydrogeology and environmental geology. The BMZ project comprised mathematical groundwater flow modeling as a major component and was jointly conducted by the Federal Institute for Geosciences and Natural Resources (BGR), the Directorate of Environmental Geology (DEG) – which became the Center for Environment Geology (CEG) in 2009 – and the Center for Groundwater Resources and Environmental Geology (CGREG) in 2012. The aim of this project was to quantify groundwater resources for urban development areas, such as Bandung, Semarang, Yogyakarta, Cilegon (all in Java), and Denpasar in Bali (Haryadi and Schmidt 1991). In 2009 the CEG also conducted a survey to establish a map of groundwater conservation zones for technical consideration when issuing groundwater permits from the production well. The Ministry of Energy and Mineral Resources (MEMR) of West Java Province in close collaboration with the Institut Teknologi Bandung (ITB – Bandung Institute of Technology) has also made several attempts to improve the water environment through development research on the establishment of groundwater simulation scenarios to plan for groundwater utility in the Bandung Basin.

Meanwhile, the CGREG, Geological Agency, MEMR, who are responsible for national groundwater management, have established technical guidelines for groundwater management to be applied by local governments.

Major responses made so far in Bandung City to improve the groundwater environment include development of a groundwater flow simulation model, formulation of a plan for groundwater utilization, and delineation of zones for groundwater conservation.

10.7.1 Development of a Groundwater Flow Simulation Model

The BGR and the DEG developed groundwater flow simulation models for Bandung with the aim of quantifying groundwater resources for urban development (Haryadi and Schmidt, 1991). A numerical model was developed and used in 1991 and reactivated in 1999 to simulate the effects of groundwater abstraction for domestic and industrial use from the aquifer system. The three-dimensional groundwater flow model covered the time span 1901 to 2000. The conceptual model and geometry were adapted directly from the 1991application. A steady-state calculation generated an initial water level distribution for 1900 leading to a plausible distribution of recharge/discharge areas. This distribution was used for long-term simulation of transient flow. Since abstraction varies from period to period, it was the only boundary condition to be changed (Schmidt et al., 2005).

Limited groundwater production data for both 1991 and 2000 were used to calibrate the model. In addition, calibration was based on single-point observations at a pumping well in order to represent regional groundwater drawdown. Consequently, calculated drawdown was smaller (Schmidt et al., 2005). Calculated drawdowns from 1901–1990, 1984–1995, and 1984–1998 are consecutively shown in Annex 10.9a. Drawdown in the last 10 years as a result of groundwater production is of the same order of magnitude as total drawdown caused by pumping from 1901 to 1990.

Comparison of calculated hydrographs with observations from Annex 10.9b makes it clear that a detailed interpretation is needed. Generally, there is good accordance between all hydrographs from wells located in the central part of Greater Bandung (PDAM observation wells, public work water supply, Annex 10.9b.i). Although most hydrographs correlate moderately well, a few do not match observations at all well. These wells are either production wells themselves or located very close to pumping wells. All are located in fast-growing industrial development areas like Padalarang, west of Bandung (Annex 10.9b.iii).

As a result of the recent financial crisis in Indonesia, there are reports that some factories have stopped groundwater abstraction. This is arguably the reason groundwater levels of the order of tens of meters have recovered. In general, there is no concern about this figure as long as such an observation refers to production wells.

Drilling is predominantly for water supply, but there a few wells that have been drilled for observation purposes. A similar situation exists at the Majalaya Wellfield in the southeast of the Bandung Basin where water for domestic and industrial supply is produced (Annex 10.9b.iv). The model simulates this wellfield by abstracting groundwater from nine cells of the model aquifer system at depths between 30 mbgl and 100 mbgl. In total, about 3.3 MCM of water are abstracted annually.

10.7.2 Formulation of a Groundwater Utilization Plan

A groundwater utilization plan was formulated by simulation runs of eight scenarios carried out at the ITB in 2002. All scenarios were run on the assumption that groundwater abstraction (Q_{abs}) from registered wells in the Bandung Basin represented only 33% of actual Q_{abs}. Three of the eight scenarios were considered to be the most important for implementation by local government as follows:

- By setting the actual Q_{abs} constant up to 2013 (Scenario 1) , the critical zone will be extended by 124 km^2 or the actual critical zone (2002) will be increased by about 56% (Annex 10.10a).
- By setting the Q_{abs} for both domestic and industrial use, increases are indicated up to 2013 (Scenario 2). The critical zone will be extended by 185 km^2 or the actual critical zone (2002) will be increased by about 84% (Annex 10.10b).
- For a recovering critical zone to become a dangerous zone in 2013, Q_{abs} needs to be constant between 2003 and 2013 in the safe zone (Scenario 3). Simulation results show that the critical zone will no longer exist. In order to reach the target, Q_{abs} in some areas will have to be reduced between 2003 and 2013 by 8% per year, with a total decrement of 80% and a decrement zone of 5% per year (total decrement 50%). Simulation results show that there will be an 85% surplus of Q_{abs} compared with total Q_{abs} for 2003 (Annex 10.10c).

10.7.3 Delineation of Groundwater Conservation Zones

Groundwater conservation zones were delineated and established in 2009 by considering such criteria as drawdown percentage, deterioration of groundwater quality, and land subsidence. The conservation zone in the groundwater basin is subdivided into a protection zone (e.g., groundwater discharge area) and utilization zone. Each zone contains requirements for groundwater utilization and conservation. Based on these criteria, six zones (Annex 10.9) were identified: damage zone, critical zone, dangerous zone, safe zone (1 and 2), and protection zone (Salahudin and Arismunandar, 2009). The damage zone can clearly be seen to occur in industrial areas, coincident with the groundwater depression cone (Annex 10.11).

10.7.4 Setting Environmental Standards and Guidelines

Technical guidelines on groundwater management were established in 2004 by the MEMR. They consist of:

Technical Guidelines on Groundwater Inventory
Technical Guidelines on Groundwater Conservation
Technical Guidelines on Groundwater Utilization
Technical Guidelines on Groundwater Permits
Special Technical Guidelines on Groundwater

In 2009 the MEMR issued Ministerial Regulation of Energy and Mineral Resources (EMR) No. 13/2008 on Technical Guidelines for the Establishment of Groundwater Basins. This regulation should be referred to by central, provincial, and city or regency governments. Presidential Decree No. 26/2011 on Establishing a Groundwater Basin in Indonesia was announced in 2011, which refers to Ministry Regulation EMR No. 13/2008. The decree should be used as the hydrogeological basis for groundwater management.

10.8 SUMMARY

Driver–pressure–state–impact–response (DPSIR) framework analysis shows that during the last two decades increasing population density, urbanization (increase in the urban population), and development of industrial areas are the driving forces behind Q_{abs} exceeding recharge in some areas of the basin. Other driving forces include a decrease in groundwater levels in industrial areas (Leuwigajah, Dayeuhkolot, Rancaekek, and Banjaran), deterioration in groundwater quality as indicated by higher salinity, and lowering of land elevation due to land subsidence in some industrial areas between 2000 and 2002 (Leuwigajah, Dayeuhkolot, and Rancaekek).

Management intervention is currently ongoing and includes establishment of a map showing groundwater conservation zones, a modeling approach (scenario Q_{abs}) toward a groundwater utilization plan, decreasing Q_{abs} for groundwater permits in dangerous and critical zones, and stopping groundwater pumping in the damage zone. In future there should also be continuous monitoring of groundwater conditions (quantity and quality) such that information on the positive impacts of management efforts already put in place in the West Java Province can be obtained.

A groundwater database and information system for local government, linked to the CGREG, should be established as soon as possible so that groundwater data and information for better management of the BSGB can be exchanged.

ACKNOWLEDGMENTS

The authors acknowledge the Director of CGREG, Geological Agency, MEMR for encouraging them to participate in the Workshop on "Enhancing the Groundwater Management Capacity in Asian Cities through the Development and Application of Groundwater Sustainability Index in the Context of Global Change." The authors also acknowledge Dr Sangam Shrestha, Associate Professor of Water Engineering and Management (WEM) at the Asian Institute of Technology (AIT) and project leader, for providing the opportunity to be part of the project.

REFERENCES

Alzwar, 1989. Geologic map of the Garut and Pameungpeuk Quadrangles. Geological Survey of Indonesia, Bandung.

ITB, 2002. Groundwater utilization plan in Bandung Basin, Final Report. Local Office of Mines and Energy of West Java Province and Bandung Institute of Technology (ITB) [in Indonesian].

Aust, H., Siebenhüner, M., Toloczyki, M., Wagner, W., Dahms, E., Geyh, M., Rosadi, D., Ruchijat, S., Soekrisno, S., Wiriosudarmo, S., Van der Wall, R., 1994. Groundwater protection and selection of waste disposal sites in the Greater Bandung area, Indonesia. Natural Resources and Development, 40: 7–243, Institute for Science Co-operation, Tübingen.

Bender, H., Djaendi, S., Wagner, R., 1981. Contribution to the hydrogeology of the Bandung Basin. Unpublished report, 98 pp., 23 figures, 2 tables, 11 annexes, BGR Archive No. 91790, Directorate of Environmental Geology and Bundesanstalt für Geowissenschaften und Rohstoffe, Bandung and Hannover.

Haryadi, T., Schmidt, G., 1991. Technical cooperation with Indonesia – groundwater quantification for four urban development areas: Groundwater Flow Simulation Model – model results: Bandung, Denpasar, Cilegon, Semarang, Yogyakarta. Unpublished report, BGR Archive No. 109220, Directorate of Environmental Geology and Bundesanstalt für Geowissenschaften und Rohstoffe, Bandung and Hannover.

Hasyim, I., 2006. Pengelolaan air tanah di Jawa Barat – Menapak harapan di masa depan, makalah Lokakarya Kebijakan Nasional Pengelolaan Air Tanah, Direktorat Tata Lingkungan Geologi dan Kawasan Pertambangan [in Indonesian].

Priowirjanto, G.H., 1985. Investigations on the hydrogeology of the Bandung basin and on the water development of Bandung City with support of mathematical models. Mitt. Ing. u. Hydrogeol., 21: 183 pp., 25 figures, 19 tables, 33 annexes, Aachen [in German].

Ruchijat, S., Sukrisna, A., 2000. Survei potensi air tanah CAT Bandung. Direktorat Geologi Tata Lingkungan, Bandung.

Salahudin, A., Arismunandar, dan, 2009. Penyelidikan konservasi air tanah Cekungan Air Tanah Bandung-Soreang. Pusat Lingkungan Geologi.

Schmidt, G., Ploetner, D., Moesta, W., 2005. Groundwater Quantification – Validation of Groundwater Models, Vol. 2. Bundesanstalt für Geowissenschaften und Rohstoffe, Bandung and Hannover.

Silitonga, P.H., 1973. Geologic Map of the Bandung Quadrangle. Geological Survey of Indonesia, Bandung.

Soetrisno, S., 1983. Hydrogeological Map of the Bandung Sheet, Java. Directorate of Environmental Geology, Bandung.

Sudjatmiko, 1972. Geologic Map of the Cianjur Quadrangle. Geological Survey of Indonesia, Bandung.

Wagner, W., Ruchijat, S., Rosadi, D., 1991. Technical cooperation with Indonesia – Project CTA 108; Environmental geology for land use and regional planning – Project Report No. 15: groundwater resources and groundwater protection in the Bandung basin. Unpublished report, 80 pp., 13 figures, 5 tables, 5 maps, 17 appendices, BGR Archive No. 109012. Directorate of Environmental Geologi and Bundesanstalt für Geowissenschaften und Rohstoffe, Bandung and Hannover.

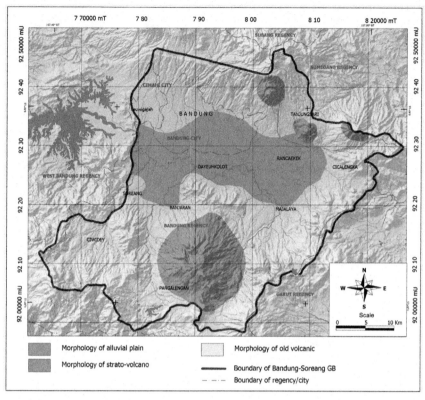

Annex 10.1 Geomorphic units of the BSGB. *Source: Ruchijat and Sukrisna (2000).*

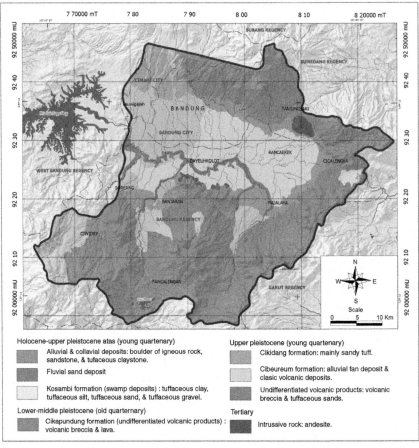

Annex 10.2 Geologic map of the BSGB. *Source: Modified from Alzwar (1989), Silitonga (1973), and Sudjatmiko (1972).*

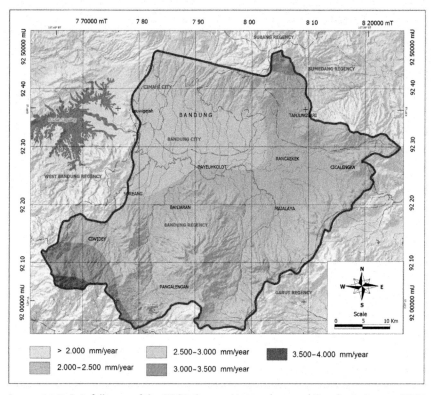

Annex 10.3 Rainfall map of the BSGB. *Source: Meteorology and Geophysic Agency (2005, Rainfall Map of Java Island, Jakarta).*

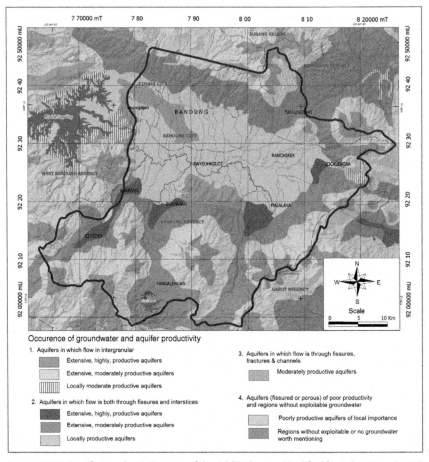

Annex 10.4 Aquifer productivity map of the BSGB. *Source: Modified from Soetrisno (1983).*

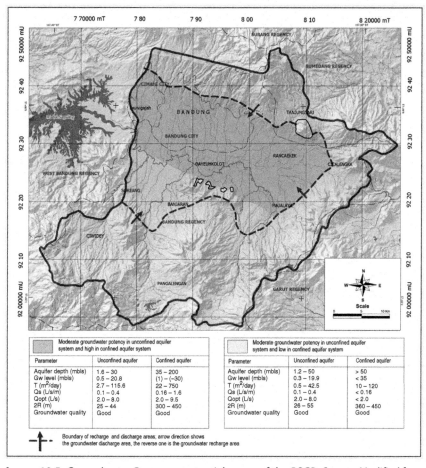

Annex 10.5 Groundwater Resource potential zones of the BSGB. *Source: Modified from Ruchijat and Sukfisna (2000).*

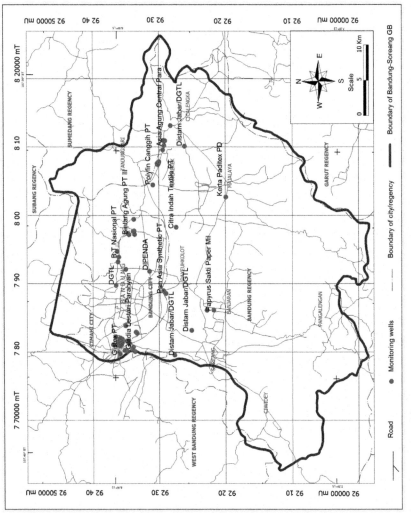

Annex 10.6 Location of groundwater-monitoring wells.

Annex 10.7 Details of groundwater-monitoring wells.

No.	Name of factory (well owner)	Address	Regency/City	Well no.	X	Y	Elevation (m)	Depth (m)	Top screen (m)	Bottom screen (m)
1	B.T.Nasional PT		Bandung City	0273140655	7,93,950	92,36,000	685	60	52	59
2	Bintang Agung PT II	Jl. Rumah Sakit Ujungberung	Bandung City	0273120654	7,97,390	92,34,650	670	90	42	54
3	Bintang Agung PT III	Jl. Rumah Sakit 114 Ujungberung	Bandung City	0273120707	7,97,225	92,34,450	675	90	72	84
4	Bintang Agung\Patal Dharma Kalimas PT I	Jl. Rumah Sakit Ujungberung	Bandung City	0273120708	7,97,235	92,33,640	675	90	61	84
5	Bintang Agung PT IV		Bandung City	0273120841	7,97,750	92,33,700	671	150	122	146
6	Bandung Synthethic Sarong Mill PT	Jl. Jend. Sudirman 823	Bandung City	0273010824	7,83,900	92,35,000	772	150	72	120
7	DGTL	Jl. Diponegoro	Bandung City	0273221549	7,89,800	92,36,450	724		36	69
8	DIPENDA	Jl. Soekarno-Hatta	Bandung City	0273090544	7,91,850	92,31,400	675	152	81	152
9	Grandtex PT	Jl. Jend. A. Yani No. 127	Bandung City	0273140653	7,94,750	92,36,250	690	130	78	126
10	IPTN	Jl. Pajajaran 154 Kec. Cicendo	Bandung City	0273193003	7,85,350	93,36,900	748	160	78	144
11	Lawe Adya Prima PT		Bandung City	0273120709	7,97,700	92,33,850	672	100	60	78
12	Siputex PT		Bandung City	0273110329	7,99,450	92,33,750	690	100	66	90
13	Tarumatex PT		Bandung City	0273150659	7,93,200	92,36,150	693	75	64	74
14	DGTL	Antapani	Bandung City	0273151534	7,92,150	92,34,950	683		117	142
15	Ayoe Indotama Textile PT	Jl. Leuwigajah 205 Ds. Utama Cimahi Selatan	West Bandung Regency	0206390845	7,81,050	92,36,400	724	176	102	172
16	Asia Agung Central Parahyangan PT	Jl. Raya Rancaekek Km 25,15	West Bandung Regency	0206100856	8,10,250	92,29,150	679	120	72	144
17	Central Georgete Nusantara PT	Jl. Cibaligo Km 7.2 Cimindi Cimahi	West Bandung Regency	0206390784	7,81,900	92,35,500	706	102	60	90
18	Citra Indah Textile PT	Jl. Raya Sapan 100 Ds. Tegalluar	West Bandung Regency	0206271533	7,98,300	92,27,350	660	147	80	116
19	Dewa Sutratex PT	Jl. Cibaligo 76, (Leuwigajah) Cimindi - Cimahi	West Bandung Regency	0206390813	7,32,200	92,36,250	650	150	102.5	137.5
20	Giri Asih Jaya PT		West Bandung Regency	0206230818	7,75,850	92,36,700	655	136	66	126
21	Gistex PT		West Bandung Regency	0206390651	7,79,750	92,35,850	725	150	78	144
22	Gladia Lestari ParahyanganPT	Jl. Nanjung 90 Ds. Lagadar	West Bandung Regency	0206240859	7,80,400	92,34,000	690	150	78	145
23	Hegar Mulya PT	Jl. Cibaligo - Leuwigajah Cimahi Selatan	West Bandung Regency	0206390843	7,82,250	92,36,500	720	150	78	120
24	Hintex Mitra Jaya PT	Jl. Cibaligo Km 2,8, Cimindi - Cimahi	West Bandung Regency	0206391522	7,81,250	92,34,450	708	150	104	146
25	How Are Yoe Indonesia PT	Jl. Raya Nanjung 206 Cimahi Selatan	West Bandung Regency	0206390847	7,80,300	92,35,150	726	150	60	138
26	Indokaha Shoes PT/Kahatex PT III	Jl. Cigondewah Blok Suci 5	West Bandung Regency	0206240814	7,82,950	92,33,450	690	120	36	99
27	Indorama PT		West Bandung Regency	0206230858	7,75,700	92,37,000	665	160	70	150
28	Kahatex PT I		West Bandung Regency	0206240705	7,82,750	92,33,250	690	127	60	126
29	Kahatex PT II	Jl. Cigondewah Blok Suci Cijerah	West Bandung Regency	0206240706	7,82,750	92,33,250	692	150	101	150
30	Kamarga Kurnia PT		West Bandung Regency	0206390821	7,81,650	92,34,650	710	200	153	183
31	Matahari Sentosa Jaya PT		West Bandung Regency	0206390848	7,81,100	92,35,600	710	195	72	174
32	Melvin PT		West Bandung Regency	0206390840	7,80,100	92,34,750	725	90	73	78
33	Nayatex Indopura PT	Jl. Raya Cicalengka Km 25, Ds. Tenjolaya, Kec. Cicalengka	West Bandung Regency	0206090810	8,13,150	92,28,200	675	90	42	86
34	Prinsa Totsuwa Jaya PT	JL Industri III 7, Cimahi	West Bandung Regency	0206390767	7,81,500	92,36,100	706	150	60	132
35	Pan Asia Synthetic PT	Jl. Moh. Toha Km. 6 Kec. Dayeuhkolot	West Bandung Regency	0206260822	7,88,600	92,28,900	665	150	72	138
36	Papyrus Sakti Paper Mills PT	Jl. Raya Banjaran Km 16,2	West Bandung Regency	0206160412	7,86,200	92,21,750	680	130	45	96
37	Pulo Mas Textiles PT	Ds. Lagadar Kec. Margaasih	West Bandung Regency	0206240846	7,80,750	92,33,850	675	148	106	142
38	Safilindo Permata PT	Ds. Sukasari Kec. Pameungpeuk	West Bandung Regency	0206131517	7,86,150	92,22,750	630	150	89	123
39	Sansan Saudaratex PT		West Bandung Regency							
40	Sapta Jaya Textile PT		West Bandung Regency	0206390725	7,81,650	92,34,800	709	120	78	108
41	Sarana Makin Mulya PT		West Bandung Regency	0206340844	7,76,200	92,40,750	710	150	90	144
42	Sinar Kontinental PT	Jl. Industri II/20 Leuwigajah	West Bandung Regency	0206390823	7,82,100	92,35,800	686	180	130	174
43	Tirtharia PT I		West Bandung Regency	0206390711	7,80,900	92,35,800	710	117	42	114
44	Tirtharia PT II	Jl. Industri III/Leuwigajah Km 8,7 Cimahi Selatan	West Bandung Regency	0206390820	7,80,100	92,33,600	706	168	150	162
45	Tridarmatex PT		West Bandung Regency	0206390657	7,81,450	92,35,500	710	110	90	110
46	Trisula Banten Textile Mill PT	Jl. Leuwigajah 142 Cimahi	West Bandung Regency	0206391523	7,81,100	92,36,900	715	246	174	218
47	Ultra Jaya PT I		West Bandung Regency	0206340762	7,77,650	92,40,350	710	65	41	275
48	Ultra Jaya PT II	Jl. Raya Padalarang	West Bandung Regency	0206340819	7,77,650	92,40,350	710	280	41	275

(Continued)

Annex 10.7 Details of groundwater-monitoring wells. *(cont.)*

			Regency							
49	Distam Jabar/DGTL	Bojong Kunci	West Bandung Regency	0206170542	7,83,200	92,25,050	666	131	57	66
50	Distam Jabar/DGTL	Moh. Toha	West Bandung Regency	0206260545	7,89,000	92,29,250	673	152	91	133
51	Distam Jabar/DGTL	Bojong Salam	West Bandung Regency	0206100540	8,10,150	92,26,050	673	155	88	127
52	Distam Jabar/DGTL	Cipedung	West Bandung Regency	0206180652	7,79,600	92,27,600	673	68	27	53
53	Kerta Paditex PD	Jl. Laswi, Ds. Sukamukti Kec. Majalaya	West Bandung Regency	0206110860	8,02,700	92,19,900	671	120	78	113
54	Coca Cola PT I	Jl. Raya Bandung - Garut Km 26 Cimanggung	Sumedang Regency	0213020267	8,11,000	92,29,250	679	110	95	104
55	Coca Cola PT II		Sumedang Regency	0213021528	8,10,950	92,29,000	679	150	72	117
54	Insan Sandang Internusa PT	Jl. Raya Rancaekek Km 22,5 Ds. Cinta Mulya Kec. Cikeruh	Sumedang Regency	0213010842	8,07,500	92,30,050	676	160	72	150
55	Kahatex PT I		Sumedang Regency	0213010718	8,07,650	92,30,000	676	70	50	72
56	Kahatex PT II		Sumedang Regency	0213010719	8,07,650	92,30,200	676	120	87	120
57	Kahatex PT III	Jl. Raya Rancaekek	Sumedang Regency	0213010811	8,07,500	92,30,150	676	150	115	140
58	Kewalram PT		Sumedang Regency	0213010658	8,09,600	92,29,350	676	110	72	110
59	Polyfin Canggih PT		Sumedang Regency	0213010002	8,04,500	92,30,850	676		150	180
60	Sunsonindo Textile PT	Jl. Raya Rancaekek Km 25,5	Sumedang Regency	0213020816	8,07,900	92,29,900	676	136	72	132

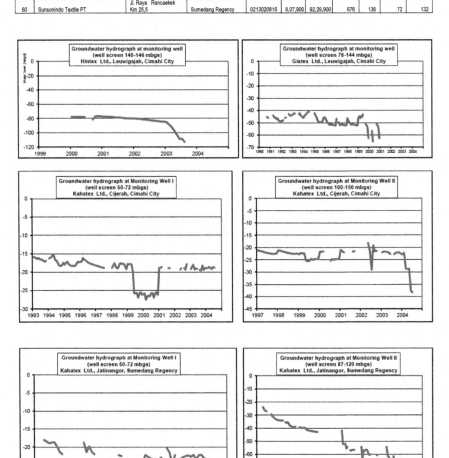

Annex 10.8 Groundwater hydrographs.

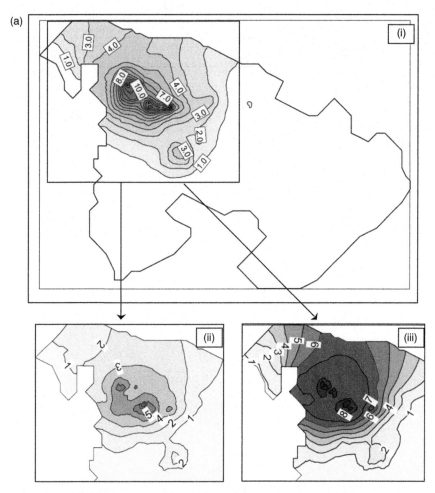

Annex 10.9 (a) Calculated drawdown in the deep aquifer as the difference in heads: (i) between 1901 and 1990 (model result, 1991); (ii) between 1984 and 1995 (Model Result 2000); (iii) between 1984 and 1998 (model result 2000) (Schmidt et al., 2005). (b) Deep-aquifer calculated groundwater heads at selected locations in the Bandung Basin (1901–2000): (i) Central Bandung area (ii) Industrial area west of Bandung (iii) Industrial area west of Bandung (iv) Other industrial areas, e.g., Majalaya (34).

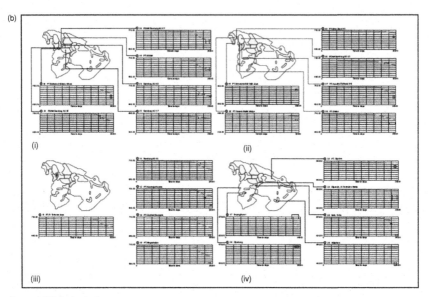

Annex 10.9 *(cont.)*

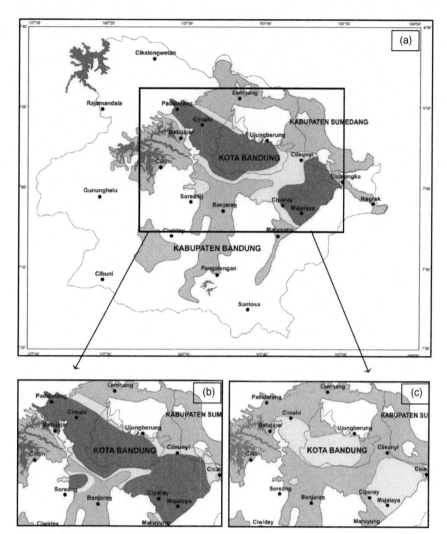

Annex 10.10 Groundwater abstraction zones in the Bandung Basin under different scenarios. *(a) Scenario 1 (b) Scenario 2 (c) Scenario 3. The scenario refers to the one developed by ITB (2002).*

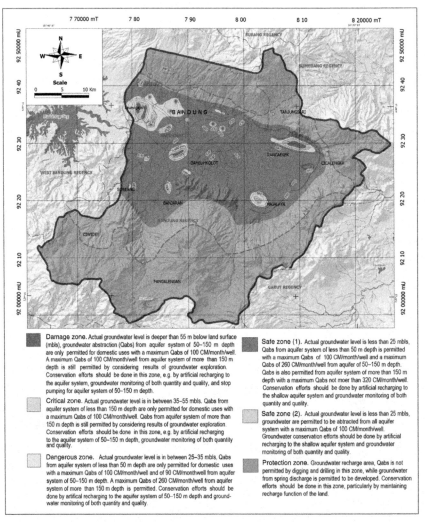

Damage zone. Actual groundwater level is deeper than 55 m below land surface (mbls), groundwater abstraction (Qabs) from aquifer system of 50–150 m depth are only permitted for domestic uses with a maximum Qabs of 100 CM/month/well. A maximum Qabs of 100 CM/month/well from aquifer system of more than 150 m depth is still permitted by considering results of groundwater exploration. Conservation efforts should be done in this zone, e.g. by artificial recharging to the aquifer system, groundwater monitoring of both quantity and quality, and stop pumping for aquifer system of 50–150 m depth.

Critical zone. Actual groundwater level is in between 35–55 mbls. Qabs from aquifer system of less than 150 m depth are only permitted for domestic uses with a maximum Qabs of 100 CM/month/well. Qabs from aquifer system of more than 150 m depth is still permitted by considering results of groundwater exploration. Conservation efforts should be done in this zone, e.g. by artificial recharging to the aquifer system of 50–150 m depth, groundwater monitoring of both quantity and quality.

Dangerous zone. Actual groundwater level is in between 25–35 mbls, Qabs from aquifer system of less than 50 m depth are only permitted for domestic uses with a maximum Qabs of 100 CM/month/well and of 90 CM/monthwell from aquifer system of 50–150 m depth. A maximum Qabs of 260 CM/month/well from aquifer system of more than 150 m depth is permitted. Conservation efforts should be done by artifical recharging to the aquifer system of 50–150 m depth and groundwater monitoring of both quantity and quality.

Safe zone (1). Actual groundwater level is less than 25 mbls, Qabs from aquifer system of less than 50 m depth is permitted with a maximum Qabs of 100 CM/month/well and a maximum Qabs of 260 CM/month/well from aquifer of 50–150 m depth. Qabs is also permitted from aquifer system of more than 150 m depth with a maximum Qabs not moer than 320 CM/month/well. Conservation efforts should be done by artificial recharging to the shallow aquifer system and groundwater monitoring of both quantity and quality.

Safe zone (2). Actual groundwater level is less than 25 mbls, groundwater are permitted to be abtracted from all aquifer system with a maximum Qabs of 100 CM/month/well. Groundwater conservation efforts should be done by artificial recharging to the shallow aquifer system and groundwater monitoring of both quantity and quality.

Protection zone. Groundwater recharge area, Qabs is not permitted by digging and drilling in this zone, while groundwater from spring discharge is permitted to be developed. Conservation efforts should be done in this zone, particularly by maintaining recharge function of the land.

Annex 10.11 Groundwater conservation zones in the BSGB. *Source: Salahudin and Arismunandar (2009).*

CHAPTER 11

Groundwater Environment in Bangkok and the Surrounding Vicinity, Thailand

Oranuj Lorphensri*, Tussanee Nettasana**, and Anirut Ladawadee**
*Department of Groundwater Resources, Thailand
**Bureau of Groundwater Conservation and Restoration, Department of Groundwater Resources, Thailand

11.1 INTRODUCTION

Bangkok is one of the megacities of Southeast Asia. It occupies the southern part of the Lower Central Plain (LCP) in Thailand (Figure 11.1), covering an area of about 10,300 km², with a total population of 11 million. Due to its fast-growing economy, rapid urbanization, poor planning, and failure of its authorities to prevent land development for various uses (e.g., residential, industrial, and commercial), the metropolitan area and its vicinity are susceptible to serious problems in water supply, sewerage, transportation, waste disposal, and other related problems.

Bangkok and its vicinity (seven provinces including Bangkok) receive water supplies from three authorities namely, the Metropolitan Waterworks Authority (MWA), Provincial Waterworks Authority (PWA), and the municipal/local authority. The MWA draws solely from surface water to supply water within its service area, which is the inner city (Bangkok, Samut Prakan, and Nonthaburi provinces). However, water supplied in the area outside the inner city (four provinces), which is the service area of the PWA and municipal/local authority, comes from both surface and groundwater.

Traditionally, water supplies for this area relied heavily on groundwater. The MWA gradually switched the source from groundwater to surface water and stopped using groundwater as a source of water in 2001. Even though water for domestic use is supplied by the MWA, groundwater is still an important source of water used to support the fast-growing economy in business and industry because of its lower cost and better quality. Groundwater abstraction in the area gradually increased and reached 2.4 million cubic metres per day (MCM/d) in 2000, which far exceeded the safe yield of 1.25 MCM/d for the area.

Groundwater Environment in Asian Cities
http://dx.doi.org/10.1016/B978-0-12-803166-7.00011-8

Copyright © 2016 Elsevier Inc.
All rights reserved.

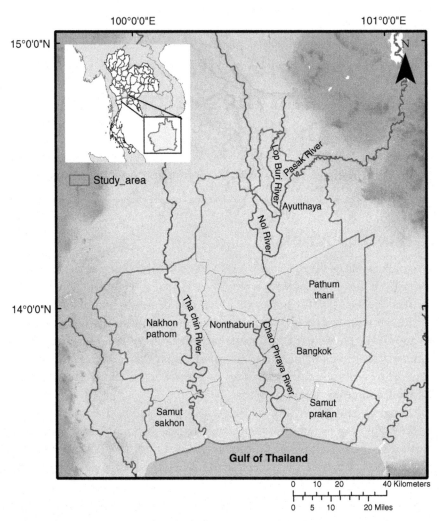

Figure 11.1 *Morphology of Bangkok and its Vicinity, Thailand.*

The overexploitation of groundwater has prominently induced land subsidence in the area. Throughout the history of groundwater management, government authorities in Thailand have put a lot of effort into the introduction of regulatory measures to control groundwater pumpage in this area, aiming to reduce the rate of land subsidence. Some of the efficient tools introduced to achieve that goal were the declaration of critical zones, introducing the conjunctive use of groundwater and surface water to business and industry, and pricing groundwater competitively for the supply of piped water to the public.

This chapter aims to compile all the available data, information, and knowledge to assess the status of the groundwater environment in Bangkok and its vicinity by analyzing: driver (D), pressure (P), state (S), impact (I), and response (R) (DPSIR), as well as their extent and interrelationships. The results will be useful for all stakeholders and policy makers, improving their understanding of the situation and serving as a basis for decision-making processes regarding the sustainable utilization of groundwater in the long term.

11.2 ABOUT THE STUDY AREA

Bangkok is located on an extremely flat, fertile, and low-lying elevation in the southern part of the LCP in Thailand. The city has been developed on the banks of the Chao Phraya River; the main river of the country, flowing through Bangkok from the northern highlands and discharging into the Gulf of Thailand 25 km south of the city (Figure 11.1). The LCP consists of the fluvial and marine deposits of the Chao Phraya Delta. It is bounded by a mountain range on the west, upper plain on the north, and the Khorat Plateau on the east. Fans and terraces occupy the west and east marginal zones of the plain. The delta that hosts Bangkok is formed by deposits of seaward tidal flat and brackish clays, marine clays, and a tidal zone. Although largely an alluvial plain, inliers of older rock occur around margins of the plain, concealed beneath a fault-bounded basin of Cenozoic nonmarine sediments (Ridd et al., 2011). Bangkok's climate is monsoon with three seasons a year: wet season from May to July with the arrival of the southwest monsoon in early May; cool season from November to February with the start of the northeast monsoon around November; and summer season from March to May. The temperature in Bangkok is high at around 30°C for most of the year but often reaching in excess of 40°C.

The total population of the study area in 2013 was about 11.3 million. With a total land area of about 10,300 km2, the average population density ranges from 300 to 3600 persons/km2 with an overall average of 1100. The population in Bangkok accounts for more than half of the total population in the study area. The total gross provincial product (GPP) of the study area in 2013 was BHT5180 billion (baht), representing approximately 44% of the country's GDP, 64% of which was generated in Bangkok (NESDB, 2013).

The study area covers the groundwater control areas of Bangkok and its six surrounding provinces, consisting of underlying unconsolidated sediment with a multiaquifer system. The top soil consists of soft to stiff, dark gray to black clay, also known as "Bangkok clay," ranging in thickness from 20 m to 30 m. Beneath the Bangkok clay layer are unconsolidated and semiconsolidated sediments intercalated by clay layers and containing large volumes of voids for water storage, forming several confined aquifers, distinguished into eight layers as follows: (i) Bangkok aquifer (BK, 50 m zone), (ii) Phra Pradaeng aquifer (PD, 100 m zone), (iii) Nakhon Luang aquifer (NL, 150 m zone), (iv) Nonthaburi aquifer (NB, 200 m zone), (v) Sam Kok aquifer (300 m zone), (vi) Phaya Thai aquifer (350 m zone), (vii) Thonburi aquifer (450 m zone), and (viii) Pak Nam aquifer (550 m zone) (JICA, 1995; Ramnarong, Buapeng, 1992) (Figure 11.2). The aquifers are highly productive, consisting of well-sorted sand and gravel. The wells tapped to these aquifers can yield over 1000 m^3/d.

11.3 DRIVERS

Population growth, urbanization, tourism, and industrialization are the main drivers of the groundwater environment in Bangkok and its vicinity. The status of indicators for drivers and other components of the DPSIR framework are tabulated in Table 11.1.

11.3.1 Population Growth

Bangkok witnessed a slow increase in population until 1975. Subsequently, in the last 40 years, the city has changed a lot due to the flow of tourists and international activities. The population of the city was only 1.5 million before Bangkok became a destination for American servicemen during the Vietnam War. The rural poor were attracted by US dollars and development began. The city grew to more than 8.5 million in 25 years. This represents nearly 15% of the country's population and is 40 times the size of any other city in Thailand (Haapala, 2002).

Trends in annual population and growth rate are shown in Table 11.2. The growth rate in Bangkok Metropolis from 1960 to 1970 was 4.8%, which increased to 5.2% between 1970 and 1980 and started to decrease over subsequent decades. While the annual growth rate in Bangkok decreased from 1980, the growth rate in its vicinity increased substantially from 1.7% in 1960 to 4.7% in 2000. The total population in both Bangkok Metropolis and its vicinity increased continuously from 3.3 million in 1960 to 11.1 million in 2010. This shows that the Bangkok suburbs are growing faster than the city center. The population densities in Bangkok Metropolis and its vicinity areas in 2010 were 3600 and 300–1800 persons/km^2, respectively.

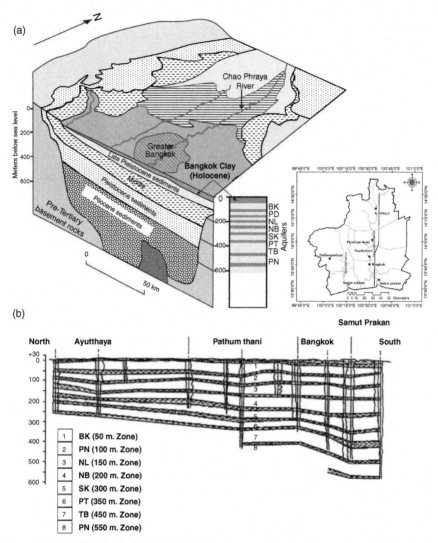

Figure 11.2 (a) Hydrogeological block diagram to show the Lower Central Plain and alluvial aquifer system (Jica, 1995); (b) Hydrogeological cross section in North-South direction to show depths of aquifer system.

11.3.2 Urbanization

Urbanization in Bangkok started in the 1960s. The rapid expansion of Bangkok and its vicinity is depicted in Figure 11.3. This is obviously attributed to the development of infrastructures such as road networks, real estate, land values, as well as an advancing economy, which has resulted in urban

Table 11.1 Status of DPSIR indicators in Bangkok and its vicinity, Thailand

	Indicators	Year 1	Year 2
Drivers	Population growth (see Table 11.2 for details)	Population: 6.6×10^6 (in 1980) Population density (persons/km^2): 782 (in 1980)	11.1×10^6 (in 2010) 3600 in the city core and 300–1800 in its vicinity (in 2010)
	Urbanization (see Table 11.3 for details)	Urban area: 1119 km^2 (in 1988)	3031 km^2 (in 2011)
	Tourism: number of tourists arriving (millions)	7.76 (in 1998)	26.74 (in 2013)
	Industrialization	Share of industry in Thailand's GDP is continuously increasing (Figure 11.6); Thai manufacturing output grew at an average annual rate of 9.7% during 1961–1985 to 13% during 1986–1996; the groundwater development pattern and depression cone in Bangkok and its vicinity also followed the pattern of industrial development in Thailand.	
Pressures	Inadequate surface water resources	Dependence on surface water has increased as indicated by increasing numbers of customers of the MWA from 1.44 to 1.75 million during 2001–2006. The MWA has increased production of surface water resources from 1482 to 1700 MCM/y. Reduced quantity of surface water in per capita terms, inadequate quality, and uneven distribution over the area has encouraged groundwater use and hence put pressure on the groundwater environment of Bangkok's aquifers.	
	Groundwater overexploitation	Groundwater abstraction: 2.3 MCM/d (in 1997) Recharge: 0.6 MCM/d (DGR, 2004) to 0.7 MCM/d (DGR, 2008)	0.836 MCM/d (in 2015)
	Land cover changes	Agricultural area: 8326 km^2 (in 1995)	4996 km^2 (in 2002)

State	Well statistics	Number of production wells: 9077 (in 2001); Well density: 0.89 wells/km²	7200 (in 2010); 0.70 wells/km²
	Groundwater abstraction	0.008 MCM/d in 1954, 0.45 MCM/d in 1982, 2.3 MCM/d in 1997, with a drop in abstraction from 1998 onward after control measures were put in place.	
	Groundwater level	2 to −77 m (in 2001)	2 to −67 m (in 2010)
	Groundwater quality	Cl⁻: 2.4–4400 mg/l (in 2001)	3.4–9200 mg/l (in 2010)
	Recharge	0.6 MCM/d (DGR, 2004) to 0.7 MCM/d (DGR, 2008)	
Impacts	Land subsidence	0.9–9 cm/y during 1978–1982	1.5–2.4 cm/y (2003–2005); 1.0 cm/y (2006–2012)
	Depletion and recovery of groundwater level	Groundwater level: 50–60 mbgl (in 1977)	20–40 mbgl (in 2013)
	Degradation in groundwater quality	Large areas of high chloride concentration in 1993 were reduced during the period of low pumpage in 2009 (Figure 11.13)	
Responses	Aquifer monitoring	Networks of 150 groundwater level–monitoring wells (Figure 11.14) and land-subsidence–monitoring wells (Figure 11.15) were established in 1978, monitoring frequency being at least twice a year for groundwater level and once for land subsidence	
	Environmental standards and guidelines	Drinking standard for groundwater established by DGR consisting of 27 parameters under the domains of physical quality, chemical quality, toxic elements, and microbiological concentrations (Table 11.6)	
	Groundwater management instruments	Management instruments in the form of regulatory measures, economic measures, and supporting measures adopted since the Groundwater Act 1977. The instruments helped to recover groundwater table and land subsidence in recent years (Section 11.7.3)	

DGR, department of groundwater resources; DPSIR, driver–pressure–state–impact–response framework; MCM, million cubic meters; mbgl, meters below ground level; MWA, metropolitan waterworks authority.

Table 11.2 Population growth of Bangkok and its vicinity, Thailand

	Total population (millions)						Annual growth rate (%)					
	1960	1970	1980	1990	2000	2010	1960 1970	1970 1980	1980 1990	1990 2000	2001 2010
(a) Bangkok metropolis	2.1	3.1	4.7	5.9	6.3	5.7	4.8	5.2	2.6	0.7	−1
(b) Vicinity (6 provinces)	1.2	1.4	1.9	2.7	3.8	5.4	1.7	3.6	4.2	4.7	4.2
(a) + (b)	3.3	4.8	6.6	8.6	10.1	11.1	4.5	3.7	3	1.7	1.09

Sources: National Statistical Office (NSO), 2010; Ministry of Interior, 2011

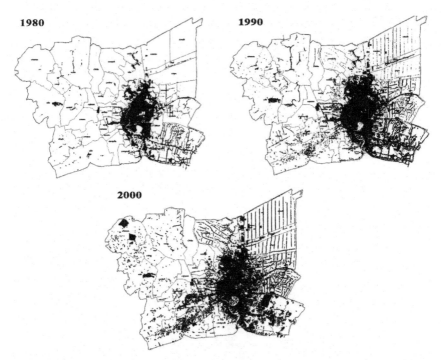

Figure 11.3 *Expansion of Bangkok and its Vicinity in 1980, 1990, and 2000. Source: Department of Public Works and Town & Country Planning (2008).*

agglomeration into surrounding areas. The urban area of Bangkok Metropolis and its vicinity increased drastically from 39% and 5% in 1988 to 63% and 23% in 2011 (Table 11.3). The Bangkok Comprehensive Plan (BMA, 2006)

Table 11.3 Urban areas of Bangkok and its vicinity

Region	Total area (km²)	Urban area (km²)				
		1988	2002	2007	2009	2011
Thailand	513,115		18,246 (3.56%)	23,729 (4.62%)	24,179 (4.71%)	
(a) Bangkok metropolis	1565	610 (39.15%)	842 (53.54%)	897 (57.33%)	993 (63.46%)	1001 (63.93%)
(b) Vicinity (6 provinces)	8750	509 (5.82%)	1110 (12.69%)	1568 (17.92%)	1605 (18.34%)	2030 (23.20%)
(a) + (b)	10,315	1119 (10.85%)	1952 (18.92%)	2465 (23.90%)	2598 (25.19%)	3031 (29.38%)

Source: Land Development Department (2014).

set targets to accommodate a population of 9.3 million in 2002, 10.2 million in 2017, and 11 million in 2022. Approximately 54.5% growth was expected between 1995 and 2015 with urban areas estimated on the basis of agglomeration instead of administrative boundaries (Figure 11.4).

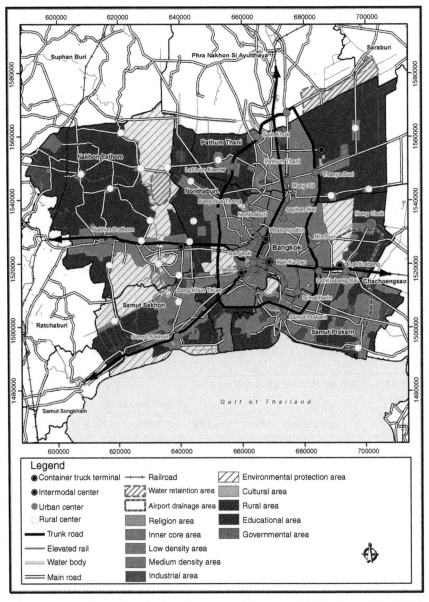

Figure 11.4 *Land Use of Bangkok, Its Vicinity, and Provinces in 2007. Source: World Bank, 2009 (modified from Department of Public Works and Town & Country Planning, 2008.)*

Bangkok and its vicinity play a fundamental role in the administration and governing of the country. Regional development in the past has enabled the area to attract various developmental activities, including infrastructures, social services, and particularly those relating to economics. The subsequent development resulted in the rapid growth of all activities, with the region being the center for settlement, industry, commerce, and services, including social services. These are key factors attracting labor, industry, and individuals from other regions (MLIT, 2013).

The urban sprawl of the area has led to inappropriate land use, causing various urban problems. There has been rapid growth within the area joining the inner city and urban fringe, creating economic, commercial, industrial, and residential centers in both vertical and horizontal directions. These types of development have problems due to insufficient services and facilities, as well as growth of urban communities along transportation routes in both the urban fringe and suburban areas. Most of these areas are developed into residential quarters, with huge department stores and industrial clusters along the main transportation routes and intersections between main roads.

11.3.3 Tourism

Thailand considers tourism to be a fast track to economic growth. Tourism requires less investment than other industries and is an effective means of creating job opportunities, and increasing local income. This sector contributed an estimated 7.3% to Thailand's GDP in 2012. If indirect benefits are also included, it accounts for 16.7% of Thailand's GDP (WTTC, 2014)

Thailand has devised a tourism-marketing approach, which encourages low, medium, and high-cost mass tourism to nearly all regions of the country. It has become one of the best known, and most sought after, international tourist destinations. On June 1, 2013, *Time* magazine reported that Bangkok was identified as the most visited city in the world by the 2013 Global Destination Cities Index (Pfotenhauer, 1994). Figure 11.5 depicts the trend in the arrival of tourists in Thailand.

Critics claim that tourism promotion in Thailand aims at quantity rather than quality. The explosion of tourism has brought an uneven distribution of financial benefits in favor of large enterprises, while costs are shouldered by local people who have no direct gain from tourist promotion. Worse still are the environmental effects of unbridled tourism development. However, most large hotels are located in the city, which already uses piped water supplies from the MWA or PWA. Therefore, pressure from the tourism industry on water usage, and groundwater in particular, may not be that large in comparison with other cities in Asia.

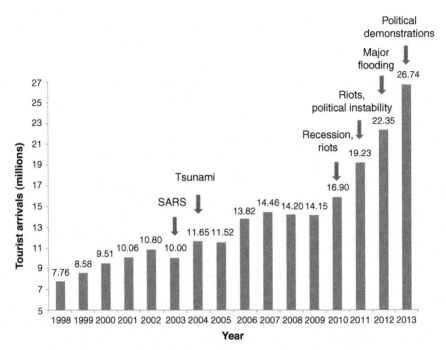

Figure 11.5 *Yearly Tourist Arrivals and Major Critical Events in Thailand from 1998 to 2013. Source: Department of Tourism (2014).*

11.3.4 Industrialization

Industrial development in Thailand has been successful for a newly industrialized country. However, industrial geographical dispersion has failed as indicated by a very high concentration of industrialization in the Bangkok Metropolis Region (BMR). The Thai government attempted to persuade business sectors to locate to the periphery or rural areas, including moving their industries from central Bangkok using several types of tax incentives, secondary city or growth pole, and supporting subregional development. However, this has not been successful and rural Thailand has yet to be industrialized (Pansuwan, 2010).

During the postwar era, manufacturing grew much faster than other sectors, resulting in its increased importance, most noticeably between 1986 and 1996. As a matter of fact, Thai manufacturing output grew at an average annual rate of 9.7% during the period 1961–1985, while from 1986 to 1996 the sector attained an annual average growth rate of around 13%. However, with the onset of the financial crisis, growth in the manufacturing sector slowed down, dropping to as low as 4.4%/y. Considering the past four decades, growth patterns of the Thai manufacturing sector can be separated into two subperiods: 1960–1985 and 1986–present day (Kohpaiboon, 2003). The purpose of such segregation is to illustrate the growth performance of different industrialization strategies between import substitution (IS) and export promotion (EP) regimes (Figure 11.6).

It is worth noting that during an IS industrialization period, the country's development started with a rapid expansion of manufacturing of textiles and clothing as well as that of transport equipment. Such trends led to the dramatic increase in the share of manufacturing from 1.7% in 1950 to 13.1% between 1976 and 1980. In order to gradually lessen successive deficits in the balance of payments between the late 1970s and early 1980s, the government shifted its industrialization strategy toward EP. In the early 1980s, the government implemented the Board of Investment (BOI) promotion scheme to partly mitigate the adverse impact of input tariffs on the international competitiveness of export-oriented industries. Under such a scheme, the BOI imposed tariff exemptions on imports over and above the usual investment promotion privileges for export-oriented activities. In the mid-1980s many East Asian investors were seeking an export base to maintain international competitiveness in their labor-intensive products, such as textiles, garments, and footwear. This was also the case with electronics and other durable consumer goods industries.

On the other hand, trade in exports of Thai-manufactured products also rapidly expanded between 1986 and 1995. Between 2001 and 2003,

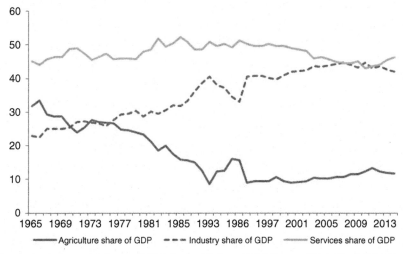

Figure 11.6 *Sectorial Share of GDP during 1965 to 2006. Source: World Bank (2014).*

manufactured products accounted for 75.2% of the country's total exports. This takes into consideration the fact that Thailand's development in manufacturing for export commenced in the late 1970s (Figure 11.6) with the production of several processed food products, especially canned pineapple, canned tuna, frozen chicken, as well as traditional labor-intensive manufactured goods (garments in particular). In 2006, the export of Thai manufactured products amounted to BHT4315 billion, which mainly constituted machinery and mechanical appliances; including computers and computer parts, vehicles and auto accessories, and electrical appliances – especially integrated circuit products.

The pattern of groundwater development in Bangkok and its vicinity also reflected industrial development. The early stage of groundwater development was concentrated in the inner zone and eastern part of Bangkok province. During the IS period, most textiles and related industries relocated in the eastern adjacent province of Samut Prakan. Subsequently, during the EP period, processed food products, canned fruit, canned tuna, frozen chicken, as well as traditional labor-intensive manufactured goods located to the western adjacent province (Samut Sakhon). The groundwater depression cone also followed this industrial development pattern.

11.4 PRESSURES

Pressures on Bangkok's groundwater environment come from inadequate surface water resources, overexploitation of groundwater resources, and land cover change.

11.4.1 Inadequate Surface Water Resources

The piped water supply in Bangkok and its vicinity is the responsibility of different agencies in each service area. The area receives its water supply from the MWA, PWA, and municipal or local authority. In addition, private water suppliers also serve some areas (World Bank, 2008). The MWA supplies water to Bangkok, Samut Prakan, and Nonthaburi provinces and the PWA to the remaining provinces. The MWA and PWA are responsible for the sourcing, production, and distribution of water. With population growth and subsequent rise in water demand, there is higher dependency on surface water as indicated by an increasing number of MWA customers (i.e., 1.44–1.75 million between 2001 and 2006). The MWA has increased production of surface water resources from 1482 to 1700 MCM/y. Increased production has reduced the availability of surface water in per capita terms.

In addition, inadequate quality of surface water has encouraged users to go for groundwater to supplement their supply. The uneven distribution of surface water is also an issue that is causing the use of groundwater to become a source of piped water for the vicinity of Bangkok. Villages without installed piped water systems, housing estates, and factories located outside the service of the MWA and PWA still depend on groundwater. Inadequate surface water is therefore putting pressure on groundwater resources.

11.4.2 Groundwater Overexploitation

Total water demand in this area for 2009 was estimated at 20.67 MCM/d and demand for groundwater was 1.5 MCM/d (Table 11.4). The total water produced by both the MWA and PWA was 6.9 MCM/d (MWA, 2014; PWA, 2013, 2014). The summation of water demand for domestic and industrial sectors, to be supplied through the piped system, was 7.04 MCM/d. Moreover, it is estimated that people outside the PWA and MWA service area still depend on groundwater over 0.5 MCM/d.

In the early stage of pumping, most of the water was drawn from the aquifer. This resulted in a lowering of the groundwater level in the aquifer, developing a hydraulic gradient between the aquifer and aquitard. This hydraulic gradient induced flow from the aquitard to the aquifer accompanied by a decrease in its hydraulic head. Such an aquifer and aquitard–dewatering process led to a concurrent reduction in pore pressure, pore volume, and land subsidence.

A groundwater modeling exercise suggested that recharge to Bangkok's aquifer system is approximately 0.6–0.7 MCM/d with the safe yield or permissible yield of this area being 1.25 m^3/d with the control groundwater level 30 meters below ground level (mbgl). Groundwater pumpage for the public water supply alone was 1.2 MCM/d in 1995 and 0.836 MCM/d in 2015. Statistics indicate the overexploitation of

Table 11.4 Water demand in Bangkok and its vicinity in 2009

| Region | Total water demand/groundwater demand (MCM/d) | | | |
	Domestic (DGR, 2009a)	Industrial (DGR, 2009a)	Agricultural (DGR, 2009b)	Total
Bangkok	1.55/0.01	0.98/0.06	0.81/–	3.34/0.07
Vicinity★	1.12/0.29	3.39/1.01	12.82/0.12	17.33/1.43
Total	2.67/0.30	4.37/1.08	13.63/0.12	20.67/1.50

★Vicinity (six surrounding provinces, currently declared as groundwater critical areas).

groundwater resources, which certainly exerts pressure on the groundwater environment.

11.4.3 Land Cover Changes

Bangkok and its vicinity is rapidly becoming urbanized as discussed in Section 11.3.2. Agricultural areas have decreased from 8326 km^2 to 4996 km^2 between 1995 and 2002, and have no doubt decreased further in recent years. Most of those areas have been converted into urban built-up sectors, creating a reduction in recharge to groundwater. Restricting inflow to groundwater imposes pressure on the groundwater environment.

11.5 STATE

The state of the groundwater environment in Bangkok and its vicinity is discussed from the perspective of well statistics, groundwater abstraction, groundwater level, groundwater quality, and recharge into the aquifers.

11.5.1 Well Statistics

Bangkok is a city with a relatively high water consumption per person per day. The MWA is responsible for supplying drinking water for domestic and industrial use. At present, the MWA and PWA use surface sources only to produce treated water. The major groundwater users apart from industrial users, are municipal and local authorities, which supply water outside the service area of the MWA and PWA. The total number of privately owned wells in Bangkok and its vicinity used by domestic, business, and agricultural sectors in 1997 and 2015 are shown in Table 11.5. This shows that the total number of working wells in Bangkok and its vicinity have decreased from 9771 in 1997 to 4240 in 2015.

Taking the statistics from 2001 and 2010, the number of production wells has decreased from 9077 to 7200 and well density from 0.89 wells/km^2 to 0.70 wells/km^2.

11.5.2 Groundwater Abstraction

Groundwater development for the public supply in Bangkok began in 1954 with an abstraction of 8360 m^3/d, increasing to 0.45 MCM/d in 1982. Including abstraction from private sectors, the figure in 1982 was 1.4 MCM/d. In 1983 the authorities designed and implemented control measures under the title "Mitigation of Groundwater Crisis and Land Subsidence in the Bangkok Metropolis." This led to a sharp drop in groundwater

Table 11.5 Well statistics (private wells) of the study area

Province	Domestic		Business		Agriculture		Total	
	Number of wells	Pumpage (m³/d)	Number of wells	Pumpage (m³/d)	Number of wells	Pumpage (m³/d)	Number of wells	Pumpage (m³/d)
					1997			
Bangkok metropolis	711	221,294	634	178,303	33	3395	1378	402,992
(a) Vicinity (6 provinces)	4033	624,364	4257	1,283,736	103	6686	8393	1,914,786
*(b)		436,954						436,954
(a) + (b)		1,061,318						2,351,740
Total	4744	1,282,612	4891	1,462,039	136	10,081	9771	2,754,732
					2015			
Bangkok Metropolis	51	1554	130	5324	9	273	190	7151
(a) Vicinity (6 provinces)	784	46,922	3179	172,861	87	40,697	4050	260,480
(b)		568,979						568,979
(a) + (b)	784	615,901	3179	172,861	87	40,697	4050	829,459
Total	835	617,455	3309	178,185	96	40,970	4240	836,610

The amount of groundwater supplied by municipal/local authorities was estimated from population.
* Population serviced by local authorities outside the MWA and PWA service area (2,844,895 persons) × rate of 200 liter/person/day is equal to 568,979 m³/d.

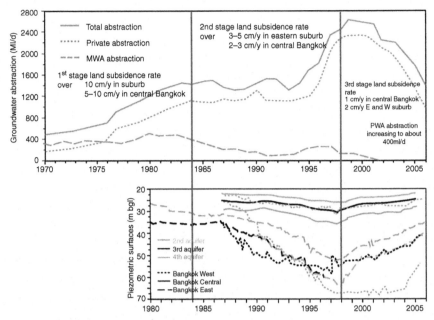

Figure 11.7 *Groundwater Abstraction and Corresponding Groundwater Levels and Land Subsidence. Source: Modified from Buapeng and Foster (2008).*

abstraction between 1985 and 1990. However, abstraction began to increase again from 1991 onward due to high economic growth. By 1997, total groundwater abstraction in the control area of the four provinces: Bangkok, Samut Prakan, Nonthaburi, and Pathumthani was 1.67 MCM/d. The total for seven provinces was 2.3 MCM/d. After the economic crisis and full implementation of economic instruments for groundwater usage, plus the groundwater conservation charge, groundwater use declined steeply (Figure 11.7).

11.5.3 Groundwater Levels

The initial groundwater levels in Bangkok were very close to the ground surface and some wells were said to be artesian. In those days, groundwater was believed to have been pumped from the shallowest good–water quality aquifer (PD aquifer) since the overlying Bangkok aquifer (BK aquifer) produced brackish to saline water.

A plot of west–east cross-sections during the period of groundwater decline and recovery is shown in Figure 11.8. In the early stages of groundwater development in around 1959, groundwater levels ranged from

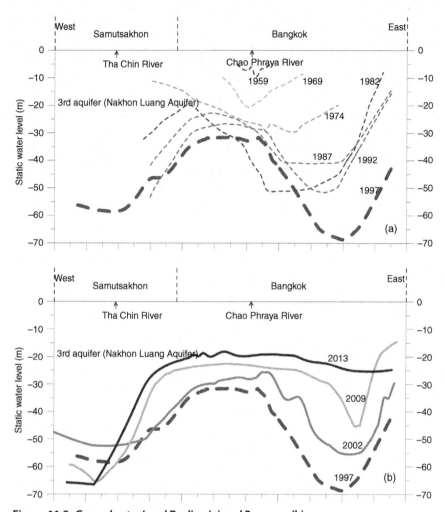

Figure 11.8 *Groundwater Level Decline (a) and Recovery (b).*

4 mbgl to 5 mbgl in eastern Bangkok to 12 mbgl in central Bangkok. After 1967, a heavy use of groundwater was observed in the eastern part of the city. By 1967, the lowest water level in the NL aquifer in central Bangkok and the eastern suburbs was 30 mbgl. Annual rates of water level decline in the NL aquifer from 1969 to 1974 were 3.6 m (i.e., 0.7 m/y) in the eastern part and 1–2 m (0.2–0.4 m/y) in central Bangkok.

From 1959 to 1982, the water level in the NL aquifer declined by 38 m in central Bangkok and 60 m in the eastern suburbs. Since 1983, the control measures on groundwater pumping together with the introduction of a groundwater tariff in 1985 have had a marked effect on groundwater use in Bangkok; the consequent decrease in groundwater withdrawal produced a rapid recovery of water levels in three aquifers (PD, NL, and NB). In central Bangkok, the public water supply produced from surface water sources replaced much of the groundwater, resulting in a continuous rise in the groundwater level. Although the groundwater crisis in central Bangkok has been improved, depression cones have developed in new areas on the outskirts of Bangkok both in the eastern province (Samut Prakan) and the western province (Samut Sakhon), which are areas of extensive industry (Figures 11.9 and 11.10).

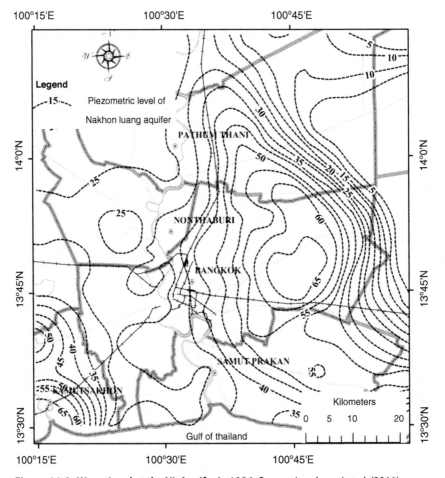

Figure 11.9 *Water Level at the NL Aquifer in 1996.* Source: Lorphensri et al. (2011).

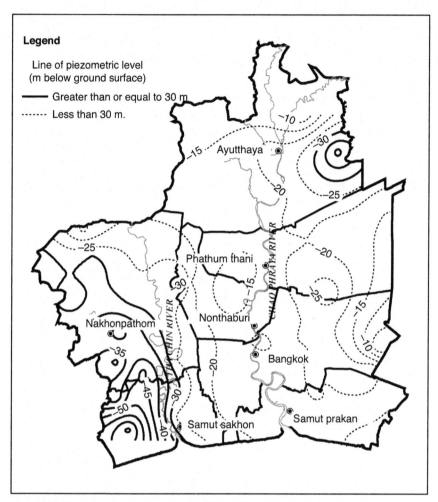

Figure 11.10 *Groundwater Level at the NL Aquifer in 2014.*

11.5.4 Groundwater Quality

The status of groundwater quality is assessed using water samples collected from monitoring and production wells. Fundamental parameters (e.g., physical and chemical) are measured in the water samples. The parameters of concern in this area are total dissolved solids, iron, and hardness. Maps of chloride concentrations and the total dissolved solids are regularly produced to monitor the water quality status of the area. Most high chloride content in the three uppermost production aquifers (PD: 100 m, NL: 150 m, NB: 200 m) occurred in areas near the river mouth and shoreline. However, inland salt water in some areas also exists. The upper 50 m deep aquifer (i.e., BK aquifer) is the only aquifer that totally contains salt water. The source of high salinity is found to be diffusion of salt from the overlying marine clay layer.

11.5.5 Recharge

The study area is totally covered by a thick marine clay layer, hence direct recharge should be minimal. However, almost no percolation of surface water or precipitation occurs due to the presence of clay layers. AIT (1981) revealed that only 3% of total recharge is through direct percolation from soil layers and surface water. This implies that almost all recharge comes through the movement of groundwater from outside the discharge areas.

Groundwater in the Bangkok aquifers is basically recharged away from abstraction areas (Figure 11.11). However, such natural recharge is limited

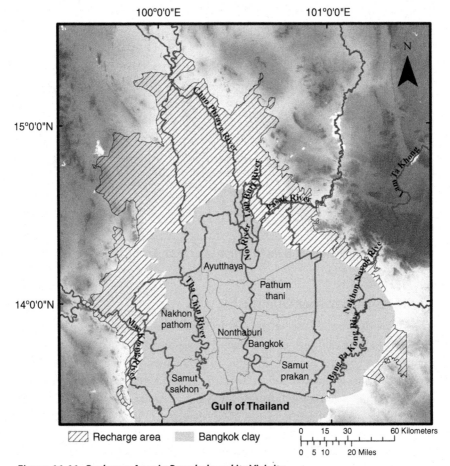

Figure 11.11 *Recharge Area in Bangkok and its Vicinity.*

because of the slow movement of groundwater. An estimated sustainable yield from all the aquifers in the Bangkok area is 0.6 MCM/d (DGR, 2004) to 0.7 MCM/d (DGR, 2008a). This means that this amount of groundwater is supposed to be recharged to maintain a balance against pumping. Estimated groundwater use in the study area is 0.836 MCM/d (Table 11.5). Comparing the amount of recharge and groundwater use shows that groundwater balance is still in deficit.

11.6 IMPACTS

Major adverse impacts of groundwater overexploitation in Bangkok's aquifers include land subsidence, depletion, and recovery of groundwater levels, together with degradation in groundwater quality.

11.6.1 Land Subsidence

Land subsidence in the area can be divided into three stages: early (1978–1981), mid (1984–2000), and recent (2000–2008) (Figure 11.12). The response of land subsidence to groundwater level changes is well described by long-term monitoring at Ramkhamhaeng University (Figure 11.12). The slope of land subsidence follows the slope of groundwater level

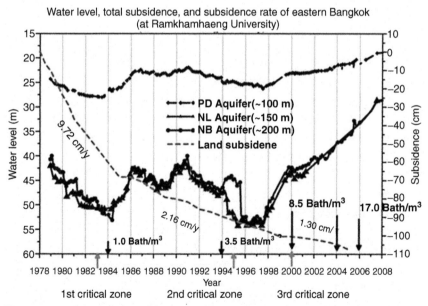

Figure 11.12 *Trends in Groundwater Levels, Land Subsidence and Chronology of Mitigation Measures at a Monitoring Station in the Eastern Suburb of Bangkok. Source: Lorphensri et al. (2011).*

(decline and recovery) (Lorphensri, 2011). In the early stage, land subsidence was over 10 cm/y in the eastern suburbs, and 5–10 cm/y in central Bangkok. Soil compression during this period in the top 50 m and in the deeper zone of 50–220 m in depth, contributed 40 and 60% to total surface subsidence, respectively (AIT, 1982). Leveling in 1982 by the Royal Thai Survey Department (RTSD) indicated that the lowest elevation in Bangkok was 4 cm below sea level at a land subsidence–monitoring station within Ramkhamhaeng University.

During the mid-stage, after introducing control measures against groundwater use in 1983, a short continuous recovery of the groundwater level was observed in central Bangkok and its eastern suburbs. This resulted in a decrease in the rate of land subsidence. The annual subsidence rate in 1989 reduced to 2–3 cm/y in central Bangkok and 3–5 cm/y in the eastern suburbs (Ramnarong and Buapeng, 1992).

From 2000 to 2014, the subsidence rate stabilized or recovered. The average subsidence rate for the entire area is 1 cm/y. The higher rate of 2 cm/y can still be found in the eastern province (Samut Prakan) and the western province (Samut Sakhon).

11.6.2 Depletion and Recovery of Groundwater Levels

The study area has experienced several stages of groundwater development. At the stage of overexploitation, the groundwater level declined to 50–60 mbgl (in 1997) at the rate of 2 m/y. After implementing control measures and expanding the services of the MWA and PWA, which use surface water, groundwater levels recovered to 20–40 mbgl in 2013 at the rate of around 3 m/y. The new concern for authorities is preparing for the possible consequences if the groundwater level continues to rise steadily.

11.6.3 Degradation in Groundwater Quality

It is noticeable that groundwater quality deteriorates during periods of high pumpage, and increasing chloride concentrations have been observed in several areas. A comparison of chloride concentrations during the years 1993 and 2009 is shown in Figure 11.13. Large areas of high chloride concentrations in 1993 have been reduced during periods of low pumpage (2009). Leakage of salt water from the uppermost layer and salt water pockets in some aquifers can be expected to diffuse or leak to the fresh water aquifer due to the past history of heavy pumpage.

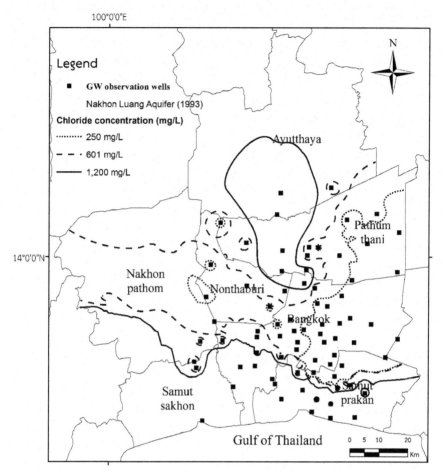

Figure 11.13 *Chloride Concentrations in 1993 and 2009.*

11.7 RESPONSES

11.7.1 Aquifer Monitoring

This consists of monitoring groundwater levels and land subsidence. Networks for groundwater level–monitoring wells (Figure 11.14) and land subsidence (Figure 11.15) were established in 1978. Groundwater levels are monitored at 150 stations (Figure 11.14), each consisting of a set of wells tapping at least three top aquifers of interest. The distance between the aquifer generally ranges from 5 km to 10 km depending on gradients of water level. Monitoring frequency is at least twice a year. For land subsidence, a network (Figure 11.15) has been established of 1 m deep benchmarks,

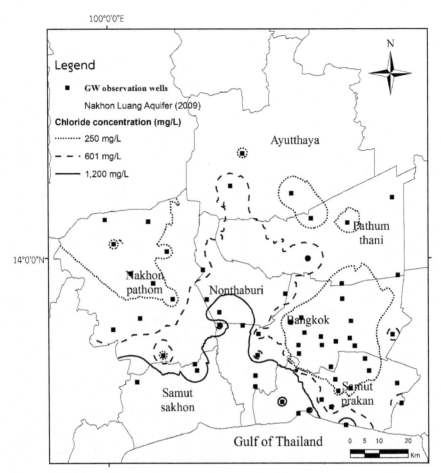

Figure 11.13 *(cont.)*

distributed evenly throughout the area. Once a year, the RTSD carries out a survey on land leveling.

11.7.2 Environmental Standards and Guidelines

The Department of Groundwater Resources (DGR) has established drinking standards for groundwater (Table 11.6). The main concerns are physical, chemical, biological, and toxic, and 27 parameters are considered for quality characterization. The Ministry of Natural Resources and Environment, the agency responsible for determining environmental standards and guidelines, has set an additional standard in order to cope with recent problems in the groundwater environment. More stringent groundwater

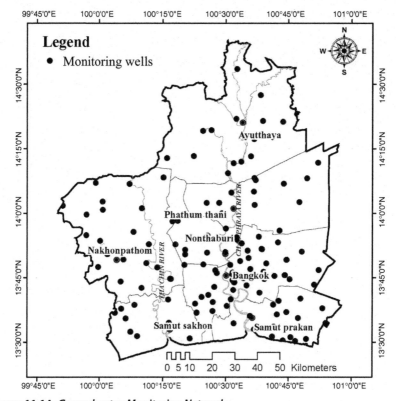

Figure 11.14 *Groundwater-Monitoring Network.*

standards have been established with consideration given to several organic constituents.

11.7.3 Groundwater Management Instruments

The Groundwater Act (1977) prescribes technical measures to protect it from pollutants and introduces standards for groundwater to be used for drinking purposes. The PWA established the Provincial Waterworks Authority Act with the intention of improving and expanding waterworks and services in provincial areas, and is also authorized to provide water supply services to areas not catered for by the MWA.

The Groundwater Act is a basic law that provides definitive terms concerning groundwater exploitation activities, such as "groundwater," "drilling," "groundwater usage," "wells," and other related terms that need legal definition. The main concept of this law is that groundwater exploitation is a public

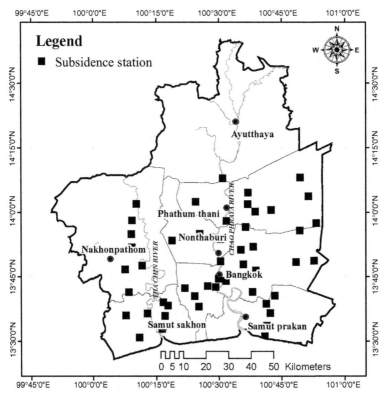

Figure 11.15 *Land Subsidence–Monitoring Network.*

matter (DGR, 2008b). Therefore, a landowner who wants to drill and exploit groundwater lying under his own land must apply for the relevant permits from the DGR. This concept may be considered an exemption to the absolute right of property owners as recognized by the Civil and Commercial Code.

In addition, the Groundwater Act requires three kinds of permit for different purposes, namely, a 1-year permit for drilling, a 10-year permit for groundwater exploitation, and a 5-year permit for discharging water into a well. In this connection, the law sets out rules and conditions for extension of permits, administrative appeals, and grounds for refusing permits or extensions, including certain measures for overseeing and control.

A second revision to the act was made in 1992, in which no pumpage areas were designated. This was in order to control water quality, prevent endangerment or deterioration of aquifers, protect natural resources and the environment, protect public health or property, and to avoid land subsidence. In addition, there are criminal sanctions for pumping groundwater

Table 11.6 Drinking standard for groundwater quality in Thailand

Physical quality

Parameter	Allowable (not exceed)	Maximum allowable
Color	5 TCU	50 TCU
Turbidity	5 NTU	20 NTU
pH	7.0–8.5	6.9–9.2

Chemical quality (mg/L)

Fe	0.5	1.0
Mn	0.3	0.5
Cu	1.0	1.5
Zn	5.0	15.0
SO_4	200	250
Cl	200	600
F	1.0	1.5
NO_3	45	45
Total hardness	300	500
Noncarbonate hardness	200	250
Total dissolved solids	750	1500

Toxic elements (mg/L)

As	0	0.05
Cn	0	0.2
Pb	0	0.05
Hg	0	0.001
Cd	0	0.01
Se	0	0.01

Microbiological water quality

Standard plate count	500 CFU/m^3	0.05
MPN of coliform organism	<2.2/100 m^3	0.2
Escherichia coli	0	0.05

CFU, colony-forming unit; MPN, most probable number; NTU, nephelometric turbidity unit: TCU, true color unit.

into a no-pumpage area and for pumping without a permit, including a procedure for the court to order an offender to return the well back to its previous condition prior to occurrence of the violation.

The third revision to the act in 2003 involved the setting up of a "Groundwater Development Fund" within the DGR to fund studies and research on conservation of groundwater and the environment. The fund is managed by a board chaired by the Director General of the DGR.

The policy/action plan for the study area is still the same as in the past few years, namely, to limit groundwater abstraction to within a "safe yield." Cooperation from the MWA and PWA is necessary to implement the policy/action plan. Domestic licences will be declined for wells when the public water supply service is extended to the area. Moreover, a certain percentage of business and industry licences will need to be switched to the public piped water supply, depending on the type of industry. The ratio of the conjunctive use of surface water to groundwater is to be maintained at between 20 and 50%.

Until recently, the DGR has adopted the policy of modernizing the well-licencing information system by introducing a geographical information system to help the licencing process. Currently, the procedure to pay a groundwater use fee has been transferred from direct payment at a provincial office to the government bank. The DGR is now in the process of decentralizing groundwater licencing and fee collection to local administration offices throughout the country. The transfer process will operate in a gradual manner. Initially, three pilot provinces were chosen for the decentralization process, with the aim of completing decentralization for all the provinces within five years. The process started by training local officers in the necessary practices and law.

As a result of the implementation of the groundwater conservation charge and expansion of the MWA and PWA services, groundwater use has been reduced below the controlled safe yield, and the rate of subsidence has reduced to 1.0 cm/y. Due to these achievements, the groundwater conservation charge has been reduced accordingly.

A groundwater conservation strategy to curb water demand takes a multiprong approach through pricing and mandatory water conservation. Pricing of water is an important and effective mechanism to encourage users to conserve water. Groundwater should be treated as an economic benefit. The water is priced not only to recover the full cost of groundwater management, but also to reflect the scarcity of this precious resource. A water conservation tax was also implemented in 1991 to further encourage water conservation. The groundwater tariffs and groundwater conservation tax were restructured over a 4-year period, starting in 1997, to reflect its strategic importance and environmental impact.

A groundwater tariff was first implemented in 1985 in the six provinces of Bangkok and its vicinity, where 1.0 baht/m^3 was charged. By 1994, the charge had increased to 3.5 baht/m^3, and the government began to charge the whole country for groundwater use. Between 2000 and 2003, the groundwater tariff was gradually increased in the Critical Zone from 3.5 baht/m^3 to 8.5 baht/m^3.

In 2003, the Groundwater Act was amended to impose a Groundwater Preservation Charge for all groundwater users in the Critical Zone. Starting at 1.0 baht/m^3 in 2004, the charge was set to increase to 8.5 baht/m^3 in 2006. The total cost of groundwater use in the Critical Zone has therefore become relatively high, but has helped in limiting the abstraction of groundwater in the area. Total groundwater charges increased from 9.5 baht/m^3 in 2004, to 12.50 baht/m^3 in 2005, and to 17 baht/m^3 for 2006 and beyond, which is deterring groundwater users in the area, especially those using large amounts such as industries. A chronological summary of the adoption of groundwater management instruments (e.g., regulatory measures, economic measures, and supporting systems) is shown in Table 11.7 and a timeline of response to land subsidence is shown in Figure 11.16.

Table 11.7 Chronology of groundwater management instruments adopted in Bangkok, Thailand (regulatory and economic)

Regulatory measures

1977	Groundwater law: the Groundwater Act was enacted in 1977 and has been amended twice (1992 and 2003). Provision for controlling of well drilling, groundwater use, discharging to well, and protection and conservation of groundwater resources in the country.
1978/1983	Designation of groundwater regions and critical zones: a groundwater region is the selected area that needs to apply for a license for groundwater use. To control groundwater use and mitigate environmental problems associated with it, areas most severely affected by groundwater-related problems such as land subsidence and groundwater depletion were designated as Groundwater Critical Zones and were subjected to more control over private and public groundwater uses.
1978	Licensing for groundwater well drilling and groundwater use: licenses were required to extract groundwater, and pumpage limits were controlled through these permits.
1985	Groundwater use metering: business and industry use of groundwater were subjected to installation of well meters. Groundwater use was charged according to meter readings.
1992/2000	Establishment of a groundwater quality standard: standards for groundwater for drinking purposes were established through the Groundwater Act. In 2000, groundwater quality standards for the conservation of environmental quality were issued.

(Continued)

Table 11.7 Chronology of groundwater management instruments adopted in Bangkok, Thailand (regulatory and economic) *(cont.)*

Economic measures

1977	Penalizing violators of regulations: violation of regulations will result in a fine of not more than THB 20,000 or imprisonment for not more than 6 months (drilling machinery and equipment will also be confiscated).
1985	Implementation of groundwater use charge: the charge was first implemented in six provinces at 1.0 baht/m^3. In 1994, the charge increased to 3.5 baht/m^3 and applied to the whole country. During 2000–2003, only in the Critical Zone, the charge gradually increased to 8.5 baht/m^3.
2003	Implementation of conservation charge: during 2003–2006, the charge for conservation started at 1–8.5 baht/m^3. In 2006, the total groundwater charge was 17 baht/m^3. These brought up the cost of groundwater use close to the public water supply. Later in 2012, after evidence that groundwater levels had persistently recovered, the conservation charge reduced from 8.5 baht/m^3 to 4.5 baht/m^3 making the total cost of groundwater use 13 baht/m^3.

Supporting systems

1978	Groundwater-monitoring system: monitoring networks were established under the study program of groundwater and land subsidence in 1978. The data collected were water level, water quality, and land subsidence in critical areas.
1982	Groundwater database: a Groundwater Database System was established electronically in 1982 and has been persistently improved through several study programs. The current status of the groundwater database is available to the public using a geographical information system and well information via web services.

11.8 SUMMARY

Groundwater development in the area has a long history, which follows the pattern of social and economic development. It is obvious that there are several impacts due to long-term groundwater development. After strict implementation of the Groundwater Act 1977 in the Bangkok area, groundwater depletion has now recovered. Land subsidence, a well-known issue for the area, is now systematically monitored. Although, land subsidence has recovered over time, there are still other impacts, which must be carefully examined, such as groundwater quality deterioration and groundwater level

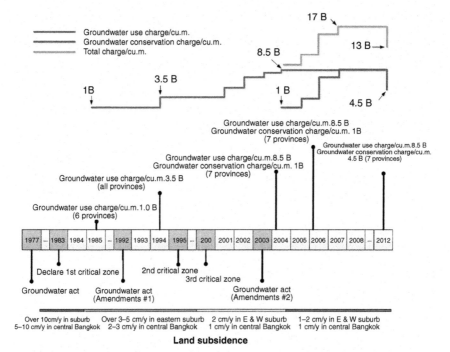

Figure 11.16 Timeline of Response to Land Subsidence in the Bangkok Aquifers.

recovery. Groundwater resources in the area are available in large quantities, and should be used wisely in the economic development of Thailand. Overall the ultimate goal of groundwater management in the area is the sustainable utilization of groundwater resources to achieve the best possible economic growth. The results of this DPSIR analysis have produced a useful knowledge base for such a purpose.

REFERENCES

AIT, 1982. Investigation of Land Subsidence Caused by Deep Well Pumping in Bangkok Area. Report NEB. Pub. 1982-002, Asian Institute of Technology.

BMA, 2006. Bangkok Comprehensive Plan 2006: Summary of Major Features, Bangkok Metropolitan Administration, The Department of City Planning.

Buapeng, S., Foster, S., 2008. Controlling Groundwater Abstraction and Related Environment Degradation in Metropolitan Bangkok-Thailand. GW-MATE Case Profile Collection No. 20, World Bank.

DGR, 2004. Effects of Groundwater Over-Pumping Mitigation: Mathematical Model Study. Department of Groundwater Resources, Kasetsart University.

DGR, 2008a. The Impact Assessment of Groundwater Conservation Tax to Groundwater Situation in Bangkok and Vicinity. Department of Groundwater Resources, Chulalongkorn University.

DGR, 2008b. The Revision of Groundwater Law for Groundwater Conservation. Department of Groundwater Resources, Environmental Law Center, Thailand Foundation.

DGR, 2009a. National Groundwater Wells Inventory. Department of Groundwater Resources.

DGR, 2009b. The Study of Conjunctive use of Groundwater and Surface Water. Pilot Project, Department of Groundwater Resources.

Department of Public Works and Town & Country Planning, 2008. Bangkok and Vicinities Regional Plan.

Department of Tourism, 2014. Thailand Tourist Statistics, http://www.tourism.go.th.

Haapala, U., 2002. Urbanization and water: the stages of development in Latin America, South-East Asia and West Africa. Master thesis, Helsinki University of Technology, 106 pp.

JICA, 1995. The Study on Management of Groundwater and Land Subsidence in the Bangkok Metropolis Area and Its Vicinity. Japan International Cooperation Agency Report submitted to Department of Mineral Resources and Public Works Department, Kingdom of Thailand.

Kohpaiboon, A., 2003. Foreign trade regimes and the FDI-growth nexus: a case study of Thailand. J. Dev. Stud. 40 (2), 55–69.

Land Development Department, 2014. Statistics of Urban Area.

Lorphensri, O., Ladawadee, A., Dhammasarn, S., 2011. Groundwater and subsurface environments. In: Review of Groundwater Management and Land Subsidence in Bangkok, Thailand, pp. 127–144, Chapter 7.

MWA, 2014. Water Production Capacity. Metropolitan Waterworks Authority. Available online at www.mwa.co.th

Ministry of Interior, 2011. Statistical Profile of Bangkok Metropolis 2011. Bureau of Registration Administration, Department of Provincial Administration.

MLIT, 2013. An Overview of Spatial Policy in Asian and European Countries. Ministry of Land, Infrastructure, Transport and Tourism, Japan. Available from: http://www.mlit. go.jp/kokudokeikaku/international/spw/general/thailand/index_e.html

NESDB, 2013. Yearly Report of Thailand Economical Statistics. Office of National Economics Social Development Board. Available from: www.nesdb.go.th

NSO, 2010, Population and Housing Census, Statistical Forecasting Bureau, National Statistical Office.

Pansuwan, A., 2010. Industrial decentralization policies and industrialization in Thailand. Silpakorn Univ. Int. J. 9–10, 117–147.

Pfotenhauer, L., 1994. Thailand's tourism industry – What do we gain and lose? TDRI Q. Rev. 9 (3), 23–26, 1994.

PWA, 2014. Plant Production Capacity. Provincial Water Works Authority.

PWA, 2013. Yearly Report 2013. Provincial Water Works Authority.

Ramnarong, V., Buapeng, S., 1992. Groundwater resources of Bangkok and its vicinity: impact and management. Proceedings of National Conference on Geologic Resources of Thailand: Potential for Future Development, Bangkok, Thailand, Vol. 2, pp. 172–184.

Ridd, M.F., Barber, A.J., Crow, M.J., 2011. Introduction to the geology of Thailand. Geology of Thailand. Geological Society, UK, pp.1–17.

World Bank, 2008. Thailand Infrastructure Annual Report 2008. World Bank, Washington, DC.

World Bank, 2009. Climate Change Impact and Adaptation Study For Bangkok Metropolitan Region, Final Report, Panya Consultant.

World Bank, 2014. World Development Indicators. Available from: http://data.worldbank. org/country/thailand

WTTC, 2014. Travel and Tourism Economic Impact, Thailand. World Travel and Tourism Council.

CHAPTER 12

Groundwater Environment in Dili, Timor-Leste

Domingos Pinto and Sangam Shrestha
Water Engineering and Management, Asian Institute of Technology (AIT), Thailand

12.1 INTRODUCTION

Dili City, the capital of Timor-Leste, is mostly dependent on the Dili aquifer as a source of water supply. The larger aquifer, also known as the Dili Groundwater Basin (DGB), is located in the downstream of the Comoro River Basin and is formed of Quaternary sedimentary deposits in the Dili alluvial plain. The texture of these sediments varies, but typically contains unconsolidated and moderate poorly sorted silts to cobbles (Wallace et al., 2012). The aquifer is bounded by the coastline in the north and mountains in other directions. There are three major well fields in the Dili aquifer, namely Comoro, Kuluhun, and Bidau, which have different hydrogeologic characteristics (JICA, 2001).

With high population growth and rapid urbanization, the DGB is experiencing a water shortage during the dry season. Water demand for various uses is increasing rapidly (e.g., agriculture, industry, tourism, and domestic) since the country started to develop in 1999. Of the total water demand of about 49,199 m³/d (or 17.9 million cubic meters, MCM) in 2010, only 14.53 MCM was fulfilled. In early 2000 there were only 14 groundwater abstraction wells in the area, which increased by more than 20 within a few years to meet 60% of the water supply needed by the city. Groundwater abstraction from the aquifer in 2010 was estimated at 27,821 m³/d (Aurecon Australia, 2012), which is equivalent to 10.15 MCM/y. The figure is close to the recharge value of 11.86 MCM/y, estimated as 19% of the recharge from the Laclo watershed area (Costin and Powell, 2006).

Recent development activities such as the construction of river banks, wetlands, and surface/subsurface drainage without considering the health of the groundwater environment, is placing stress on the aquifer by limiting recharge in the wet season. Due to increasing groundwater abstraction, the deep aquifer downstream of the Comoro River has experienced cases

Groundwater Environment in Asian Cities
http://dx.doi.org/10.1016/B978-0-12-803166-7.00012-X

Copyright © 2016 Elsevier Inc.
All rights reserved.

of saltwater intrusion. Failure by the National Directorate of Water and Sanitation Services (DNSAS or NDWSS) to supply adequate quantities and quality of water on time, has encouraged the public to look for alternative sources of supply to meet their daily needs (Aurecon Australia, 2012). The DGB, therefore, needs to take immediate action to restore and conserve the groundwater environment to ensure its sustainable use for various economic activities.

Nevertheless, management has yet to respond, probably because of inadequate understanding of the DGB, and related statistics and their parameters (e.g., hydrogeology, basin discharge, location, etc.), groundwater quality across the basin, groundwater potential zones, and recharge areas. Several studies conducted since the early 1990s on the Dili aquifers were aimed at groundwater development rather than management. These have certainly generated a knowledge base of the aquifer system such as geological formations (Prasetyadi and Harris, 1996; Thompson, 2011), groundwater quality (MoH, 2010; Wallace et al., 2012), and hydrogeology (JICA, 2001; Wallace et al., 2012). However, the results of the studies are either scattered or not analyzed coherently enough to generate information relevant to policymakers. As such, the contents of this chapter place emphasis on the development of a systematic knowledge base of the groundwater environment in Dili by analyzing driver (D), pressure (P), state (S), impact (I), and response (R) and their interrelationships. This knowledge base is expected to be useful for a diverse range of stakeholders – including policymakers – to understand the current status of the groundwater environment in Dili and develop strategies for sustainable utilization of groundwater resources.

12.2 ABOUT THE CITY

12.2.1 Physiography and Climate

Dili City is located at 8°34'0S' and 125°34'0'E and covers an area of 48 km². It lies on the northern coast of Timor island (Figure 12.1). The city is the only urban area in the Dili District, which covers an area of 170 km² and supports a population of nearly 0.15 million (Aurecon Australia, 2012). The district extends from coastal roads and beaches to the rugged mountainous terrain in the south. The elevation of the city varies from 0 to 60 meters above mean sea level (masl). The city is divided into the subdistricts of Nain Feto, Vera Cruz, Dom Aleixo, and Cristo Rei and is divided into 26 sucos, which are headed by an elected Village Chief. Eighteen out of the 26 sucos of the four subdistricts are categorized as urban.

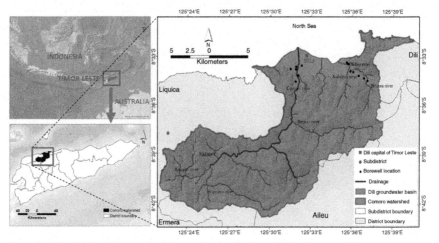

Figure 12.1 *Surface Watershed, Groundwater Basin, and the Plain in Dili.*

The city is situated in a tropical hot climate region. The watershed is influenced by the northern monomodal rainfall pattern having four to six months of wet season from December to May (Wallace et al., 2012). An analysis of climatic data over the past three decades suggests that daytime temperatures throughout November generally reach up to 29°C, dropping down to around 25°C at night (Fenco Consultants, 1981). In recent years (2003–2013), the highest temperature recorded was 36°C in November 2011 and the lowest was 14°C in August 2013 at Comoro Climate Station (Pinto et al., 2015). Average monthly precipitation is recorded as 75 mm. Annual precipitation ranges from 940 mm/y to 1761 mm/y as shown in Figure 12.2.

12.2.2 Hydrogeology

Hydrogeologically, Timor-Leste is divided into three major rock categories: (i) intergranular aquifers at Dili, (ii) fissured aquifers (karst) at Baucau, and (iii) localized aquifers (fractured) at Aileu. As per Wallace et al. (2012), the intergranular aquifers are composed of sediments with groundwater filling the spaces between sediment grains. The type of sediments that form an aquifer influence the properties of the aquifer and its ability to store and transmit groundwater. For this reason it is important to understand the sediments themselves. The sediments in the Dili aquifer have been mapped as Quaternary alluvium. The texture of these sediments varies, but is typically unconsolidated, with moderately poor silts and cobbles. Much of the alluvium is undifferentiated Quaternary sediments, but in many areas the

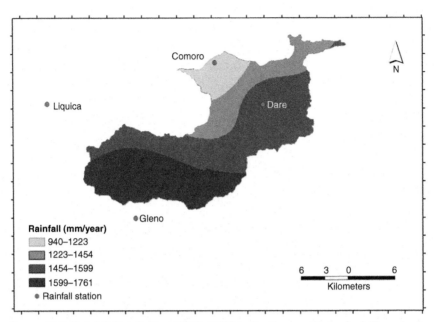

Figure 12.2 *Rainfall Distribution (2003–2013) in the Comoro Watershed, Timor-Leste.*
Source: Pinto et al. (2015).

alluvium forms part of defined units, such as the Suai Formation, Ainaro Gravels, and Dilor Conglomerate. It rests over all other older rocks and is present throughout the Timor-Leste river valleys and coastlines.

A significant portion of water resources comes from groundwater. The DGB starts from the coast with elevations of 0–60 masl and covers an area of 26 km², about 12.2% of the Comoro watershed area (i.e., 250 km²) (Figure 12.1). Groundwater is abstracted from a thick intergranular alluvial aquifer dominated by silt clay and an intermittent thin clay layer underlain by a saprolite layer and hard rock (Figure 12.3a). It indicates that the city's aquifer receives direct and indirect recharge from rainfall. The clay layer is more than 50 m thick in the central part of the Comoro wellfield. The Comoro River comprises a large amount of aquifer sediment and therefore has a high groundwater potential zone within the watershed (Figure 12.3b). Other rivers such as the Bidau and Kuluhun also contribute sediment to the aquifer. Those rivers serve as rechargeable areas for the deeper part of Dili's aquifer. The distribution of sediment in shallow and deeper parts of the aquifer and intermittent clay layers is irregular with discontinuities occurring both vertically and horizontally. The Dili aquifer is highly heterogeneous and likely to have zones of preferential flow. Rather than containing large areas of homogeneous sand, the aquifers have many clay lenses, which require more

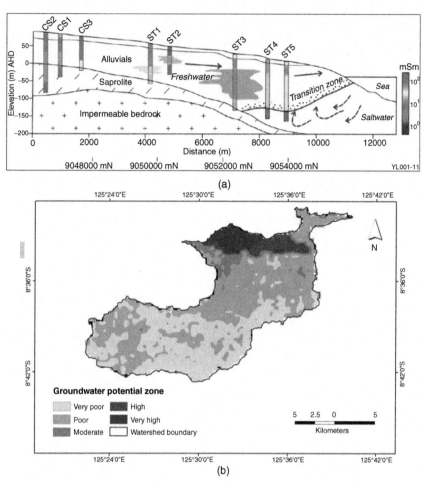

Figure 12.3 (a) Interpreted Geological Section at the Comoro Wellfield; (b) Groundwater Potential Zone in the Comoro Watershed. *Sources: (a) Furness (2012) and (b) Pinto et al. (2015).*

detailed investigation to determine the actual architecture of water storage (Wallace et al., 2012). Overall, the greatest portions of aquifer storage are located in the western and eastern parts of the basin. They are exploited largely by the NDWSS or DNSAS through two wellfields and also by private wells.

12.3 DRIVERS

12.3.1 Population Growth

Dili City has been the center of state administration, economics, education, and politics since it was a Portuguese colony in 1515. It was taken over by the Indonesian military in 1975, until Timor-Leste returned to independence in 2002. The population has steadily increased since then, mostly as a result of urbanization. However, there is no estimate of the total population within the DGB. This study uses an approach of household density inside and outside the DGB within the Dili District to estimate the population. An analysis using Google Earth revealed that 90% of the population within the Dili District can be estimated as being the population within the DGB. Using this approach, the total population within four subdistricts of Dili District was converted into the population of the DGB (Table 12.1). Results showed that the population from 2004 to 2010 increased from about 0.15 to 0.20 million within the DGB (Table 12.2). This resulted in an increase in population density from 1187 persons/km^2 to 1402 persons/km^2 (Table 12.2). In 2010, with an estimated groundwater abstraction of 27,821 m^3/d (Table 12.2), and a total population of nearly 0.2 million, per capita groundwater abstraction became 140 liters per capita per day (lpcd). Considering the rate of abstraction as constant, an increase in nearly 0.05 million people from 2004 to 2010 is equivalent to 7000 m^3/d (approx.) of groundwater abstraction. This clearly shows how population growth is exerting pressure on the groundwater environment of Dili City. If abstraction continues at the same rate, in 2030 an estimated 0.41 million people (NSD, 2010) would pump 57,400 m^3/d or 20.95 MCM/y (approx.) of groundwater, much higher than the estimated recharge of 11.8 MCM/y. This would result in groundwater mining, with abstraction exceeding recharge.

12.3.2 Urbanization

Dili was established in 1520 by the Portuguese, who made Dili the capital of Portuguese Timor in 1769. It was proclaimed a city in January 1864. However, urbanization only gained momentum from the early 1980s, when the Indonesian Government started to develop urban infrastructures such as roads and water supply systems. Development was further continued by Timor-Leste in 1999 as the capital of the country. Exploiting the weaknesses of inadequate planning, urban areas are expanding randomly in an unplanned way. The government's failure to meet public expectations, together with the subsequent military crisis in 2006, led more rural youth to migrate to Dili City for better jobs. Within six years, the urban population rose from 0.164 million to 0.221 million in four subdistricts, which accounts for 35% of the population in Dili City (Census Atlas, 2013). This has led to an increase in population density for the urban areas during 2004–2010 (Table 12.2).

Table 12.1 Population changes within DGB

| District | Subdistrict | Area (km²) | Basin population | | Population density | | Change 2004–2010 | | |
| | | | 2004 | 2010 | 2004 | 2010 | Population | Population (%) | Density |
		DGB	DGB	DGB	DGB	DGB	DGB		DGB
Dili	Cristo Rei	4	30,736	49,442	7684	12,361	18,707	62%	4677
Dili	Dom Aleixo	16	58,117	94,639	3632	5915	36,522	61%	2283
Dili	Nain Feto	3	26,928	23,933	8976	7978	−2995	112%	−998
Dili	Vera Cruz	3	31,733	30,614	10,578	10,205	−1120	104%	−373
Total		**26**	**147,514**	**198,627**					

DGB, Dili groundwater basin.

Table 12.2 Status of DPSIR indicators in Dili, Timor-Leste

Indicators		2004	2010
Drivers	Population growth (within GW basin; area: 26 km²)	Population: 0.15 million people Population density: ~1187 persons/km²	0.20 million people ~1402 persons/km²
		Annual growth rate during 2004–2010: 4.2%; Census Atlas (2013)	
	Urbanization	Urbanization is taking place but no official data are available as it is unplanned and unregulated	
	Tourism	Tourist arrival: 14,000 (in 2006). Number of hotels: 3 (until 1991)	51,000 (in 2011) 14 hotels (1990–2014)
Pressures	Inadequate surface water resources	Increased development activities, urbanization, rising economic activities, and quality of life are water demanding and have reduced per capita surface water availability	
	Land cover change	There is evidence of increasing urban development and a decrease in open land and agriculture during 1990–2014, but quantitative data are not available	
	GW overexploitation	Abstraction: NA Recharge: 11.80 MCM/y (estimated as 19% of Laclo watershed recharge)	27,821 m³/d (or 10.15 MCM/y) 11.80 MCM/y
	Salt water intrusion	Lindsay Furness in 2011 reported high content of salts in the downstream of the Comoro River with an average depth 140 m. There is evidence that high abstraction without proper management will lead to salt water intrusion.	

State	Well statistics	—	Total: 17; production wells: 14; abandoned wells: at least 3
	GW abstraction	Abstraction: NA Recharge: 11.80 MCM/y Rainfall: 1761 mm/y = 440.25 MCM/y (within Comoro watershed)	27,821 m³/d (or 10.15 MCM/y) 11.80 MCM/y 1761 mm/y = 440.25 MCM/y (within Comoro watershed)
	GW level	In general, water level rises from November/December and declines from June/July; (i) in latitude direction: decreasing toward southern part, shallower toward the northern coastline; (ii) in longitude direction: no visible trend.	
	GW quality	GW is characterized by PH (6.2–8.9), TDS (8511 mg/L), iron (11.3 mg/L), turbidity (13.2 NTU), sulfate (65.1 mg/L), fluoride (0.2 mg/L), nitrate (12.5 mg NO_3/L), and arsenic (0 mg/L)	
	Recharge	11.80 MCM/y (estimated as 19% rainfall of the Laclo recharge area)	11.80 MCM/y
Impacts	Depletion in GW level	NA	6.0 m at GW borewell Comoro A; 1.8 m at borewell Comoro D; and 10.94 m at borewell Asgor
	Decline in production capacity of wells	Production capacity of wells during five years (2009–2013) declined by: 7884 m³ in Comoro A, 50 m³ in Comoro D, and 75,000 m³ in Cendana	
	Land subsidence	No monitoring of land subsidence so far in the area	
	Public health	60% of water in Dili's water supply system comes from GW. However, WHO (2010) reports higher concentration of nitrate in Dili's GW. It therefore needs monitoring on a regular basis. The report also recommended testing GW sources in close proximity to mechanical workshops for petroleum products.	

(Continued)

Table 12.2 Status of DPSIR indicators in Dili, Timor-Leste (cont.)

	Indicators	2004	2010
Responses	GW monitoring	Monitoring began in 2000; however, a military crisis in 2006 caused loss of all the data.	Continuous monitoring since 2012 under the Department of Water Resources Management, National Directorate for Water Control and Water Quality, Ministry of Public Work.
	Constitutional and legal provisioning	Article 61 of the country's constitution mentions environment conservation as a duty of everyone. Ministry of Public Work drafted a bill for Water Resources Law through NDWRM (2010–2014). The law is expected to minimize negative impacts on the groundwater environment.	
	Dili Urban Water Supply Sector Project	The proposed project aims to improve water supply management in Dili; improve relationship of DNSAS with the community; focus on five areas to achieve the desired level of service; and prepare a strategic plan for Dili's water supply system. These activities will ultimately reduce pressure on the groundwater environment.	

DNSAS, Dirasaun Nasional Sistema Agua no Saneamentu; DPSIR, driver–pressure–state–impact–response framework; GW, groundwater; MCM, million cubic meters; latitude, north–south; longitude, east–west; NA, not available; NDWRM: National Directorate of Water Resources Management; NTU, nephelometric turbidity unit; TDS, total dissolved solids.

Annual average rainfall data rainfall distribution of Comoro watershed (GIS data, 2008–2013).

Sources: Rainfall data (Pinto et al., 2015).

Considering per capita water demand as 150 lpcd for urban areas (Aurecon Australia, 2012) and 30–60 lpcd for rural areas (Water Aid, 2009), urban people use 90 lpcd more than their rural counterparts. An increase of 0.057 million in the urban population from 2004 to 2010 is equivalent to 8519 m^3/d water demand. The direct impact of urbanization in Dili City is equivalent to the abstraction of 5111 m^3/d of groundwater. Apart from the volume of abstraction, alluvial deposits in the Dili City area are very important recharge for Dili's aquifer system. Urbanization in those areas limits the supply of water to aquifers as a form of recharge and therefore affects the groundwater environment. In addition to quantity, a large volume of waste generation in urban areas, along with improper management, adversely impacts groundwater quality. Thus, urbanization is exerting pressure on Dili's groundwater environment in many ways.

12.3.3 Tourism

As the capital city of the country, Dili plays hosts to many tourists each year. The arrival of foreign tourists in Dili from 2006 to 2011 increased from 14,000 to 51,000 (Table 12.2). With the increase in tourism, there are more hotels and restaurants (Timor-Leste International Tourism, 2012). There are more than 14 hotels in the DGB, ranging from two to four stars, which rely mostly on groundwater as a source of water supply to meet their high demands. The majority of hotels are connected to the DNSAS supply, of which 60% comes from groundwater. Most hotels have drilled their own borewells to supplement their water in the absence of a DNSAS supply. However, accurate monitoring of groundwater abstraction for hotels is not available as the government still provides water free of charge to consumers.

12.4 PRESSURES

12.4.1 Inadequate Surface Water Resources

Insufficient surface water to meet the city's water demand is the major source of pressure on the groundwater environment. Expansion of social systems, urbanization, rising economic activities, and an improving quality of life are water intensive and therefore reduce per capita surface water availability. At the same time, uncontrolled disposal of the city's waste has deteriorated the quality of surface water sources. For that reason, DNSAS has introduced groundwater in Dili City's supply system since the 1980s and this has been continued by the Timorese Government after its separation from Indonesia in 1999. Currently, groundwater contributes around

60% to the total DNSAS supply in the dry season. Inadequate supply (i.e., irregular, poorly managed, and often polluted) from the DNSAS has led industry, the private sector, institutions, individuals, and communities to increase their own water supply by pumping huge quantities of groundwater. A lack of regulation on groundwater abstraction has exerted more pressure on the city's groundwater environment.

12.4.2 Land Cover Change

The land cover of the basin has changed toward nonagricultural in the last two decades. Even though no accurate data on agricultural land are available, the best possible estimate suggests a decrease to below 40% between 1990 and 2014. Several agricultural areas which existed in the 1990s in the eastern part of the basin (such as Caicoli, Bairo Pite, and Hudi-Laran) are now utilized for horticulture and aquaculture. Similarly, agricultural land in the western part of the basin (such as Manleoana, Fomento, Kampung Baru, Beto, and Don Bosco) has now mostly converted to urban land and the greater wetlands in the Caicoli area have disappeared. Wetland areas such as Kaikoli, Aimutin, and Biro Pite with a high potential to recharge groundwater, particularly in the rainy season, have been reclaimed and changed into residences, hotels, and government offices. Such a change in land cover has affected groundwater recharge. In addition, the conversion of rural land into urban development has led to increased pumping in many areas, and more importantly, created extensive pollution of both surface streams and the groundwater aquifer.

12.4.3 Groundwater Overexploitation

These days, groundwater is accessible with inexpensive technology. Therefore, various agencies in need of water (such as DNSAS, hotels, industry, government buildings, private institutions, and hospitals) are continuously pumping groundwater from Dili's aquifers. According to the data for 2010, groundwater abstraction is estimated at 27.821 m^3/d or 10.15 MCM/y, which is very close to the estimated recharge of 11.80 MCM/y (Table 12.2). If the abstraction continues at the same rate, it is expected to reach 22.3 MCM/y in 2030; almost double the recharge volume. In the absence of appropriate management interventions, this is likely to result in groundwater mining and its associated consequences.

12.4.4 Salt Water Intrusion

The DGB is geologically vulnerable to salt water intrusion, which is to a certain extent due to the aquifer being attached directly to sea water, posing

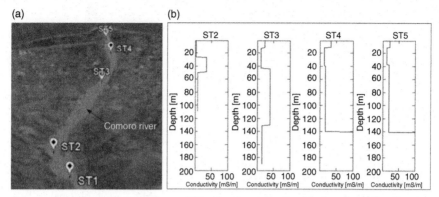

Figure 12.4 *Electrical Sounding within the Groundwater Basin, along the Comoro River.* (a) Top Elevation and (b) Long Sections of Electrical Sounding at ST2 to ST5. *Source: Furness (2012).*

groundwater overexploitation. Between 2010 and 2012 a project was carried out by the National Directorate for Control and Quality of Water (DNCQA) in cooperation with the Australian Government to monitor salt water intrusion into the Dili aquifer. The results as reported by Lindsay Furness in 2011 suggested the presence of high concentrations of salt water downstream of the Comoro River, with an average depth of 140 m (Figure 12.4 a,b). The evidence indicated that high abstraction without proper management leads to salt water intrusion toward inland areas in the southern part of the Dili aquifer. This exerts pressure by degrading the quality of groundwater, which then requires treatment before it can be used for various purposes.

12.5 STATE

12.5.1 Well Statistics

About three to five deep wells were initially drilled in the Dili aquifer, probably during the 1980s by the Indonesian Government, to supply drinking water to the community. After independence in 1999, more wells were drilled under the Democratic Republic of Timor-Leste (RDTL) in cooperation with international agencies such the Asian Development Bank (ADB) and Japan International Cooperation Agency (JICA). Those wells now belong to the DNSAS. From early 2000, private hotels and industries started to drill their own wells to meet increasing water demand, due to an inadequate and irregular water supply from DNSAS. Apart from hotels and industry, the DNSAS also introduced groundwater into its water supply

Table 12.3 Total boreholes in the basin

No.	Borehole	Borehole cashing (mm)	Q (l/s)	Borehole depth (m)	Operated (h)	Remark
1	Comoro A	200	20	80	16	Yield decline
2	Comoro B1	200	30	85	16	Normal
3	Comoro B2	150	22	70	9	Normal
4	Comoro C	150	17	70	12	Normal
5	Comoro D	200	28	70	24	Yield decline
6	Comoro E	200	13	80	12	Normal
7	Kuluhun A	200	26	80	22	Normal
8	Kuluhun B	200	26	80	24	Normal
9	Becora 2	200	26	70	13	Normal
10	Becora 1	150	8	70	24	Normal
11	Cendana	150	8	70	12	Normal
12	Bidau 1	200	17	70	24	Normal
13	Bidau 2	200	8	70	24	Normal
14	Bidau 3	150	2	70	8	Normal
15	Asgor	200	15	80	16	Yield decline
16	Mascarhinas	200	15	80	11	Normal
17	Marconi	200	15	70	16	Normal
18	Exs. Hely Port A	200	15	80	13	Normal
19	Exs. Hely Port B	200		80		Not operated yet
20	Becora Sede	150		80		Not operated yet
21	Becora Ailok	150		80		Not operated yet
22	Becora Merkado	150		80		Not operated yet

Source: DNSAS record.

system at almost the same time. The DNSAS gradually increased the number of borewells, and by 2010 there were 18 reported by Aurecon Australia (2012). By 2013, there were 22 borewells. Table 12.3 highlights their key characteristics. An analysis of trends in the drilling of borewells indicates that the majority of wells were drilled between 2008 and 2013 to abstract groundwater from the Dili aquifers. In 2012 and 2013, seven deep borewells were drilled by the DNSAS to increase volume to its water supply system. Even though borewells are increasing year-by-year, city dwellers are still struggling to access safe drinking water over a 24 h period.

12.5.2 Groundwater Abstraction

Groundwater was first abstracted from the Dili aquifer in the 1980s. However, significant abstraction occurred only after 1999. Abstraction by the DNSAS alone (excluding abstraction by hotels, industry, and the private sector) is 10.15 MCM/y. Abstraction from many shallow wells by the private sector is unquantified due to the lack of a well registration system. Total abstraction from all sectors is predicted to be more than 12.0 MCM/y.

12.5.3 Groundwater Level

Reliable information on groundwater levels in shallow and deep aquifers is limited and varied, due to poorly spaced monitoring wells and a discontinuity in monitoring. The data in 2012 show that the static water level decreases toward the foothills in the southern part and is shallower toward the northern coastline.

In an east–west direction, there is no visible trend and levels vary randomly. For example, in the Comoro wellfield, the groundwater level at Comoro-B fluctuates between 6.68 m and 7.54 m and gradually increases toward the south at AS (Asgor), where it fluctuates between 18.25 m and 31.30 m (Figure 12.5). The general trend of groundwater flow is toward the northern coastline. The trend is similar for all seasons (premonsoon, monsoon, and postmonsoon). In general, water levels rise for one to two months after November/December and decline after June/July. All borewells are categorized as deep with a depth from 70 m to 80 m. Hydrographs from 2009 at selected wells in Comoro A and B and AS show fluctuations in monthly groundwater levels from January to December (Figure 12.6); reflecting the general trend in groundwater recharge. In general, water levels start rising after November/December and decline after June/July.

12.5.4 Groundwater Quality

Groundwater quality is not being monitored routinely in the Dili aquifer. The latest study of groundwater quality in this aquifer was made from November 2009 to April 2010 as part of a national program to gather the necessary information to develop water quality–monitoring guidelines, including water quality standards. Groundwater quality in the Dili aquifer, based on an analysis of 14 borewells, can be characterized as turbid (in general) with very high concentrations of total dissolved solids, iron, and nitrate. Details are provided in Table 12.4. Some samples showed the presence of fluoride but these did not exceed the recommended value. Arsenic was not found in groundwater or other sources (Water Quality Report, 2010).

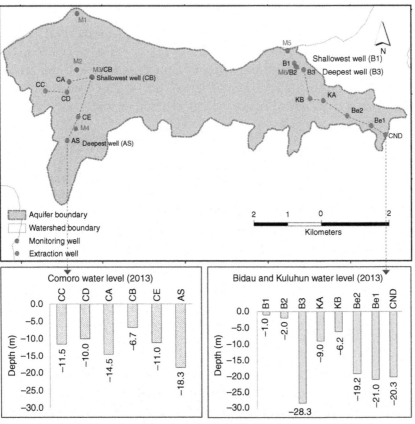

Figure 12.5 *Static Water Level Trends at Comoro Wellfield – Comoro C (CC), Comoro D (CD), Comoro A (CA), Comoro B (CB), Comoro E (CE), and Asgor (AS) – and at Bidau Wellfield – Bidau 1 (B1), Bidau 2 (B2), Bidau 3 (B3), Kuluhun A (KA), Kuluhun B (KB), Becora 1 (Be1), Becora 2 (Be2), and Cendana (CND) – for premonsoon (August) of 2009 to 2013. Monitoring wells: M1, M2, M3, M4, M5, and M6. Source: DNSAS records.*

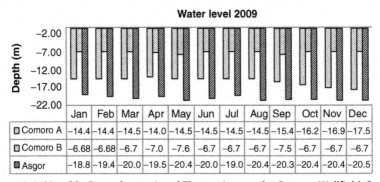

Figure 12.6 *Monthly Groundwater-Level Fluctuations at the Comoro Wellfield. Source: DNSAS record.*

Table 12.4 Groundwater chemistry and recommended guidelines in selected districts, including 14 DNSAS borewells in Dili, Timor-Leste

Parameter	Lowest value	Highest value	Average	Timor-Leste – recommended value
pH	6.2	8.9	–	6.5–8.5
TDS (mg/L)	40.3	60,300	8510.7	600 (WHO)
Temperature (C)	19.6	30.9	27.2	
Turbidity (NTU)	0.2	87.9	13.2	5
Iron (mg/L)	0	288	11.3	0.3
Sulfate (mg/L)	0	650	65.1	250
Fluoride (mg/L)	0	0.98	0.2	1.5
Nitrate (mg NO_3/L)	0	50	12.5	10
Arsenic (mg/L)	0	0	0	0.01

DNSAS, National Directorate of Water and Sanitation Services; NTU, nephelometric turbidity unit; TDS, total dissolved solids.

12.5.5 Recharge

Groundwater recharge is the amount of water that infiltrates through the ground surface and is stored underground for a certain amount of time. Recharge is a complicated phenomenon, particularly in respect of recharge into deep aquifers. As a tropical hot climate, precipitation is the only means of groundwater recharge in the Dili aquifer. The DGB receives a large amount of precipitation from the Comoro watershed, about 1,761 mm/y. Recharge is limited or controlled by small sand and gravel deposits and discontinuities bounded by irregular impermeable clay. Although there is no estimation, the Railaco Valley in the upper part is probably contributing a significant proportion of recharge to the Dili aquifer due to its flat topography and permeable soil type (sandy and loose sediment). The preliminary estimate of recharge into the Dili aquifer is made up of 19% recharge into the Laclo watershed (i.e., 62.4 MCM/y), which accounts for 11.80 MCM/y. This recharge is very close to the existing groundwater abstraction of 10.15 MCM/y.

12.6 IMPACTS

12.6.1 Depletion in Groundwater Levels

Historical data on groundwater levels are not available in the area. The monitoring data for recent years show that the water level has gone down by 6 m (17.5–23.5 m), 1.8 m (7.63–7.40 m), and 10.94 m (20.36–31.30 m) in the Comoro A, Comoro D, and Comoro AS wells, respectively. Between

Table 12.5 Aquifer depletion at selected locations during the dry season

Location	WID	Previous water level (mbgl)		Current water level (mbgl)		Decline (m)
		Base year	SWL	Current year	SWL	
Comoro WF	Comoro A	2008	17.50	2013	23.50	6.00
	Comoro B	2008	7.63	2013	7.40	0.23
	Comoro C			2013	14.49	
	Comoro D	2008	10.00	2013	11.80	1.8
	Asgor	2008	20.36	2013	31.30	10.94
Bidau WF	Bidau 1	2011	1.50	2012	6.60	5.1
	Bidau 2			2013	3.33	
	Bidau 3			2013	29.60	
	Becora 1			2013	13.25	
	Becora 2 (Cipl)			2013	31.80	
	Cendana			2013	25	
Kuluhun WF	Kuluhun A			2013	10.50	
	Kuluhun B			2013	7.42	

mbgl, meters below ground level; SWL, static water level; WF: wellfield; WID, well identification number.

2008 and 2013 these borewells were heavily pumped in the western part of the basin, which caused a decline in groundwater (Table 12.5, Figure 12.7). There is relatively more depletion in the southern part of the Comoro wellfield due to the high surface elevation and longer recovery time needed to return to the normal water level compared with a lower elevation near to the northern coastline, which is affected by the seawater tide after abstraction. At Comoro A, it is related more to the aquifer's capacity to respond to pumping rates and recharge restoration from the Comoro River. Figure 12.7 shows that water levels in the borewells vary considerably. The decreasing and increasing water levels also directly respond to the seasonal rainfall in November/December and the dry season of May/June.

12.6.2 Decline in the Production Capacity of Wells

Some wells have not shown water level depletion in the period from 2008 to 2013, probably because the pumps did not run continuously for 24 h and/or as a result of poor maintenance leading to nonoperation of the borewells over a long period of time. In such cases, a decline in the production capacity of wells has indicated that groundwater is certainly depleting. At least three borewells, namely, Comoro A, Comoro B, and Cendana, have shown declines in well yield from 2008 to 2013 (Table 12.6).

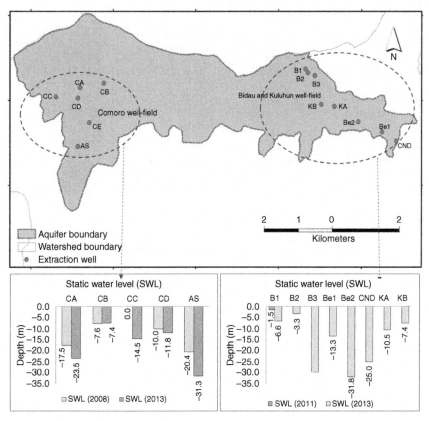

Figure 12.7 *Decline in Static Water Levels over Time at Different Locations within the Aquifer.*

12.6.3 Land Subsidence

Land subsidence is a serious global issue and is mostly affected by groundwater abstraction. However, in Timor-Leste, so far there is no research focusing on land subsidence. Land subsidence may occur as a result of the overexploitation of groundwater, geothermal fluids, oil, and gas; hydrocompaction of sediments, and oxidation and shrinkage of organic deposits (Nurhamidah et al., 2011). Even though there is no research, land subsidence is predicted for the DGB due to the tremendous change in land cover and high abstraction of groundwater.

12.6.4 Public Health

Water contamination is a serious public health issue in the basin. In 2010, the World Health Organization (WHO) and Ministry of Health Timor-Leste

Table 12.6 Well yield (m³/y) from 2008 to 2013 in wellfields in Dili City

SN	Wellfield	WID	Areas within GW basin	Previous discharge		Q (m³/y) in 2013	Decline in Q (m³/y)
				Year	Q (m³/y)		
1	Comoro A	CA	Comoro	2008	265,414	257,530	7884
2	Comoro B	CB	wellfield	2008	107,945	107,895	50
3	Comoro C	CC		2008	467,015	467,015	
4	Comoro D	CD		2008	48,501	48,501	0
5	Comoro E	CE		2008	65,727	65,727	
6	Asgor	AS		2008	0	178,821	
7	Bidau 1	B1	Bidau and	2008	855,008	855,008	
8	Bidau 2	B2	Kuluhun	2008	0	32,903	
9	Kuluhun A	KA	wellfields	2008	369,308	369,308	
10	Kuluhun B	KB		2008	546,774	546,774	
11	Becora 1	Be1		2008	0	52,998	
12	Becora 2	Be2		2008	189,057	189,057	
13	Cendana	CND		2008	136,539	61,539	75,000

SN, serial number; WID, well identification number; GW, groundwater; Q, discharge.
Source: DNSAS record.

(MoHTL) reported that about 70% of the sources tested across the country, including 14 wells in the DGB are found to be microbiologically contaminated. This is of great concern as the contaminated water may result in many diseases such as diarrhea, cholera, and typhoid (Figure 12.8). Results from the water quality–monitoring system clearly show microbial contamination of water. Much of the stored water in homes is found to be contaminated, and therefore a campaign for household water treatment and safe storage is required to prevent water-borne diseases across the country. The results of bacteriological tests on groundwater in the DGB are presented in Table 12.7.

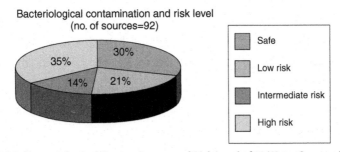

Figure 12.8 *Bacteriological Contaminants and Risk Level of 92 Water Sources in the Entire Country Including the DGB, Timor-Leste. Source: Water Quality Report (2010).*

Table 12.7 Bacteriological test results of the groundwater in the DGB

No	Date	Sampling point	Type of source	Total coliform (CFU/100 mL) 0 recommended
1	12/12/2009	Culu Hun A	Well	0
2	12/12/2009	Culu Hun B	Well	0
3	12/12/2009	Comoro B	Well	0
4	12/12/2009	Comoro D	Well	0
5	12/12/2009	Comoro E	Well	0
6	18/02/2010	Bidau Masau	Well	TNC
7	18/02/2010	Lab. Nas	Well	14
8	18/02/2010	Cacaulidun	Well	0
9	18/02/2010	Kampung Alor	Well	TNC
10	18/02/2010	Kampung baru	Well	TNC
11	18/02/2010	Matadoru	Well	TNC
12	18/02/2010	Lahene	Well	0
13	18/02/2010	Laboratory DNSAS	Well	TNC
14	18/02/2010	MDS	Well	TNC

CFU, colony-forming unit; DGB, Dili groundwater basin; DNSAS, National Directorate of Water and Sanitation Services; TNC, too numerous to count.
Source: Water Quality Report (2010).

12.7 RESPONSES

Since 1999, the Government of Timor-Leste, in cooperation with many related stakeholders and international organizations, has made numerous efforts to improve the water environment through development programs, organizational adjustments, and research activities. Some of the notable initial activities on the water resources sector are: rehabilitation and new river intake for the Dili Water Supply Sector Project (JICA, 2010); Dili Urban Water Supply Sector Project (Asian Development Bank, 2007); construction and rehabilitation of damaged water supply systems; building surface water intake in the Comoro River upstream to collect river flow to minimize groundwater pumping, and building a water pond in Tasi Tolu and a solid waste plant in Tibar to minimize groundwater contamination. All these responses are helping to minimize pressure on groundwater resources directly or indirectly.

12.7.1 Groundwater Monitoring

Groundwater monitoring began in 2000; however, a military crisis in 2006 resulted in the loss or damage of all the collected data. Subsequently, monitoring recommenced in 2010. Between 2010 and 2012 the Government of

Timor-Leste, with support from Australian groundwater experts through the BESIK project, constructed six groundwater-monitoring wells in Dili (Figure 12.6) to observe groundwater levels and quality. Due to a lack of maintenance and management, only two of six monitoring wells (M3 and M6, see Figure 12.6 for their location) are still in use. Data are recorded using an automatic system at a frequency of one hour and are downloaded at intervals of three months. Although groundwater-level monitoring has been initiated, it is inadequate. Spatial coverage of monitoring, inclusion of groundwater quality in the monitoring system, discontinuance of ground-water production from monitoring wells, and monitoring of the aquifer at various depths are some of the concerns requiring attention from the relevant authorities.

12.7.2 Constitutional Provisions

As a young country, Timor-Leste has shown concern and sensitivity toward environmental issues since the restoration of its independence on May 20, 2002. Thus, recognizing the quality of its environment as an integral and essential part of the quality of life of all Timorese, the Constitution of the Democratic Republic of Timor-Leste provides in Article 61, that people are not only entitled to ecologically balanced environmental health to maintain human life, but that everyone also has a duty to conserve and pro-tect the environment in the interests of future generations. From 2010 to 2014, the Ministry of Public Works developed draft water resources legisla-tion through the National Directorate for Water Resources Management (NDWRM). However, the bill is still under process and will be imple-mented after obtaining approval from the National Parliament. Chapter V of the water resources bill mentions user entitlement, and Articles 21 to 34 refer to regulating the use of water from surface or groundwater resources. The contents of these articles contain water abstraction entitlements, gen-eral obligations to obtain such entitlement, licences for private abstraction of public water, concessions for private abstraction of public water, and borewell construction, etc. Strict implementation of these legal provisions will help to minimize negative impacts on the groundwater environment and promote sustainable use of the resource.

12.7.3 Dili Urban Water Supply Sector Project (DUWSSP)

To overcome the issues of inadequate water (both in terms of quantity and quality) and reduce pressure on the groundwater environment by means of lowering groundwater abstraction, the DUWSSP is supported by the ADB and JICA. The ADB is responsible for rehabilitation of the existing water supply system in the basin and JICA is responsible for improving the surface water facility from upstream (outside the basin) as a supplement to groundwater supply within the basin. The proposed project aims to improve water supply management in Dili, improve the community relationship with the DNSAS, focus on five areas to achieve the desired level of service, and prepare a strategic plan for the Dili water supply system.

12.8 SUMMARY

The major drivers exerting pressure on Dili City's groundwater environment are population growth, urbanization, and tourism. Due to pressures such as inadequate surface water, land cover changes, and groundwater overexploitation, groundwater has depleted as indicated by groundwater levels and well yield. For example, in the Comoro A well, in the Comoro wellfield, water levels from 2008 to 2013 have decreased from 17.50 mbgl to 23.50 mbgl and well yield has decreased by 7884 m^3. Well yields have also decreased in other wells such as Comoro D (by 50 m^3) and Cendana (by 75,000 m^3) during the same period. Even though stress on the aquifer is increasing, neither functional regulatory interventions nor institutional responsibility for aquifer monitoring and management is put in place. Even though this chapter has developed a knowledge base of the Dili groundwater environment, more studies are needed to understand the aquifer storage capacity, flow paths, interaction between surface and groundwater, and identification of recharge areas and recharge amounts in the DGB, and these are some of the key issues to be addressed in future.

REFERENCES

Asian Development Bank (ADB), 2007. Proposed ADB Fund Grant RDTL: Dili Urban Water Supply Sector Project. Asian Development Bank, Dili.

Aurecon Ausralia, 2012. ADB grant Dili Urban Water Supply Sector Project design and construction supervision consultancy services. Ministry of infrastructure, Republic Democracy of Timor-Leste.

Costin, G., Powell, B., 2006. Situation analysis report: Timor Leste. In: Brisbane (Ed.), Australian Water Research Facility. International Water Center, Brisbane, Australia.

Census Atlas, 2013. Timor-Leste Population and Housing Census 2010, vol. 15.

Furness L., 2012. Construction of four groundwater monitoring wells in Dili, Timor Leste. National Directorate for Water Resources, Water and Climate Adaptation.

Fenco Consultants, 1981. Sumbawa Water Resources Development Planning Study. Hydrology Report, Technical Report No. 2, Directorate General of Water Resources Development, Ministry of Public Works, Jakarta.

JICA, 2001. Japan International Cooperation Agency: the study on urgent improvement project for water supply system in Timor Leste.

JICA, 2010. Project for Urgent Improvement of Water Supply System in Bemos-Dili.

MoH, 2010. Water quality study. Democratic Republic of Timor Leste, Ministry of Health Environmental Health Devision.

NSD, 2010. Preliminary results from the 2010 census. National Statistics Directorate, Dili, Timor-Leste.

Nurhamidah, Nick van de G., Hoes, O., 2011. Subsidence and deforestation: implications for flooding in delta's southeast and East Asia. ISBN 978-983-42366-4-9.

Prasetyadi, C., Harris, R.A., 1996. Hinterland structure of the active arc-continent collision, Indonesia: constraint from the Aileu complex of Timor Leste. Proceedings, IAGA Annual Convention 25, 121–129.

Pinto, D., Shrestha, S., Babel, M.S., Ninsawat, S., 2015. Delineation of groundwater potential zones in the Comoro watershed, Timor Leste using GIS, remote sensing and Analytic Hierarchy Process (AHP) technique. Appl. Water Sci.

Thompson, 2011. Soil and geology in Timor Leste.

Timor-Leste International Tourism, 2012. Available from: http://www.indexmundi.com/facts/timor-leste/international-tourism

Wallace, L., Sundaram, B., Brodie, R.S., Marshall S., Dawson, S., Jaycock, J., Stewart, G., Furness, L., 2012. Vulnerability assessment of climate change impact on groundwater resources in Timor Leste. Australia Government Departement of Climate Change and Energy Efficiency, 55 pp.

Water Aid, 2009. Water Aid in Timor-Leste. Country Strategy 2010–2015.

Water Quality Report, 2010. Democratic Republic of Timor-Leste, Ministry of Health Environmental Division.

CHAPTER 13

Groundwater Environment in Ho Chi Minh City, Vietnam

Bui Tran Vuong, Phan Nam Long, and Le Hoai Nam
Division of Water Resources Planning and Investigation for the South of Vietnam (DWRPIS), Ministry of Natural Resources and Environment of Vietnam

13.1 INTRODUCTION

Ho Chi Minh City (HCMC) is situated on the banks of the Saigon River in the southern part of Vietnam (Figure 13.1). The city with an area of 2095 km² is home to more than 7.95 million residents (2014), accounting for approximately 0.6% of the total area and 8.79% of the total population of the whole country. The metropolitan area is the most important economic center in Vietnam, contributing approximately 20% to the country's GDP. Characterized by a tropical monsoon climate, the city has two seasons in a year: the rainy season runs from May to October, and the dry season from November to April of the following year. It receives an average annual rainfall of 1946 mm, of which about 1130 mm (calculated at Tan Son Hoa Station, during the period 1997–2008) is lost in the form of evaporation.

According to Decision No 729 on Approval of the Water Supply Planning of HCMC to 2025 (The Prime Minister, 2012), water demands in 2015 will be 2.750 million cubic meters (MCM)/d, of which the domestic sector will account for approximately 52%, industry 6%, other sectors 12%, and losses 30% of the total demand. In 2025, the demand will reach 3.57 MCM/d and domestic, industry, other sectors, and losses will account for 53, 7, 16, and 24%, respectively. Management of water is the responsibility of the People's Committee of HCMC (PC). The mandate of PC is listed in Appendix 13.1.

Water supply in the city is derived from both surface and groundwater sources. Until now, groundwater has played an important role in meeting the city's water demand. In 2012, the total amount of water supply in HCMC was about 1.5 MCM/d, of which 44.67% was produced from groundwater. It is planned to reduce the amount of groundwater abstraction for the water supply to 0.44 MCM/d (or 15.5% of the total water

Groundwater Environment in Asian Cities
http://dx.doi.org/10.1016/B978-0-12-803166-7.00013-1

Copyright © 2016 Elsevier Inc.
All rights reserved.

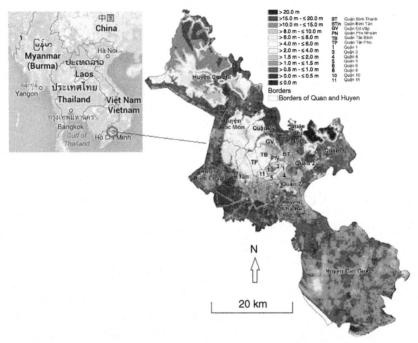

Figure 13.1 *Location and Ground Surface Elevation of HCMC in Vietnam* (Google Map and Bui Tran Vuong, 2010).

demand) by 2015 and to 0.10 MCM/d (or 2.7% of the total water demand) by 2020 (The Prime Minister, 2012).

Groundwater has been used in HCMC since 1920. The rapid increase of groundwater use started in 1990 when the economic policies of Vietnam were instituted. High industrialization and urbanization resulted in a rapid increase in water demand. Surface water production and development have so far been unable to meet the increasing demand. Besides, water is free of charge and uncontrolled abstraction has significantly increased rates and volumes of groundwater pumping in HCMC. Groundwater is under threat due to salt-water intrusion, reduction in groundwater levels, and contamination. The excessive groundwater abstraction has also led to land subsidence at some areas in HCMC. Subsidence at the rate of a few centimeters per year is measured at heavily pumped groundwater stations (Trung et al., 2008).

Several studies were carried out in order to better understand the groundwater condition of HCMC (e.g., Tin, 1988; Dung, 1991; Luan, 1993; Hung, 2001; Nga, 2006; Vuong, 2010). However, these focused particularly on the characteristics of aquifer systems and possibilities for groundwater development rather than on proper management, and they have not shed any light on the analysis of the groundwater environment by considering both natural and social systems together. This case study aims to synthesize the situation in HCMC to better understand the groundwater environment and the origin of its stresses, current state, expected impact, and responses needed to restore it back to health.

13.2 ABOUT THE CITY

HCMC began as a small fishing village commonly known as Prey Nokor. In 1623, King Chey Chettha II of Cambodia (1618–28) allowed Vietnamese refugees fleeing the Trịnh-Nguyễn civil war in Vietnam to settle in the area of Prey Nokor and to set up a custom house there. In 1698, Nguyễn Hữu Cảnh established Vietnamese administrative structures in the area. Conquered by France and Spain in 1859, the city was influenced by the French during their colonial occupation of Vietnam. In 1929, Saigon had a population of 0.124 million. In 1931, a new region called Saigon-Cholon was formed. In 1956, following South Vietnam's independence from France in 1955, the region of Saigon-Cholon became a single city called Saigon. The Viet Minh led by Ho Chi Minh (HCM) proclaimed independence for Vietnam in 1945. In 1954, the Geneva Agreement partitioned Vietnam along the 17th parallel (Bến Hải River), with the Communist Việt Minh, under HCM, gaining complete control of the northern half of the country, while the Saigon government continued to govern the State of Vietnam in the southern half of the country. South Vietnam fought against the Communist North Vietnamese during the Vietnam War. On April 30, 1975 Saigon fell and the war ended with a communist victory. In 1976, upon the establishment of the unified Communist Socialist Republic of Vietnam, the city of Saigon (including Cholon), the province of Gia Định, and two suburban districts of two other nearby provinces were combined to create HCMC in honor of the late communist leader HCM.

HCMC, with coordinates $10°10'–10°38'N$ and $106°22'–106°54'E$, is located in the southeastern region of Vietnam; 1760 km south of Hanoi (Figure 13.1). The average elevation of the entire city ranges from 0.5 to 1.0 m above mean sea level (masl). There are six distinct aquifers in HCMC (Figure 13.2), namely Upper Pleistocene (qp_3); Upper–Middle Pleistocene (qp_{2-3}); Lower Pleistocene (qp_1); Middle Pliocene (n_2^2); Lower Pliocene and Upper Miocene (n_1^3) (Figure 13.2). The aquifer materials consist of alluvial

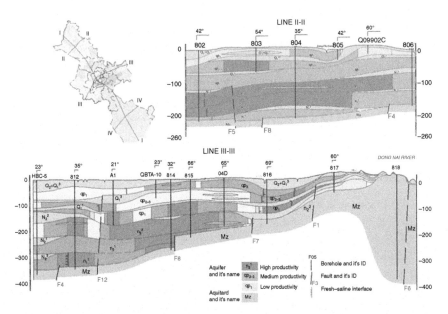

Figure 13.2 *Aquifer System in HCMC, Vietnam. Source: Vuong (2010).*

and marine deposits (gravel, sand, silt, clay, and peat). The areal extent and main parameters of the aquifer system are presented in Table 13.1. The aquifers contain both fresh and saline water.

The groundwater reserves in the city's aquifers are tabulated in Table 13.2. Total groundwater reserve (Q_{kt}, m³/d) was calculated as a sum of natural dynamic reserve and static groundwater reserve as detailed in Appendix 13.2. The natural dynamic groundwater reserve contains all components of water flow in the aquifers (recharge from rainfall, and surface water bodies, as well as horizontal and vertical infiltration between aquifers). The static groundwater reserve, on the other hand, is composed of two components, namely gravity reserves and elastic reserves. Gravity reserve is the amount of water storage in an intergranular or fractured aquifer. Elastic reserve is the amount of water that is discharged by reducing one unit head of groundwater.

13.3 DRIVERS

The major drivers exerting pressure on HCMC's groundwater environment are population growth, urbanization, and the increase in tourism. The change in drivers in two time periods are tabulated in Table 13.3 and discussed in subsequent sections.

Table 13.1 The main parameters of the aquifer system in Ho Chi Minh City

Parameters	Aquifer					
	qp_3	qp_{2-3}	qp_1	n_2^2	n_2^1	n_1^3
Distribution area (km²)	1983	2020	2042	2012	1634	682
Saline GW area (km²)	1199	1190	1157	912	980.6	454.2
Fresh GW area (km²)	784	830	885	110.0	653.2	228
Average thickness (m)	22.6	27.2	27.1	37.6	33.8	25.0
Head above the top of aquifer (m)	8.83	35.59	72.44	110.60	168.80	276.99
Specific yield (μ)	0.255	0.261	0.258	0.258	0.257	0.249
Specific storage ($\mu^\star$)	0.00901	0.00227	0.00133	0.00066	0.00371	0.00371

GW, groundwater.
Source: Vuong (2010).

13.3.1 Population Growth

HCMC has been the center of administration, economics, education, and politics for many years. As a result, the population is ever increasing mainly due to migration. Between 2006 and 2010, HCMC's population increased from 6.5 million to 7.4 million within the groundwater basin (Table 13.3). There was a corresponding increase in population density within the groundwater basin from 3095 persons/km² to 3522 persons/km² (Table 13.3). Considering the case of 2006, with an estimated groundwater abstraction of 525,000 m³/d (Table 13.3) and a population of 6.5 million, per capita groundwater abstraction becomes 81 liters/capita/day (lpcd). Thus, an increase of 0.9 million people from 2006 to 2010 is equivalent to 72,400 m³/d of groundwater abstraction. This clearly shows how population expansion imposes pressure on the groundwater environment.

Table 13.2 Groundwater (GW) reserves of Ho Chi Minh City (m^3/d)

STT	Components of GW reserves	Aquifer						Sum
		qp_3	qp_{2-3}	qp_1	n_2^2	n_2^1	n_1^3	
1	Dynamic reserves (dry season)	75,986	109,041	120,587	165,192	41,533	13,911	526,250
2	Dynamic reserves (rainy season)	77,100	112,100	124,092	165,896	41,801	15,551	536,540
3	Static reserves	78,812	157,086	212,609	372,146	239,569	73,106	1,133,331
3.1	Gravity reserves	75,821	152,139	204,556	364,314	198,65	49,676	1,045,159
3.2	Elastic reserves	2,991	4,947	8,053	7,832	40,919	23,43	88,172
4	GW reserves (in dry season)	154,798	266,127	333,196	537,338	281,102	87,017	**1,659,581**
5	GW reserves (in rainy season)	155,912	269,186	336,701	538,042	281,370	88,657	**1,669,871**

Source: Vuong (2010).

Table 13.3 Status of DPSIR indicators in Ho Chi Minh City, Vietnam

Component	Indicator	2006	2010	Data sources
Drivers	Population growth Population (million people)	6.5	7.4	*HCMC Statistics Book* (2010)
	Population density (persons/km²)	3095	3522	*HCMC Statistics Book* (2010)
	Annual growth rate (2006–2010)		3.23%	*HCMC Statistics Book* (2010)
	Urbanization Urban area (km²)	371 (during 1990–2002)	424 (during 2002–2010)	Son et al. (2012)
	Urban population (million people)	4.56 (in 2002)	6.14 (in 2010)	*HCMC Statistics Book* (2010)
	Tourism Tourist arrivals (million people)	2.8 (in 2005)	9.2 (in 2011)	
Pressures	Inadequate water delivery pipeline system	In several districts (Go Vap, Binh Tan, Thu Duc, Binh Chanh, and Hoc Mon), new industrial and residence zones, there are no surface water delivery pipeline systems to take water from surface water supply station.		
	Land cover change Urban area	371 km² (in 1990–2002)	424 km² (in 2002–2010)	Son et al. (2012)
	Nonagricultural area	252 km² (in 2002)	285 km² (in 2010)	Son et al. (2012)
	Overexploitation of GW resources (GW abstraction as % of recharge)	99.81	127.38	Calculated by Vuong (2010) based on data for various sources

(Continued)

Table 13.3 Status of DPSIR indicators in Ho Chi Minh City, Vietnam *(cont.)*

Component	Indicator	2006	2010	Data sources
State	Well statistics Total wells	95,828 (in 2006)	257,254 (in 2012)	Nga (2006), Tuyen (2014)
	GW abstraction Abstraction (MCM/d) Recharge (MCM/d)	0.525 0.526	0.670 (in 2012) 0.526	Nga (2006), Tuyen (2014) Nga (2006), Tuyen (2014) Chan (2009)
	GW level	There is an increasing trend in the average depth of groundwater level below the surface in all aquifers		Vuong (2010)
	GW quality	Groundwater quality in all aquifers is very good, only needs simple treatment before using. Groundwater quality is characterized by high pH value, ammonia, iron, and hardness.		Vuong (2010)
	Recharge (MCM/d)	0.526	0.526	Chan (2009)
Impacts	Decline in GW level (m)	Decline in GW level in 10 years (1999–2009) for each aquifer is as follows:		
		In aquifer qp_3 In aquifer qp_{2-3} In aquifer qp_1 In aquifer n_2^2 In aquifer n_2^1 In aquifer n_1^3	0.07–2.74 0.81–0.74 0.95–19.05 2.42–12.99 3.43 10.5	Vuong (2010) Vuong (2010) Vuong (2010) Vuong (2010) Vuong (2010) Vuong (2010)
	Land subsidence	Several areas with a total area of 115 km^2 have land subsidence with magnitude ranging from 5 mm/y to 15 mm/y		Trung et al. (2008)

			Nga (2006), Tuyen (2014)
	Public health	GW meets about 33% of the HCMC's water demand. Only one tenth of total wastewater is treated each day. Groundwater quality is under threat because of solid waste disposal and effluent discharge into rivers and open ground. Bacterial indicators, which cause water-borne diseases, have already been detected in groundwater samples.	
Responses	Initiation of GW monitoring	Groundwater levels in HCMC's aquifer system have been monitored since the 1990s. At present, there are 45 monitoring wells belonging to the National Monitoring Network and 28 monitoring wells belonging to Ho Chi Minh's People's Committee monitoring network.	
	Environmental standards and guidelines	Environmental standards and guidelines of Vietnam on GW resources is fairly complete. It helps HCMC's authorities to improve the GW environment. Decree No. 201/2013/NĐ-CP; Circular No. 13/2014/ TT- BTNMT, No. 27/2014/TT-BTNMT, Circulars No. 19/2014/TT-BTNMT, Circulars No. 15, 16 and 17/ TT-BTNMT; Decision No 14, 15, and 16/2008/QĐ-BTNMT; Decision No 53/ 200-QĐ/BCN… are related to groundwater.	

DPSIR, driver–pressure–state–impact–response framework; GW, groundwater; HCMC, Ho Chi Minh City; MCM, million cubic meters.

13.3.2 Urbanization

The urban population has increased from 4.56 million (1990–2002) to 6.10 million (2002–2010). This has led to an increase in urban areas from 371 km² in 2002 to 424 km² in 2010 (Figure 13.3).

Considering water demand as 120 lpcd for urban areas and 80 lpcd for rural areas, urban people use 40 lpcd more than rural people. Therefore, a 1.54 million increase in the urban population from 1990 to 2010 is equivalent to 63,160 m³/d of water demand. Considering 30% of that comes from the groundwater reserve, the direct impact of urbanization would be equivalent to 18,950 m³/d abstraction of groundwater.

Apart from the increase in groundwater abstraction, there is a higher volume of solid waste, hazards from their disposal, and leachage from those sites contributes to the degradation of groundwater quality. A reduction in the permeable area is a result of the increase in built-up areas, and apparently limits the supply.

13.3.3 Tourism

Tourism has increased from 2.8 million to 9.2 million from 2005 to 2011 (Table 13.3). Assuming each tourist stays 3 days in HCMC and water

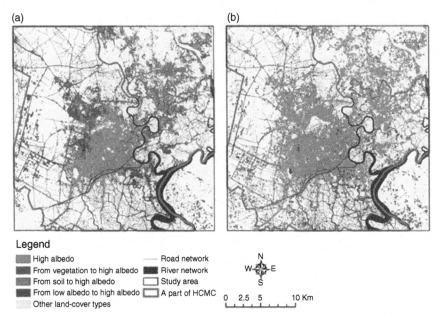

(a) (b)

Legend

High albedo	Road network
From vegetation to high albedo	River network
From soil to high albedo	Study area
From low albedo to high albedo	A part of HCMC
Other land-cover types	

N
W–E
S

0 2.5 5 10 Km

Figure 13.3 *Increase in Urban Area.* (a) 1990–2002 and (b) 2002–2010. *Source: Son et al. (2012).*

demand is 120 lpcd, a 6.4 million increase in tourists over 6 years (2005–2011) is equivalent to 2.3 MCM or 1050 m^3/d of water consumption. This is a direct pressure from the expansion of the tourism industry on groundwater resources.

13.4 PRESSURES

Pressure on HCMC's groundwater environment comes from an inadequate water delivery pipeline system, overexploitation of aquifers, and a decrease in rechargeable areas (i.e., limiting supply to the aquifers).

13.4.1 Inadequate Water Delivery Pipeline System

In several districts (Go Vap, Binh Tan, Thu Duc, Binh Chanh, and Hoc Mon) with new industrial and residence zones, there is no surface water delivery pipeline system to take water from the surface water supply station. This is a major source of pressure on the groundwater environment. In these districts and areas, groundwater is the only source for domestic and industrial use, and until there is a new pipeline system, people have no choice but to drill into a well and abstract groundwater to meet their own water demand. Uncontrolled drilling makes the situation of groundwater pollution more serious.

13.4.2 Land Cover Change (Decrease in Recharge Areas)

Land cover of HCMC has changed toward nonagricultural in the last two decades. The nonagricultural area has increased from 251 km^2 in 2002 to 285 km^2 in 2010 (Table 13.3). The urban area has also increased from 371 km^2 to 424 km^2 (Son et al., 2012). These data indicates that it is likely that recharge areas have decreased along with the change in land cover, adversely affecting supplies to the groundwater reserve. In addition, conversion of rural land into urban land has led to increased pumping in many areas and, more importantly, extensive pollution of both surface streams and shallow aquifers due to the direct disposal of municipal solid waste and wastewater in rivers and flood plains.

13.4.3 Overexploitation of Groundwater Resources

After groundwater became accessible with the advent of affordable technology, different agencies in need of water (such as water supply companies, hotels, industry, government and private institutions, hospitals, housing companies, and individuals) continued to pump groundwater.

It is very difficult to gain firm control over groundwater abstraction because there are too many illegal drilling activities, and those responsible for water resources management in the city do not have enough human resources to inspect these illegal activities. As a result, in 2006, the amount of groundwater abstraction was 0.525 MCM/d or 99.8% of groundwater recharge. It increased in 2012 to 0.670 MCM/d or 127.38% of groundwater recharge (Table 13.3). These figures suggest that the city's aquifer system is already overexploited.

13.5 STATE

The state of the groundwater environment in HCMC is described using the following indicators: number of wells, volume of groundwater abstraction, groundwater levels, groundwater quality, and annual recharge into the aquifers.

13.5.1 Well Statistics and Groundwater Abstraction

An investigation by the Department of Natural Resources and Environment (DNRE) of HCMC in 2006 indicated that there were 95,828 abstraction wells in HCMC (Nga, 2006), of which, 78,752 are in the Pleistocene aquifer (qp_3 and qp_{2-3}); 17,010 in the Middle Pliocene aquifer (n_2^2), and 5 abstraction wells are in the Lower Pliocene aquifer (n_2^1). The total amount of groundwater abstraction is 524,456 m^3/d, mainly from the Pleistocene aquifers (277,585 m^3/d) and Middle Pliocene aquifer (245,315 m^3/d). The number of the abstraction wells and volume of groundwater pumped increased to 257,254 and 669,091 m^3/d, respectively, in 2012 (Tuyen, 2014). More details on abstraction well statistics and the volume of abstraction is provided in Table 13.4.

13.5.2 Groundwater Level

Table 13.5 shows the average depth of the groundwater level below the surface in the dry season (April) and rainy season (October) for 1999 and 2009 at monitoring wells in aquifer qp_3. Note that the minus sign in the column entitled "difference" means an increase in groundwater level.

Table 13.6 shows the average depth of the groundwater level below the surface in the dry season (April) and rainy season (October) for 1999 and 2009 at monitoring wells in aquifer qp_{2-3}. It is obvious that the depth of groundwater level below the surface increases at almost all monitoring wells. The increases are from 0.81 m to 20.74 m, and the maximum increase is up to 20.74 m.

Table 13.4 Number of abstraction wells and total volume of groundwater abstraction in 2012

No	In aquifer	Number of abstraction wells	Total amount of GW abstraction (m³/d)
1	qp_3	300	64,232
2	qp_{2-3}	442	226,507
3	qp_1	137	102,121
4	n_2^2	206	129,960
5	n_2^1	38	19,745
6	Small wells	256,131	126,526
	Sum	257,254	669,091

GW, groundwater.
Source: Tuyen (2014).

Table 13.5 Average depth to the groundwater level below the surface after 10 years (1999–2009) in aquifer qp_3

	The average depth of groundwater level below the surface (m)					
ID	April 1999	April 2009	Difference	October 1999	October 2009	Difference
Q011020	4.05	6.00	1.95	1.70	1.63	−0.07
Q01302A	2.44	3.05	0.61	1.76	1.33	−0.43
Q01302B	4.93	5.64	0.71	3.59	3.16	−0.43
Q01302C	5.77	6.08	0.31	4.32	3.77	−0.55
Q01302D	0.67	1.12	0.45	0.44	0.44	0.00
Q01302E	1.12	1.18	0.06	0.82	0.45	−0.37
Q01302F	0.69	0.99	0.30	0.48	0.61	0.13
Q09902A	10.63	10.39	−0.24	9.80	7.73	−2.07
Q09902B	4.26	4.16	−0.10	2.43	0.74	−1.69
Q09902C	8.73	8.39	−0.34	7.15	5.24	−1.91
Q09902D	6.34	5.98	−0.36	4.24	2.08	−2.16
Q09902E	8.88	8.32	−0.56	7.33	4.59	−2.74
Q804020	3.56	4.09	0.53	2.51	1.11	−1.40
Q808020	3.57	7.30	3.73	2.41	2.52	0.11

Source: Vuong (2010).

Table 13.7 shows the average depth of the groundwater level below the surface in the dry season (April) and rainy season (October) for 1999 and 2009 at monitoring wells in aquifer qp_1. It is obvious that the depth of groundwater level below the surface increases at almost all monitoring wells. The increases are from 0.95 m to 19.05 m.

Table 13.6 Average depth to the groundwater level below the surface after 10 years (1999–2009) in aquifer qp$_{2-3}$

	The average depth of groundwater level below the surface (m)					
ID	April 1999	April 2009	Difference	October 1999	October 2009	Difference
Q00202A	2.93	4.4	1.47	2.54	3.85	1.31
Q003340	4.57	7.34	2.77	2	1.49	−0.51
Q007030	4.94	8.26	3.32	1.09	3.35	2.26
Q011340	14.9	27.26	12.36	13.74	25.63	11.89
Q019340	14.58	35.27	20.69	13.83	34.57	**20.74**
Q808030	4.21	9.46	5.25	3.03	8.2	5.17
Q822030	2.02	2.83	0.81	1.91	2.9	0.99

Source: Vuong (2010).

Table 13.7 Average depth to the groundwater level below the surface after 10 years (1999–2009) in aquifer qp$_1$

	The average depth of groundwater level below the surface (m)					
ID	April 1999	April 2009	Difference	October 1999	October 2009	Difference
Q00204A	2.64	4.06	1.42	2.26	3.61	1.35
Q004030	5.63	15.74	10.11	5.15	15.52	10.37
Q015030	11.78	27.83	16.05	11.69	27.65	15.96
Q017030	10.79	15.36	4.57	9.23	13.02	3.79
Q821040	2.05	2.99	0.94	2.24	3.04	0.8

Source: Vuong (2010).

Table 13.8 shows the average depth of the groundwater level below the surface in the dry season (April) and rainy season (October) for 1999 and 2009 at monitoring wells in aquifer n_2^2. It is obvious that the depth of groundwater level below the surface increases at almost all monitoring wells. The increases are from 2.42 m to 12.99 m.

There is only one monitoring well in aquifer n_2^1. The depth of groundwater level below the surface increases from 3.43 m to 3.70 m in the rainy and dry seasons, respectively.

Figure 13.4 shows a plot of the average depth in groundwater level below the surface for April and October in the period 1999–2009 at all monitoring wells in HCMC. It is clear that there is an increasing trend in the average depth of groundwater level below the surface.

Table 13.8 Average depth to the groundwater level below the surface after 10 years (1999–2009) in aquifer n_2^2

	The average depth of groundwater level below the surface (m)					
ID	April 1999	April 2009	Difference	October 1999	October 2009	Difference
Q011040	15.36	28.35	12.99	14.74	27.81	13.07
Q80404T	8.34	10.76	2.42	7.57	10.49	2.92
Q808040	5.32	16.45	11.13	No data	No data	No data
Q822040	2.02	2.02	0	No data	No data	No data
Q80404Z	8.61	12.04	3.43	No data	No data	No data
Q808050	5.44	16.26	10.82	No data	No data	No data

Source: Vuong (2010).

13.5.3 Groundwater Quality

Several studies from 1988 onward (Tin, 1988; Dung, 1991; Luan, 1993; Hung, 2001; Vuong, 2010) have discussed the variation in groundwater quality for different aquifers. These studies indicated that groundwater quality in all aquifers is very good; only simple treatment is necessary before use.

Groundwater quality in HCMC aquifers is generally characterized by high concentrations of pH, ammonia, iron, and hardness. In the Upper Pleistocene aquifer (qp_3): in comparison with QCVN 02:2009/BYT (the national standard for drinking water sources), from 126 water samples analyzed, 31 and 60% do not meet the standard for pH value and ammonia, respectively. In the Upper–Middle Pleistocene aquifer (qp_{2-3}): 5, 10.3, and 5% of 126 water samples analyzed have not met the standard for ammonia, iron, and hardness as $CaCO_3$, respectively. In the Lower Pleistocene aquifer (qp_1): 30.77 and 22.03% of 60 water samples analyzed have not met the standard for iron and pH value, respectively. In the Middle Pliocene aquifer (n_2^2): 16.83, 77.5, and 39.33% of 101 samples analyzed have not met the standard for pH value, ammonia, and iron, respectively. In the Lower Pliocene aquifer (n_2^1): 18.18 and 7.5% of 11 samples analyzed have not met the standard for pH value and ammonia, respectively. In the Upper Miocene (n_1^3) groundwater quality is quite good, with all components meeting the national standards.

13.5.4 Recharge

HCMC receives a large amount of precipitation (1600–1800 mm/y) (Table 13.3), but aquifer recharge is limited by a widely spread silty clay

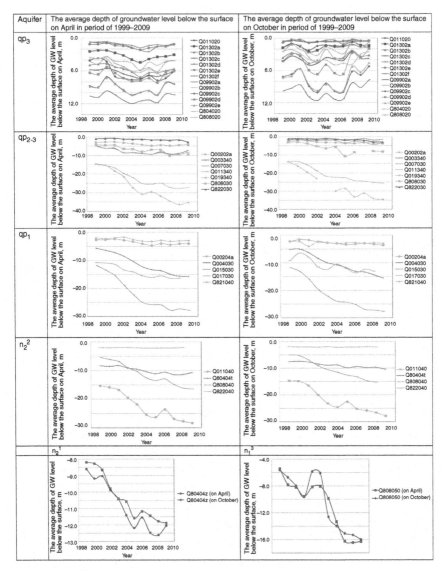

Figure 13.4 *Average Depth of Groundwater Level below the Surface for April in the Period 1999–2009 at Monitoring Wells in HCMC.*

layer 10–18 m thick on most of the surface in HCMC. This prevents the easy access of percolating rainwater to the aquifers.

Rechargeable areas are in the northern and northwestern part of the city (Figure 13.5) and cover an area of 435 km². Major rechargeable areas are Cu Chi, Thu Duc, and a part of the Hoc Mon District. A modeling

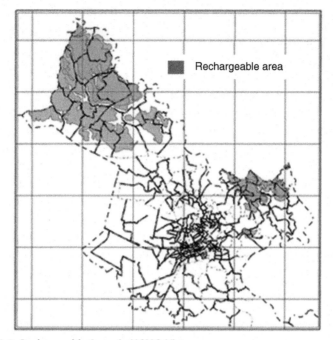

Figure 13.5 *Rechargeable Areas in HCMC, Vietnam.*

study (Chan, 2009) estimated that recharge into aquifers varies from 526,250 m³/d to 536,540 m³/d depending on whether it is a rainy or dry season.

The recharge to aquifer system is from direct infiltration of rainfall in the rechargeable area and from the Saigon River in the northern part. The minimum recharge is 536,540 m³/d and the recharge is limited to 27.6% of rainfall. These figures reveal the prospects for further research to explore possibilities for storing the remaining rainfall in HCMC's aquifer system.

13.6 IMPACTS

The impacts of overexploitation of groundwater resources in HCMC are reflected in the decline of groundwater levels and land subsidence.

13.6.1 Decline in Groundwater Level

Groundwater level in all aquifers shows a decreasing trend (Figure 13.4). A decline in groundwater levels in each aquifer during the 10-year period, from 1999 to 2009, is shown in Table 13.9.

Table 13.9 Decline in groundwater levels in 10 years from 1999 to 2009

Aquifer	Decline in GW level
Aquifer qp_3	0.07–2.74 m
Aquifer qp_{2-3}	0.81–0.74 m
Aquifer qp_1	0.95–19.05 m
Aquifer n_2^2	2.42–12.99 m
Aquifer n_2^1	3.43 m
Aquifer n_1^3	10.5 m

Source: Vuong (2010).

13.6.2 Land Subsidence

As a result of groundwater mining from deep aquifers and subsequent lowering of the piezometric head, the overlying aquitard (clay and silt layers) and deep aquifer may consolidate due to a decrease in pore water pressure. This may result in subsidence or settlement of the ground surface.

A study using the Interferometric Synthetic Aperture Radar (InSAR) technique to monitor land surface subsidence was carried out by the HCMC Center for Geoinformatics in 2008. The results of this study are shown in Figure 13.6. It shows that a total of 115 km² surface areas have land subsidence with magnitudes ranging from 5 mm/y to 15 mm/y.

13.6.3 Public Health

The amount of 669,091 m³/d groundwater abstraction meets one third of HCMC's water demand. However, its quality is under threat because of solid waste disposal and effluent discharge into rivers and open ground. Bacterial indicators, which cause water-borne diseases, have already been detected in groundwater samples.

About 2 MCM of wastewater is discharged every day from a population of over 7 million. Until now, there have been only three wastewater treatment plants: Tan Qui Dong having a treated capacity of 500 m³/d, Binh Hung Hoa having a capacity of 26,500 m³/d, and Binh Hung with a capacity of 144,000 m³/d. Therefore, only 180,000 m³/d of wastewater is treated daily; this is less than one tenth of the total wastewater generated in the city.

Most of the solid waste and wastewater from urban areas is being discharged directly into rivers and/or open ground. Such practices are responsible for the deterioration in surface and groundwater quality especially at shallow depth. Bacterial indicators capable of spreading water-borne diseases are observed in groundwater samples, and are the most serious threat to public health caused by insanitary sewage disposal.

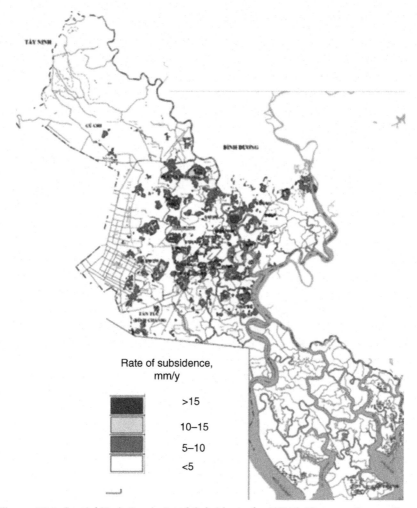

Figure 13.6 *Spatial Variation in Land Subsidence for HCMC, Vietnam. Source: Trung et al. (2008).*

13.7 RESPONSES

The Government of Vietnam, in close collaboration with local authorities and stakeholders, has made several attempts to improve the groundwater environment through the formulation of laws, organizational adjustments, and research activities aimed at groundwater management, protection, abstraction, and use. These are discussed using two indicators: formulation of a legal system relating to groundwater management and the initiation of groundwater monitoring.

13.7.1 Initiation of Groundwater Monitoring

Groundwater levels in HCMC's aquifer system have been monitored since the 1990s. At present, there are two monitoring networks in HCMC (Figure 13.7); one belongs to the national monitoring network and the other to the Ho Chi Minh's People Committee (PC).

There are 38 monitoring wells under the national monitoring network, which monitor groundwater for the period 1995–2014. Monitoring frequency is once in every 5 days for groundwater levels and twice a year for water quality. The 38 wells represent various aquifers; 2 are in qh, 14 in qp_3, 8 in qp_{2-3}, 5 in qp_1, 5 in n_2^2, 2 in n_2^1, and 2 in n_1^3 (see Appendix 13.3).

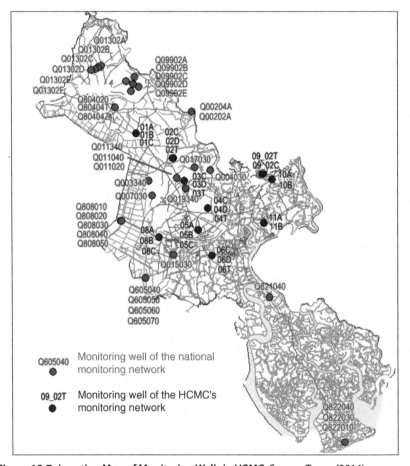

Figure 13.7 *Location Map of Monitoring Wells in HCMC. Source: Tuyen (2014).*

Under the local monitoring network there are 28 wells that have monitored groundwater for the period 2000–2006. Observation frequency is the same as that of the national monitoring network. Of the 28 wells, 9 are in qp_3, 5 in qp_{2-3}, 6 each in qp_1 and n_2^2, 1 in n_2^1, and 1 in n_1^3.

Although, the density of the monitoring wells in HCMC is sparse and the local monitoring network is not continuously operated, the data from the two monitoring networks provide reasonable information for decision makers and HCMC's authorities to recognize the adverse impacts of groundwater overexploitation. In HCMC's water supply planning to 2025, it is proposed to reduce the amount of groundwater abstraction from 0.67 MCM/d in 2010 to 0.4 MCM/d in 2015 and 0.1 MCM/d in 2025 (The Prime Minister, 2012).

13.7.2 Environmental Standards and Guidelines

The legal system of Vietnam consists of legal documents with a hierarchy ranging from high to low enforcement. The legal documents are divided into laws/ordinances and secondary regulations. The laws/ordinances are legal documents passed by the National Assembly, the highest constitutional body of the Socialist Republic of Vietnam. The secondary regulations are issued by state organizations (administrative and judicial) as legal documents ranking lower than the laws/regulations (Loan, 2010).

The current legal system focusing on the groundwater sector in Vietnam consists of the following:
1. Laws and resolutions of the National Assembly
2. Ordinances and resolutions of the Standing Committee of the National Assembly
3. Orders and decisions of the State President
4. Decrees of the Government
5. Decisions of the Prime Minister
6. Circulars from Ministers or heads of ministry-equivalent agencies
7. Legal documents of People's Councils and People's Committees

Several important legal documents related to groundwater are described below in chronological order.

On February 17, 2014 the Minister for Natural Resources and Environment (MONRE) issued the Circular No. 13/2014/TT-BTNMT to stipulate in detail the content, terms of reference, products of investigation, and assessment of groundwater resources.

On May 30, 2014 the Minister for MONRE issued the Circular No. 27/2014/TT-BTNMT to stipulate the registration of groundwater exploitation: application forms for the grant, extension, adjustment, and regrant of water resources permissions; contents of the project and a final report in a dossier for the grant, extension, adjustment, and regrant of groundwater abstraction permissions.

On November 27, 2013 the Vietnamese Government issued Decree No. 201/2013/NĐ-CP to stipulate detailed regulations on some articles in the Law of Water Resources. This decree stipulates for public consultation in water resources exploitation and use; wastewater discharge into water sources; a baseline investigation of water resources; licensing of water resources; funds for the provision and transfer of water resources exploitation rights; river basin organization, coordination, and supervision of water resources exploitation; protection and prevention to combat and remediate the adverse impacts caused by water in river basins.

On July 18, 2013 the Minister for MONRE issued Circular No. 19/TT-BTNMT on the Technical Guide to Groundwater Resources Monitoring. This circular regulates the monitoring frequency, factors, equipment, maintenance, data processing, and promulgation of monitoring data.

On June 21, 2013 the Minister for MONRE issued the Circulars Nos. 15/ TT-BTNMT, 16/ TT-BTNMT, and 17/ TT-BTNMT for building a series of groundwater resources maps (SGWRMs) using a scale of 1/50,000, 1/100,000, and 1/200,000 respectively. The SGWRMs consist of the groundwater resources practice material maps, groundwater quality maps, and groundwater resource quality maps. This circular regulates the basic content and format of the SGWRMs.

In 2012 the National Assembly promulgated the Law on Water Resources. The law regulates the management, protection, extraction, use, and prevention to combat and overcome the consequences and harmful effects caused by water within the territory of the Socialist Republic of Vietnam.

According to the law, the Vietnamese Government is carrying out the unified state management of water resources in general and groundwater resources in particular. The Ministry of Natural Resources and the Environment (MONRE) is implementing state management of water resources. Ministries and ministerial level agencies implement state management of water resources according to assignments given by the Government. People's Committees (provincial level) are responsible for state management of water resources within their localities. The Department of Natural Resources and the Environment (DONRE) is the institution and/or authority responsible for monitoring, managing, and regulating groundwater use, with a special responsibility for recording well installation or utilization.

The regulation and distribution of groundwater resource for usage purposes must be based on the planning of the river basin. Organizations and individuals have the right to exploit and use groundwater resource for purposes of living, agricultural, forestry, industrial production, and other purposes. Organizations and individuals that exploit and use water resources must get permission from the relevant state agencies, except for the exploitation and use of groundwater on a small scale within the family.

Organizations and individuals have rights to groundwater: to exploit and use water resources for domestic, production, business, and other purposes in accordance with the Law on Water Resources and other relevant laws; to benefit from water resource exploitation and use; to have their rights and legitimate interests protected by the state; to use data and information on water resources in accordance with this law and other relevant laws; to conduct water through adjacent land plots which are managed and used by others in accordance with laws.

Organizations and individuals that exploit and use water resources for domestic use are *not* subject to registration or licensing as long as use is at the household scale; small scale; for salt production; for cultural, religious, or scientific research activities; for fire extinguishment, controlling of pollution incidents and epidemics, and other emergency cases. Organizations and individuals that exploit and use water resources, but do not belong to the above-mentioned cases must get permission from the authorized State agencies before the decision of investment.

The Ministry of Natural Resources and provincial-level People's Committees shall carry out the granting, renewing, adjusting, suspending, and revoking of the licenses on water resources.

On December 31, 2008 the Minister for MONRE issued Decision No. 16/2008/QĐ-BTNMT for the national technical regulation of underground water quality.

On December 31, 2008 the Minister for MONRE issued Decision No. 15/2008/QĐ-BTNMT on groundwater protection. This decision stipulates about prohibited, limited zones to build new groundwater exploitation wells – groundwater protection for drilling, excavation, field tests, groundwater exploitation, construction, and other groundwater-related activities.

On September 4, 2007 the Minister for MONRE issued Decision No. 14 /2007/QĐ-BTNMT stipulating procedures, order, terms of reference for treatment, and filling in of abandoned boreholes made in groundwater research, investigation, assessment, dewatering, and excavation.

On September 14, 2000 the Minister for Industry (MOI) issued Decision No. 53/ 200-QĐ/BCN for the regulation of mapping hydrogeological maps on a scale of 1/25,000 and 1/50,000. This guideline regulates in detail every single phase of hydrogeological mapping – from preparation of the proposal and details of the hydrogeological mapping project, to implementation of works on the approved project, and reporting and publishing the hydrogeological maps.

Other guidelines also issued in 2000 by the MOI are Regulations on Pumping Tests, Regulations on Drilling, Regulations on Well Logging, Regulations on Groundwater Sampling and Analysis, Regulations on Vertical Electrical Sounding, Regulations on Land Surveys, and Regulations on Groundwater Monitoring. These regulations stipulate in detail the procedures, order, terms of reference, and use of products.

The legal system of Vietnam on groundwater resources is fairly comprehensive. It is of considerable help to HCMC's authorities in improving the groundwater environment.

13.8 SUMMARY

DPSIR-based evaluation shows that population growth, urban population increase, and tourism are responsible for groundwater abstraction exceeding recharge. Consequently, the groundwater level in all six-production aquifers has decreased dramatically, raising concerns about the risk of land subsidence in areas with high percentages of compressible clay and silt layers. New industrial and residence zones are being constructed in places where a water pipeline system is not yet available, so people have to drill a well and abstract groundwater to meet their own water demand. Uncontrolled drilling makes the situation of groundwater pollution more serious. This is a major source of pressure on the groundwater environment. In additional, nonagricultural and rural areas have increased, creating a corresponding decrease in recharge areas, which adversely affect supplies to the groundwater reserve. In response to those situations, HCMC has a comprehensive legal system in place for groundwater, as well as a groundwater-monitoring network. Such regulations and restrictions help the decision makers and HCMC's authorities to recognize adverse situations in the groundwater environment and plan for future reduction of groundwater abstraction. Therefore, in the near future, it is hoped that the groundwater environment will be greatly improved.

ACKNOWLEDGMENTS

The authors acknowledge the Asia Pacific Network for Global Change (APN); Asian Institute of Technology (AIT), Thailand; Institute for Global Environmental Strategies (IGES), Japan; and International Research Center for River Basin Environment (ICRE), University of Yamanashi, Japan for supporting this project. They also acknowledge Dr Sangam Shrestha and Dr Vishnu P. Pandey for their valuable inputs in improving the quality of this chapter.

REFERENCES

Chan, N.D., 2009. Application of modeling method to assess groundwater reserves of Ho Chi Minh City and neighboring areas.

Dung, N.Q., et al., 1991. Primary Groundwater Exploration in Hoc Mon–Cu Chi, Ho Chi Minh City. Report.

Hung, Ð.T., et al., 2001. Groundwater Exploitation and Utilization Planning in Ho Chi Minh City. Report.

Loan, N.T.P., 2010. Legal Framework of the Water Sector in Vietnam.

Luan, L.Q., et al., 1993. Groundwater Exploration in Binh Chanh Area. Report.

Nga, N.V., 2006. State of groundwater management in HCM City. Fifth Research Meeting on the Sustainable Water Management Policy, November 27, 2006, Ho Chi Minh City.

Son, N.T., et al., 2012. Urban growth mapping from Landsat data using linear mixture model in Ho Chi Minh City, Vietnam.

The Prime Minister, 2012. Decision No. 729 on Approval of Water Supply Planning of Ho Chi Minh City to 2025.

Tin, D.V., 1988. Hydrogeological Mapping on the Scale of 1/50,000 of Ho Chi Minh City. Report.

Trung, L.V., et al., 2008. Measuring ground subsidence in Ho Chi Minh City using differential INSAR techniques.

Tuyen, P.V., 2014. Prohibited and limited groundwater exploitation zoning.

Vuong, B.T., 2010. Hydrogeological and Engineering Geological Maps on the Scale of 1/50,000 of Ho Chi Minh City. Report.

APPENDIX 13.1 RESPONSIBILITIES OF PEOPLES' COMMITTEE (PC) ON WATER RESOURCES

- To issue authorization/license and arrange the implementation of the legal documents on water resources;
- To prepare, approve, publish and arrange the implementation of the water resources plans; baseline survey plans, regulation and allocation of water resources, and restoration of the polluted and depleted water sources;
- To delineate and publish the prohibited and restricted areas for groundwater exploitation; the areas requiring artificial groundwater recharge;

publish the minimum flow, thresholds for groundwater exploitation as authorized; the areas for prohibited or temporarily prohibited exploitation of sand, soil and minerals in rivers; and publish the list of lakes, ponds, swamps where filling-up is not allowed;

- To arrange the response to and overcoming of water pollution incidents; oversee, detect and take part in the handling of water pollution incidents of the international water sources as authorized; set up and manage the corridors for the protection of water resources and the sanitary protection areas of domestic water sources; ensure domestic water supply in case of drought, water shortage, or water pollution incidents;
- To propagate, disseminate, and educate legislations on water resources;
- To issue, renew, adjust, suspend and revoke licenses on water resources and allow the transfer of the rights to exploit water sources as authorized; guide the registration of water resources exploitation and use;
- To arrange the implementation of baseline survey and water resources monitoring as decentralized; report to the Ministry of Natural Resources and Environment the results of baseline survey on water resources, the situation of management, exploitation, use and protection of water resources, the prevention, combat against and overcoming of the harms caused by water in the locality;
- To develop database; manage, and store data and information on water resources;
- To inspect, check, settle disputes, deal with violations to legislations on water resources.

APPENDIX 13.2 CALCULATION OF GROUNDWATER RESERVES

$$Q_{kt} = \alpha_1 Q_{tn} + \alpha_2 \frac{V_{tn}}{t} \tag{1}$$

where Q_{kt} = groundwater reserves (m³/d); Q_{tn} = natural dynamic groundwater reserves (m³/d); V_{tn} = static groundwater reserves (m³/d); α_1, α_2 = coefficients of groundwater reserves utilities; t_{kt} = abstraction time (= 10^4 d).

Natural dynamic groundwater reserve is composed of all components of water flow in the aquifers (recharge from rainfall and surface water bodies, horizontal and vertical infiltration from one aquifer to other aquifers). The amount of the natural dynamic groundwater reserve in this article was calculated using the groundwater flow model of Ngo Duc Chan, (2009).

Static groundwater reserve has two components: gravity reserves and elastic reserves.

"Gravity reserves" are the amount of water storage in interganular or fracture aquifer and are calculated by the following formula:

$$Q_{dt} = \alpha\mu\frac{mF}{t_{kt}} \qquad (2)$$

where Q_{dt} = gravity reserves (m^3/d); α = coefficients of gravity reserves utilities (= 0.35); μ = specific yield; m = aquifer thickness (m); F = distribution area of aquifer (m^2); t_{kt} = abstraction time (= 10^4 d).

"Elastic reserves" are the amount of water that is discharged by reducing one unit head of groundwater and are calculated by the following formula:

$$Q_{dh} = \mu^{\star}\frac{FH}{t_{kt}} \qquad (3)$$

where Q_{dh} = elastic reserves (m^3/d); $\mu^{\star}$ = storage coefficient; F = distribution area of aquifer (m^2); H = average groundwater head above the top of aquifer (m); t_{kt} = abstraction time (= 10^4 d).

APPENDIX 13.3 DETAILS OF GROUNDWATER-MONITORING WELLS

Monitoring wells under the national monitoring network in HCM City:

No.	ID	Depth (m)	VN 2000 coordinates X	VN 2000 coordinates Y	Absolute elevation (m)	Aquifer
1	Q00202A	52.0	1214,361	679,458	1.94	qP_{2-3}
2	Q00204A	66.0	1214,361	679,462	2.15	qP_1
3	Q003340	132.5	1200,776	670,715	5.36	qP_{2-3}
4	Q004030	101.0	1202,837	683,115	2.38	qP_{2-3}
5	Q007030	126.0	1198,000	671,368	3.71	qP_{2-3}
6	Q011020	36.0	1201,413	676,370	7.91	qP_3
7	Q011040	225.0	1201,398	676,357	8.05	n_2^2
8	Q011340	130.0	1201,404	676,357	8.10	qP_{2-3}
9	Q01302A	23.0	1222,917	660,855	13.64	qP_3
10	Q01302B	24.0	1223,059	661,071	15.48	qP_3
11	Q01302C	22.5	1223,276	661,408	14.73	qP_3
12	Q01302D	23.0	1222,762	660,654	11.34	qP_3

(Continued)

No.	ID	Depth (m)	VN 2000 coordinates X	Y	Absolute elevation (m)	Aquifer
13	Q01302E	22.5	1222,591	660,200	10.78	qp_3
14	Q01302F	22.0	1222,371	659,354	8.71	qp_3
15	Q015030	165.0	1186,363	675,665	1.24	qp_1
16	Q017030	124.5	1203,416	679,924	8.66	qp_1
17	Q019340	134.0	1199,313	678,122	2.84	qp_{2-3}
18	Q09902A	26.4	1221,066	668,034	15.13	qp_3
19	Q09902B	25.8	1220,157	666,294	11.69	qp_3
20	Q09902C	25.0	1219,640	667,702	15.54	qp_3
21	Q09902D	27.6	1218,261	667,285	12.22	qp_3
22	Q09902E	30.5	1219,130	668,868	14.22	qp_3
23	Q804020	22.0	1215,192	664,143	10.23	qp_3
24	Q80404t	124.6	1215,189	664,144	10.29	n_2^2
25	Q80404Z	159.0	1215,197	664,152	10.30	n_2^1
26	Q808010	33.0	1192,962	665,306	1.23	qh
27	Q808020	63.0	1192,970	665,306	1.36	qp_3
28	Q808030	126.0	1192,981	665,317	1.24	qp_{2-3}
29	Q808040	235.0	1192,979	665,312	1.40	n_2^2
30	Q808050	295.0	1192,974	665,315	1.31	n_1^3
31	Q821040	152.1	1178,132	694,960	1.01	qp_1
32	Q822010	10.0	1149,819	709,889	2.56	qh
33	Q822030	123.0	1149,768	709,882	2.47	qp_{2-3}
34	Q822040	197.0	1149,819	709,896	2.56	n_2^2
35	Q605040	142.0	1181,820	670,001	2.13	qp_1
36	Q605050	190.0	1181,819	669,994	2.15	n_2^2
37	Q605060	230.0	1181,825	669,986	2.15	n_2^1
38	Q605070	300.0	1181,826	669,991	2.15	n_1^3

Monitoring wells under the local monitoring network in HCM City:

No	ID	Depth (m)	VN 2000 coordinates X	Y	Absolute elevation (m)	Aquifer
1	01A	37	1210,094	668,205	9.85	qp_3
2	01B	122	1210,094	668,203	9.84	qp_1
3	01C	190	1210,096	668,204	9.88	n_2^2
4	02C	40	1205,251	675,621	6.39	qp_{2-3}
5	02D	165	1205,253	675,622	6.39	n_2^1
6	02T	110	1205,253	675,620	6.39	n_2^2
7	03C	107	1200,832	677,879	5.94	n_2^2
8	03D	162	1200,832	677,877	5.94	n_2^1
9	03T	38	1200,834	677,876	5.94	qp_3

(Continued)

No	ID	Depth (m)	VN 2000 coordinates		Absolute elevation (m)	Aquifer
			X	Y		
10	04C	128	1195,427	682,576	9.69	qp_1
11	04D	184	1195,428	682,573	9.69	n_2^2
12	04T	41	1195,426	682,575	9.69	qp_3
13	05A	44.5	1191,161	680,669	4.78	qp_3
14	05B	145	1191,159	680,670	4.78	qp_1
15	05C	208	1191,159	680,668	4.78	n_2^2
16	06C	57	1186,257	683,392	1.62	qp_3
17	06D	155	1186,255	683,392	1.62	qp_3
18	06T	236	1186,259	683,392	1.62	n_2^2
19	08A	65	1189,797	672,743	1.6	qp_3^3
20	08B	133	1189,795	672,743	1.6	qp_{2-3}
21	08C	174	1189,796	672,745	1.6	qp_1
22	09-2C	12	1202,193	693,639	18.93	qp_3
23	09-2T	70	1202,215	693,575	18.93	qp_1
24	10A	30.5	1201,170	695,532	29.05	qp_3
25	10B	60.5	1201,162	695,533	29.05	qp_{2-3}
26	11A	85	1192,529	693,879	3.6128	qp_1
27	11B	46	1192,527	693,879	3.62	qp_{2-3}
28	M1A	17	1192,176	693,834	3.67	qp_3

CHAPTER 14

Groundwater Environment in Yangon, Myanmar

Khin Kay Khaing

Department of Geography, University of Yangon, Myanmar

14.1 INTRODUCTION

The provision of a reliable supply of safe, potable, and adequate water is of prime concern to the future development of Yangon City in Myanmar. The development of economic activities and population growth in Yangon has led to an increased exploitation of natural resources including water. Today, with the extension of the city area, upgrading of existing housing estates, construction of new housing projects, shopping centers, supermarkets, hospitals, and hotels, water demand is increasing in the city. Yangon City Development Committee (YCDC), which controls the water supply infrastructure, currently takes water from tubewells and four reservoirs (i.e., Hlawga, Gyobyu, Phugyi, and Ngamoeyeik). The YCDC cannot supply water to the whole city, so residents have been relying on groundwater as an alternative source.

According to available statistics, 10.5% of the annual water supply to YCDC in 2003 came from groundwater, which increased to 12.5% in 2014. According to computational estimates, 53.5% of the city's total amount of current water use is from groundwater. In actual practice, the amount is expected to be more than that figure because a significant amount of water is wasted through leakage in the supply system. It is therefore essential to obtain a detailed knowledge and evaluation of groundwater conditions in order to plan and manage the future water supply and demand in Yangon. This chapter attempts to develop a better understanding of Yangon City's groundwater environment by identifying driver (D), pressure (P), state (S), impact (I), and response (R) and analyzing their extent and interrelationships.

14.2 ABOUT THE CITY

Yangon City is situated at the confluence of the Yangon and Bago rivers on the eastern margin of the Ayeyarwady Delta. The city lies between 16°44′ and 17°2′N and 96°0′ and 96°21′E (Figure 14.1).

Groundwater Environment in Asian Cities
http://dx.doi.org/10.1016/B978-0-12-803166-7.00014-3

Copyright © 2016 Elsevier Inc.
All rights reserved.

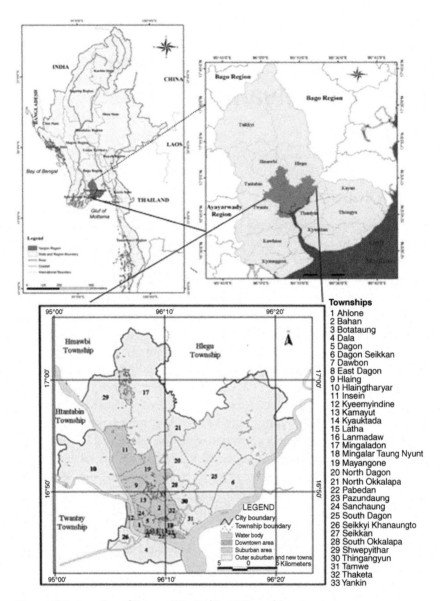

Figure 14.1 *Location of Yangon City and its 33 Townships.*

Yangon City was the capital of Myanmar from the colonial era until the last decade, before the capital relocated to Nay Pyi Taw in 2007. Originally, the city was designed for a population of 36,000. In the past 125 years, the city and corresponding built-up areas have increased steadily with the population influx. An estimated 4.94 million people are living in the present city area of 948 km². The city is composed of 33 townships as administrative units (Figure 14.1). The YCDC is responsible for administering social, economic, commercial, recreational, environmental, cultural, and infrastructural development activities within the city.

The center of the city has ridges formed by the southern narrow spur of Bago Yoma with an average height of 30 m and fault ponds with an artificial dam, namely Kandawgyi Lake, Inya Lake, and Hlawga Lake. Both the eastern and western parts of the central ridges are typically flat and formed by delta deposits (Figure 14.2). Most parts of the city area are within an elevation range of 3–6 m above mean sea level (masl). There are several earlier studies on the geological formation and hydrological condition of Yangon City (e.g., Oldham, 1893; Foy, 1903; Banerji, 1924; Coggin-Brown, 1926; Leicester, 1959; Win Naing, 1972, 1996, 2009; Maung Maung, 1996; Ahleinmar Htwe, 2000). The bedrock beneath the city can be classified into three main groups based on geological age. In order from oldest to youngest, (Figure 14.2), these are as follows: (i) Oligocene–Miocene of the Pegu group; (ii) Pliocene series of the Irrawaddy group; and (iii) Pleistocene–Holocene group of quaternary deposits. The availability of groundwater mainly depends on the group of bedrock units. In the city area, groundwater is being abstracted from valley fill deposits and Irrawaddy sand rocks. Four main groundwater zones can be classified on the basis of geological structure and composition: good potential zone, fair-to-good potential zone, poor-to-

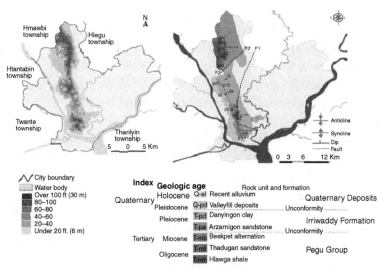

Figure 14.2 *Relief and Geology of Yangon City. Sources: Topographic map and Win Naing (1972) and further upgraded by Maung Maung (1996) and Win Naing (1996).*

fair potential zone, and poor potential zone (Khin Kay Khaing, 2011). Win Naing (1996) has summarized the topographic features, underlying material, and potential groundwater condition, as shown in Table 14.1.

Two distinct wet and dry seasons, with an annual rainfall of around 2500 mm are beneficial to the city, providing the opportunity for rain water to both fill reservoirs and recharge groundwater. Generally, the temperature in April is high, with a maximum monthly temperature of 38.1°C recorded in April 2010. A minimum monthly temperature of 16°C was recorded in January 2010 (Figure 14.3). However, since 1970, there has been a greater variation in the total amount of annual rainfall and intensity due to a decreasing number of rainy days (Figure 14.4). This process may affect the infiltration rate and replenishment of groundwater resources in the city's aquifers.

Presently, the YCDC can serve 58% of its total population with the total amount of 727.4 million cubic meters (MCM/d) (636.5 MCM/d from reservoirs and 90.9 MCM/d from tubewells) (YCDC, 2014). The remaining residents have to use water from their own tubewells. It is estimated that 53.5% of the city's total water use is derived from groundwater. However, with the exception of some townships, the city has no regulating body to control the construction of tubewells and the rate of groundwater withdrawal. As a result, the number of groundwater abstraction wells is increasing steadily in an uncontrolled manner.

Groundwater development for water supply in the Yangon area began in the early 1890s, and by the 1930s there were estimated to be 400 wells (Leicester, 1959). The existence of brackish water intrusion from the Yangon River and bacterial contamination in many tubewells during that time gained the interest of the Geological Survey of India and resulted in the Underground Water Act (Burma Act No. IV of 1930) (Leicester, 1959). This act required the licensing of all new wells, enabled a supervising agency to regulate new well construction practices, use of wells, and to some extent groundwater withdrawals. Following implementation of the act, development of groundwater resources was slow between 1930 and 1945 (Win Naing, 1972). However, there was so much organizational restructuring due to area growth, rendering the 1930 Groundwater Act on well construction practice in Yangon ineffective after 1945. As a result, the number of tubewells was increased. Even with the increased reservoir supply, the gap between water demand and supply became wider because water demand increased continuously due to population growth and the water-intensive lifestyles of the growing urban population. Yangon City authority is continuously trying to supply more water to its residents.

Table 14.1 Major topographic features, underlying material, and potential groundwater condition

Terrain	Topography	Geology	Groundwater condition
Tidal flat and channel	Low-lying area, along the river and creek, usually flooded and swampy area	Recent water-laid sediment of clay, silt, and decayed organic matter of alluvium	Not aquifer
Low-lying flat plains	Relatively low-level lands, far away from eastern part of the Shwedagon–Mingaladon ridge	Mixed with silt and clay of alluvium unit Low-level areas near the eastern part of ridge consist of outwash material from Azarnigon sand and Pegu group	Usually impervious, lower horizon, yield small amount of groundwater Yield is moderate
Terrace	Low elevated flat land at the southern part of the Shwedagon–Mingaladon ridge with more or less gentle slope	Interbedded sands and fine-to-course gravel of valley fill deposit, sometimes containing silt and clay	Yield is very high water quality is good as good aquifer
Ridge	Slightly to moderately rolling plain, highest elevation at Shwedagon Pagoda, more or less gentle slop Hillrock at the North of Yangon near Hlawga Lake with steep slope	Mainly Danyingon clay and Arzanigon sand rock of Irrawaddy formation from the anticlinal ridge of Thingangyun and Shwedagon ridge, synclinal structure in South Okkalapa Sandstone, shale, and siltstone of Pegu group	Good aquifer having at least three water-bearing horizons. In some places having high iron content due to literitization Yield small amount due to well-consolidated unit. Water quality is poor

Source: Modified from Win Naing (1996).

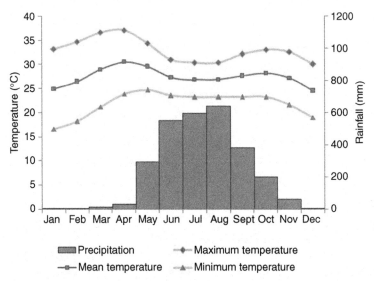

Figure 14.3 *Climograph of Yangon City. Source: Based on the data from Department of Meteorology and Hydrology, Kaba-Aye, Yangon.*

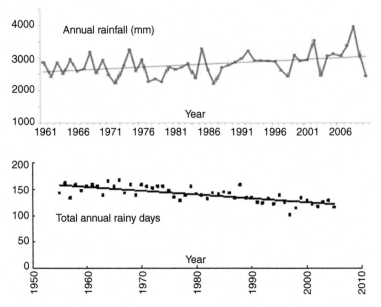

Figure 14.4 *Trends of Average Annual Rainfall and Rainy Days. Source: Based on the data from Department of Meteorology and Hydrology, Kaba-Aye, Yangon.*

14.3 DRIVERS

Population growth, urbanization, and tourism are the main drivers exerting pressure on Yangon's groundwater environment.

14.3.1 Population Growth

The population of Yangon City is continually growing by both natural increase and migration from other parts of the country. In 1990, the city's population was estimated to be about 2.9 million. Three years later, it stood at about 3.1 million. In 2000, the population increased to over 3.5 million, which further rose to 4.1 million with a density of over 5200 persons/km^2 in 2003. In 2010, it further increased to nearly 4.3 million with a population density of over 6000 persons/km^2, and again increased to 4.93 million in 2013. This city has the highest population concentration in Myanmar. Table 14.2 depicts the increasing trend in the population of the city.

The YCDC tried to expand its supply by using more water from both surface and groundwater sources. Before 2000 the amount of daily water supplied by YCDC was 0.39 MCM, which was taken from three reservoirs only. In 2003, YCDC's total piped water distribution became 0.43 MCM. YCDC abstracted groundwater to increase supply from 2000 to 2003. During the period from 2000 to 2006 the annual abstraction of groundwater by YCDC-owned tubewells was 16.6 MCM, and the city's total annual supply was 157.6 MCM from both reservoir water and groundwater. However, about half the water supplied by YCDC is lost owing to weaknesses in the distribution system. Although per capita water consumption varies according to ease of water access and living standards, the average water consumption per capita can be estimated as 160 liters/capita/day (lpcd) for the city as a whole. According to the YCDC, only

Table 14.2 Estimated population of Yangon City

Year		Population number	Net increase number
1990		2.90 million	–
1993		3.09 million	0.19 million
2000		3.55 million	0.46 million
2003		4.11 million	0.56 million
2010		4.34 million	0.23 million
2013		4.93 million	0.58 million

Source: Khin Kay Khaing (2006) and Khine Moe Nyunt (2013).

37% of the population in 2003 could be served by their water supply system. Some townships in the YCDC area did not have reservoir water and used mostly groundwater. Therefore, 63% of the residents in 2003 used water from private sources, mostly groundwater from their own tubewells. Hence, citywide groundwater abstraction (not including commercial and industrial uses) could be estimated at 151.2 MCM by private facilities and 16.6 MCM by the YCDC.

In accordance with the increased water demand of an increasing population, the YCDC expanded its capacity to supply water. At present, 58% of the total population is served by YCDC's water supply systems. However, the annual amount of groundwater withdrawal from YCDC has increased to 33.2 MCM. Although the amount of supply increased from 157.6 MCM to 265 MCM with the augmentation of water from a new reservoir, 42% of the total population has to use private sector groundwater. With the growth of economic activities, and commercial and industrial water consumption coming mostly from groundwater, the environment is under excessive pressure. Water supply and groundwater abstraction are shown in summarized form in Table 14.3.

Table 14.3 Water supply and groundwater abstraction of Yangon City

Indicators	Year 2003	Year 2013
Total population	4,111,524	4,934,766
Daily per capita average use	160 Litre	160 Litre
Annual total distribution from YCDC	157.6 MCM/year	265 MCM/year
Wastewater from inefficient supply system	50% of total amount of supply	40% of total amount of supply
Population served by YCDC	37% of total population	58% of total population
Daily groundwater withdrawal by YCDC	0.045 MCM/year	0.09 MCM/year
Total annual withdrawal by YCDC	16.6 MCM/year	33.2 MCM/year
Population unserved by YCDC	63% of total population	42% of total population
Groundwater withdrawal by private sector (domestic only)	151.2 MCM/year	121 MCM/year
Annual groundwater withdrawal (YCDC + private sector)	167.8 MCM/year	154.2 MCM/year

MCM, million cubic meters; YCDC, Yangon City Development Committee.

14.3.2 Urbanization

Urbanization is the process by which the population of a city grows. Yangon City stood as the capital of the country for years, and it is still a center for social and economic development. It is also the largest urban area in Myanmar. Considering the city's land use pattern, the whole area within YCDC's boundary can be regarded as urban. The significant expansion of urban areas can be seen in Table 14.4 and Figure 14.5.

Expansion of the urban area of the city has been more rapid during recent decades. The increase in area was over twofold during a 20-year period (311 km^2 from 1983 to 1993 and 215 km^2 from 1993 to 2003). In 2014, the total area of the city reached 947 km^2 and is expected to increase further to 154 km^2 in the next 10 years. YCDC's current piped water supply cannot

Table 14.4 Urban expansion in Yangon

Year	Area (km^2)
1963	164.6
1973	208.6
1983	346.0
1993	578.0
2003	793.0

Source: Zin New Myint (2011).

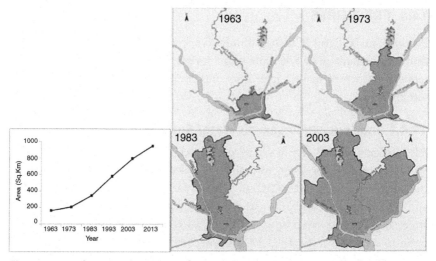

Figure 14.5 *Urban Area Expansion of Yangon City within the Yangon Region from 1963 to 2003. Source: Zin New Myint (2011).*

cope with this expansion since both reservoir water and the pipe network extension is insufficient. Therefore, people from these areas have to rely on groundwater, causing the number of privately owned tubewells to mushroom during this period. According to the YCDC, in 1996, the total number of privately owned tubewells was 69,172 for domestic use and 345 for industrial use. In 2003, YCDC had 215 tubewells, rising to 443 in 2013. However, none of the private well owners have recorded the amount of extraction. Without the figures from YCDC's tubewells, the daily amount of withdrawal from groundwater is unknown. What is certain, though, is increased amounts of water will be needed from these tubewells as a result of urbanization.

14.3.3 Tourism

Tourists to Yangon City increased from 0.12 million to 0.16 million in the fiscal years from 2000 to 2001 and 2009 to 2010, increasing further to 0.36 million in 2013. As the number of tourists increases, related facilities to serve them also increases. Therefore, with the increase in tourism, the number of hotels, motels, guesthouses, inns, and restaurants has also been increasing. In 2012, there were 204 hotels of various levels and this figure rose to 300 in 2013. All use water partly from the YCDC supply, but mostly from groundwater from private facilities. Tourism is therefore becoming a driver of increased groundwater exploitation in the city.

14.4 PRESSURES

Pressure on Yangon City's groundwater environment is being exerted through its inadequate reservoir water supply, land cover change, and subsequent reduction in recharge areas, and overexploitation of groundwater resources.

14.4.1 Inadequate Reservoir Water Supply

All residents in the city want to use piped water served by the YCDC. However, except for seven townships downtown, in 2003 the amount of the population serviced by YCDC was very low. Only the downtown area and parts of old suburban areas have a distribution system where every street has a piped water supply. Almost all the population in certain expansion areas was not served. In 2003, according to YCDC data, 75–100% of the population relied on groundwater in 16 townships, 50–74% in five townships, 25–49% in two townships, and 1–24% in six townships

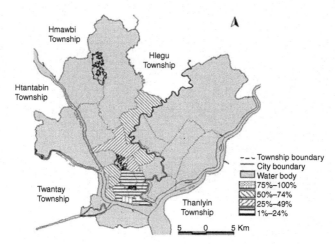

Figure 14.6 *Percentage of Population Dependent on Groundwater from Private Tube-wells. Source: Khin Kay Khaing (2006).*

(Figure 14.6). Only four townships did not use groundwater as they had a regular supply from YCDC piped water. Since the majority of people were relying on groundwater, the drilling of tubewells mushroomed in Yangon City.

With the construction of a new reservoir in 2013, the YCDC has increased its water storage capacity to serve up to 58% of the total population. However, the area covered by the piped network is much less than this figure and maintenance is inadequate. Leakage of piped water due to pipework falling into disrepair has created an abundant wastage of water. The combination of water demand for special economic activities and domestic water demand means that a significant amount of water is still needed to meet the YCDC's supply requirements, thereby exerting pressure on groundwater resources.

14.4.2 Land Cover Change (Decrease in Recharge Area)

Land cover change is expected to affect groundwater resources through infiltration. With the accelerated extension of urban areas, land cover in the city has changed from agricultural and forest cover into built-up areas. According to the YCDC, built-up areas have increased from 360 km² in 2000 to 492 km² in 2012. Land cover in the city for 2013 is shown in Figure 14.7. With the increase in urban areas, runoff patterns and infiltration changes result in a significantly adverse impact on groundwater recharge. Rechargeable areas in the city have decreased from 72% in 2002 to 63% in 2013.

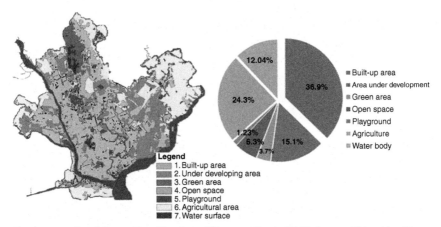

Legend
1. Built-up area
2. Under developing area
3. Green area
4. Open space
5. Playground
6. Agricultural area
7. Water surface

Figure 14.7 *Land Cover Classification of Yangon City in 2013. Source: Khine Moe Nyunt (2013) and YCDC (2013).*

Table 14.5 Land cover categories of Yangon City

Land cover categories	2002 (%)	2012 (%)	2013 (%)
Built-up area	28	34	37
Underdeveloped area	18	14	15
Playground	1	1	1
Agriculture	32	29	24
Open space	6	6	6
Green area	4	4	4
Water surface	12	12	12

Source: Khine Moe Nyunt (2013) and YCDC (2013)

With the continuous increase in urban built-up areas and development of pavements, undeveloped areas (agriculture and trees) have decreased from 2002 to 2013 (Table 14.5).

14.4.3 Overexploitation of Groundwater Resources

Both YCDC-owned and private tubewells are abstracting groundwater as and when necessary. Hospitals, hotels, government buildings, private institutions, housing projects, and gated communities also use groundwater as a reliable alternative source of water. Almost all industrial and commercial activities have used groundwater for a considerable time. For example, 18 purified drinking water plants having a daily production rate of 147.4 m³ between 1995 and 2000 used only groundwater as their raw source. The number of plants and production rates in recent years has increased to 51 and 604 m³/d

Table 14.6 Estimation of recharge in Yangon City

Total area	794.5 km^2 (in 2013)
Recharge area (63% of total area)	500.53 km^2
Total annual rainfall depth	2500 mm/y
Average recharge rate	12% of average annual rainfall/y
Average annual recharge depth	300 mm/y
Recharge volume	150.2 MCM (recharge depth × recharge area)

MCM, million cubic meters.

in 2005, with 56 and 713.5 m^3/d in 2009 (Cho Win Kyaw, 2012). Therefore, purified drinking water plants alone are abstracting about 0.3 MCM/y of groundwater in the area. If we record abstraction by all type of users (e.g., domestic, commercial, and industrial), the total is expected to be more than twice the current level of domestic water consumption in the city.

Overexploitation can generally be detected by comparing rates of recharge and withdrawal. Here, the total recharge rate is estimated using a water balance approach. The average annual rainfall of Yangon City is 2500 mm based on data from 1961 to 2010. There are no records of runoff within the city area. However, Win Naing (1972) estimated potential evaporation to be 38.2% of the average annual rainfall and then reasonably expressed the rainfall–runoff coefficient as 50% for the calculation of groundwater recharge in the Yangon area. Although infiltration may change in accordance with land use, land cover, and the intensity of rainfall, the expected percolation or recharge to groundwater is estimated as 12% of the total annual rainfall in the area. This percentage is equivalent to an annual recharge of 300 mm/y (or 150.2 MCM/y) (Table 14.6), which is less than that abstracted for domestic use only of 154.2 MCM/y (Table 14.3). It is almost certain that current socioeconomic activities will be further intensified and therefore groundwater abstraction will continue, worsening the groundwater overexploitation situation.

14.5 STATE

The state of groundwater from the perspective of well statistics, volume of groundwater abstraction, groundwater level, quality, and annual recharge is difficult to express in Yangon City due to the lack of adequate monitoring. There is no organization to regulate groundwater levels and the volume of groundwater abstraction. The following subsections discuss the situation based on the best available information.

14.5.1 Well Statistics and Amount of Groundwater Abstraction

A review of groundwater development history reveals an increasing number of tubewells and groundwater withdrawal with the growth of the city population. According to a literature survey, the total number of tubewells in the city area during 1893–1901 increased from 17 to 56. In 1920, Malcolm Perine Engineers, estimated the total groundwater abstraction from those tubewells at 1.5 million gallons a day (MGD), which increased to 2.8 MGD in 1924, 5.4 MGD in 1930, and 12.7 MGD in 1972 (Table 14.7). The increasing trend in the amount of tubewells and groundwater abstraction is summarized in Table 14.7. It is interesting to note that out of a total of 69,172 wells, 68,823 were for domestic use and the remainder for industrial use in the YCDC jurisdiction (YCDC, 2001). The actual figure is likely to be much higher than that reported, because there are many unauthorized tubewells in the townships of Yangon City.

The number of tubewells in the YCDC also increased from 217 in 2001 to 443 in 2014, resulting in an increase in groundwater withdrawal from 45,455 m³/d in 2003 to 90,922 m³/d in 2014. According to available records, the dependency of YCDC's supply system on groundwater increased from 10.5% in 2003 to 12.5% in 2014.

Table 14.7 Increasing trend in the number of tubewells and corresponding groundwater abstraction

Year	Number of tubewells	Total groundwater abstraction (MGD) by public and private wells	Reference
1893	17		Oldham (1893)
1901	56		Foy (1903)
1920	56	1.5	Malcolm Pirnie Engineers (1958)
1924	139	2.8	Banerji (1924)
1930	400	5.4	Leicester (1959)
1958	600	5.5	Malcolm Pirnie Engineers (1958)
1972	900	12.7	Win Naing (1972)
1986	2700		Tin Myo Win (1995)
1993	9340		Tin Myo Win (1995)
1996	69,172		YCDC (2001)

MGD, million gallons a day.

14.5.2 Groundwater Level and Quality

There are two major aquifers, Irrawaddian and valley fill deposits, serving as sources of groundwater in the Yangon area. The typical groundwater type in the Irrawaddian aquifer of the Yangon area is HCO_3–Na–Ca, whereas that of the valley fill deposit is Cl–HCO_3–Na–Ca/Mg, and the transitional HCO_3–Cl–Na–Mg/Ca between Irrawaddian and valley fill deposits (Win Naing 1972). Magnesium and sulfate are strongly associated in the Irrawaddian aquifer and the major type of groundwater is Ca–Mg–NaCl (Win Naing, 2009). There is no continuous monitoring of groundwater quality in the Yangon City, and therefore the exact information is not available. Only certain individual institutions are measuring groundwater quality, and on that basis, users accept it as being of potable quality. Groundwater quality varies spatially within the city. Although water quality from the two groundwater aquifers is generally acceptable, Win Naing's paper of 1995 expresses the need for enforcement of regulations governing the construction and operation of tubewells in Yangon by highlighting the results of the chemical analysis of 400 samples collected from private tubewells, and the bacteriological analysis of 970 samples from the same sources by the National Health Laboratory in 1989. According to the analysis, 120 contained high iron content, 22 contained high chloride, and 2 were classified as not potable. In addition, bacteriological analysis showed 42% were unsatisfactory and *Escherichia coli* was detected in 30 samples. In the Yangon area, 56 water samples from tubewells were analyzed by using the Wagtech Arsenator system, of which 6 water samples showed an arsenic content of 4–11 parts per billion (ppb), less than the maximum permissible arsenic content of the World Health Organization (WHO) standard for drinking water. Therefore, the groundwater from the Irrawaddian aquifer is safe for drinking (Win Naing, 2009). However, regular spatial monitoring is needed in Yangon City.

14.6 IMPACTS

The impacts on the groundwater environment are discussed using the following indicators: depletion of groundwater reserves and land subsidence.

14.6.1 Depletion of Groundwater Reserves

The decline in groundwater levels and production capacity of wells can be considered parameters for indicating the depletion of groundwater reserves. Since 1972, no systematic records of groundwater levels have existed in the YCDC area. An estimate in 1972 suggested that groundwater elevation may range from about 1.5 m near the Hlaing River to more than 10 m along the eastern boundary of valley fill deposits with water table gradients of 7–10 m/km (Win Naing, 1972). Therefore, the area near Hlaing River is sensitive to saline water intrusion if overabstraction occurs. According to the same study, the annual fluctuation in the groundwater table for the whole Yangon City ranged from 1.5 m to 5 m in 1972. By conducting open discussions with some of the experienced tubewell drillers, it seems that the annual fluctuation in the area is currently 10 m. In 1972, the complete replenishment of groundwater occurred during the months of June to October (Win Naing, 1972). However, today's situation of overabstraction casts doubt on the complete replenishment of withdrawals.

Most groundwater exploitation occurs in fair-to-good potential zones in the aquifer. Domestic and industrial users are mostly located in here. The accelerated increased abstractions gradually lower the water level. Some abandoned tubewells are also situated in this area. On the basis of abstraction activities, the amount of groundwater level lowering and decreasing capacity is significant, especially in the dry season. In the dry period, all users need more water and their abstraction increases during the same period, so the depression cone at all abstraction sites is unpredictably lower. Therefore, the abstraction time to satisfy the usual amount required takes much longer in the dry period than in the wet, due to the significant lowering of water levels, which causes a reduction in both production capacity and rate.

14.6.2 Land Subsidence

There are no records concerning land subsidence in Yangon City due to the lack of research on the topic so far. If the current situation of overexploitation continues for a long period, land subsidence is likely to occur in the area. A timely initiative is required to monitor land subsidence in order to ensure any potential impact is minimal.

14.7 RESPONSES

Fortunately, Yangon City has good access to large reservoirs and an annual rainfall of 100 inches. Existing reservoirs are adequate to meet domestic water demand, provided that the pipe networks and/or distribution systems are efficient. However, due to inefficient or leaky pipe networks, the amount of supply that reaches domestic users is inadequate and unreliable, and therefore, they rely heavily on groundwater resources. More than one half of water consumption city-wide comes from groundwater. Serious public health issues concerning groundwater pollution have not yet occurred in the groundwater history of Yangon City. However, some areas with "good-to-fair" and "fair-to-poor" potential groundwater zones are experiencing deterioration in groundwater quality even though existing tubewells are quite deep, in a range of 400–750 ft. Electrical conductivity (EC) in some places within the Dagon Myothit and Hlaingtharya townships is higher. There are cases of tubewells being abandoned after about 5 years of operation due to saline water intrusion. Drilling companies/technicians have experienced the need to drill deeper than before to abstract groundwater. Furthermore, there is no space to drill more tubewells in downtown areas for two reasons: (i) the area is near to the Yangon River which can cause brackish water and salt water intrusion into the aquifer; and (ii) high population density and the high building density can cause land subsidence if more groundwater is abstracted. Some townships in inner suburban areas have responded by specifying limits when drilling tubewells.

Seasonal lowering of the groundwater table has been experienced in recent decades. However, in the absence of a groundwater-monitoring system, exact quantitative data are not available. Such a system is essential to monitor the groundwater in different aquifer layers beneath the city. It is likely that the establishment of a single institution to monitor, manage, and regulate groundwater resources, will result in the initiation of a well-equipped groundwater-monitoring program in the near future.

14.8 CONCLUSIONS AND RECOMMENDATIONS

Like other urban areas, the water supply in Yangon City is significantly reliant on groundwater. Even though the YCDC, as the state-owned water supply authority, has increased the supply of piped water from MCM 157.6 to 265 MCM during 2003–2013, this still falls short of the total water demand of the city. Various drivers – population growth, urbanization, and tourism – are driving toward a significant withdrawal of groundwater. Total abstraction exceeds total recharge annually, resulting in groundwater mining in the city. Transformation of agricultural land into built-up areas is resulting in the decrease of rechargeable areas from 72% to 66% during 2002–2012, and then to 63% in 2013. The pressures on the groundwater environment due to various drivers and modification of the state of the groundwater environment can be measured by the increasing number of tubewells and the lowering of the groundwater table. Any significant responses for maintaining groundwater levels, identifying and conserving groundwater recharge areas, specifying groundwater abstraction rates, in particular areas based on groundwater potential, etc., are yet to be put in place. Groundwater monitoring should be started as soon as possible since it generates the fundamental information necessary to design and implement management strategies. Some awareness-raising programs aimed at promoting a better understanding of the situation in Yangon City's aquifers may help the authorities concerned to put "groundwater conservation and management" at the top of their agenda.

REFERENCES

Ahleinmar Htwe, 2000. Environmental potential maps of Yangon area. Unpublished MSc thesis, Department of Geology, Yangon Technological University.

Banerji, S.K., 1924. A list of 139 tubewells situated between Insein and Syriam. Unpublished report, Government of Burma.

Foy, E.G., 1903. Report on tube-wells in Rangoon. Unpublished report, Government of Burma.

Coggin-Brown, J., 1926. The study of underground water supply of Rangoon. General report for 1926, Geological Survey of India.

Cho Win Kyaw, 2012. Geographical analysis on production, distribution and consumption of purified drinking water in Yangon city. Unpublished PhD dissertation, submitted to Department of Geography, University of Yangon, Ministry of Education.

Khin Kay Khaing, 2006. Assessment of groundwater resource and sustainable use in Yangon City. Unpublished PhD dissertation, submitted to Department of Geography, University of Yangon, Ministry of Education.

Khin Kay Khaing, 2011. Groundwater utilization and availability in Yangon City. Universities Research Journal, vol. 4, No. 5, Department of Higher Education (Lower Myanmar) and Department of Higher Education (Upper Myanmar), Government of the Union of Myanmar, Ministry of Education.

Khine Moe Nyunt, 2013. Interaction of land and man in Yangon City. Unpublished research paper, Zoning & Land Use Sector, Urban Planning Division, City Planning & Land Administration Department, YCDC.

Leicester, P., 1959. Geology and underground water of Rangoon (with special reference to tubewells). Rangoon, Union of Burma.

Malcolm Pirnie Engineers, 1958. Report on water supply and transmission for Rangoon, Burma.

Maung Maung, 1996. Urban geology of western Yangon area. Unpublished MSc thesis, Department of Geology, University of Yangon.

Oldham, R.D., 1893. On the alluvial deposits and subterranean water supply of Rangoon. Record of Geological Survey of India, XXVI, 64–70.

Tin Myo Win, 1995. Geographical study of water supply in Greater Yangon. Unpublished MA thesis, Department of Geography, University of Yangon.

Win Naing, 1972. The hydrogeology of Greater Rangoon. Unpublished MSc thesis, Department of Geology, Arts and Science University, Rangoon.

Win Naing, 1996. Urban geology of eastern Yangon. Unpublished MSc thesis, Department of Geology, University of Yangon.

Win Naing, 2009. Evaluation of Irriwaddian aquifer in Yangon area. Unpublished PhD dissertation, submitted to Department of Geology, University of Yangon, Ministry of Education.

YCDC, 2001. The development and management of water supply system of Yangon City. Unpublished interim report submitted to Yangon City Development Committee by JICA study term.

YCDC, 2014. Strategic urban development plan of the Grater Yangon. Presentations at Expert meeting of City Planning & Land Administration Department, Yangon City Development Committee.

Zin New Myint, 2011. Processes and patterns of urban development of Yangon City. Unpublished departmental research paper, Department of Geography, University of Yangon.

SECTION IV

Groundwater Environment in Central and East Asia

CHAPTER 15

Water Environment in Central and East Asia: An Introduction

Sangam Shrestha*, Vishnu Prasad Pandey*,**,
and Binaya Raj Shivakoti†
*Water Engineering and Management, Asian Institute of Technology (AIT), Thailand
**Department of Civil Engineering, Asian Institute of Technology and Management (AITM), Nepal
†Water Resources Management Team, Natural Resources and Ecosystem Area, Institute for Global
Environmental Strategies (IGES), Japan

15.1 PHYSIOGRAPHY AND CLIMATE

The Central and East Asian subregion consists of 10 countries covering an area of 15.45 million km^2 (WB, 2015), within a geographical boundary of 18°N to 56°N and 46°E to 146°E (Figure 15.1). It is an extremely large subregion of varied geography, including high-mountain passes, vast deserts, and treeless grassy steppes. The subregion stretches from the border with Russia in the north to the High Himalayas in the south, and from the Caspian Sea in the west to the East China Sea and Pacific Ocean in the east. Central Asia was the crossroads of the Great Silk Road, which saw the intermingling of cultures over hundreds of years, and it was also a cradle of civilization, leaving achievements in arts and architecture, science, philosophy, and literature for centuries to come. Despite the many-shared cultural, historical, and environmental features among the countries of the subregion, it is characterized by great contrasts in climate and topography. There are large variations among the subregional countries in terms of geographical area, climate, population size and density, and per capita income (Table 15.1).

The climate varies from temperate with colder winters and warm summers in East Asia, to moderate in most parts of Central Asia. Air temperature fluctuations in Central Asia are more severe because it is not buffered by a large body of water. In the winter, the cold, dry heavy air over Central Asia flows outward toward the sea; in the summer months, the warm air over the Central Asian landmass rises and cools, and moist air from the ocean flows back bringing rainfall over

Groundwater Environment in Asian Cities
http://dx.doi.org/10.1016/B978-0-12-803166-7.00015-5

Copyright © 2016 Elsevier Inc.
All rights reserved.

Figure 15.1 *Location of Countries and Case Study Cities in Central and East Asia.*

Table 15.1 Annual water withdrawals by sector and sources in Central and East Asian countries

	Annual water withdrawal				With-drawal as % of TRWR	Source of fresh water	
	Agriculture sector		Total withdrawal			Groundwater	
Country	Volume (BCM)	% of total	Volume (BCM)	m³/ person		Volume (BCM/y)	% of total with-drawal
China	358.0	64.6	554.1	406	19.5	101.40	1.8
Japan	54.6	60.7	90.0	713	20.9	9.37	1.8
Kazakhstan	14.0	66.2	21.1	1299	18.4	1.03	0.5
Kyrgyzstan	7.1	88.7	8.0	1560	32.6	0.31	0.4
Mongolia	0.2	43.9	0.6	197	1.6	0.44	8.0
North Korea	6.6	76.3	8.7	359	11.2	NA	NA
South Korea	16.0	62.7	25.5	549	36.5	3.72	NA
Tajikistan	10.4	90.9	11.5	1616	51.1	2.26	2.0
Turkmenistan	26.4	94.3	28.0	5753	NA	0.31	0.1
Uzbekistan	50.4	90.0	56.0	2100	NA	5.00	0.9
Central and East Asia	543.7	67.7	803.4			123.84	

BCM, billion cubic meters; NA, not available.
Source: FAO (2015).

the land. This phenomenon brings most of the rainfall during the warm summer months, especially in the Central and Eastern parts of the sub-region. The subregion receives a sizable amount of rainfall, which varies from 78.6 billion cubic meters (BCM) per year in Turkmenistan to 6192 BCM/y in China (Table 15.2).

Table 15.2 Water resource availability in Central and East Asian Countries in 2012

Country	P (BCM/y)	TRWR BCM/y	TRWR m³/person/y	TRGWR (BCM/y)
China	6192.0	2840.0	2005	828.8
Japan	630.4	430.0	3382	27.0
Kazakhstan	681.2	108.4	6593	33.9
Kyrgyzstan	106.6	23.6	4257	13.7
Mongolia	377.0	34.8	12,258	6.1
North Korea	127.0	77.2	3099	13.0
South Korea	127.6	69.7	1415	13.3
Tajikistan	98.5	21.9	2669	6.0
Turkmenistan	78.6	24.8	4727	0.4
Uzbekistan	92.2	48.9	1689	8.8
Central and East Asia	8511	3679.2		951.0

BCM, billion cubic meters; P, precipitation; TRWR, total renewable water resources; TRGWR, total renewable groundwater resources.
Source: FAO (2015).

15.2 SOCIOECONOMIC AND ENVIRONMENTAL ISSUES

This subregion is home to some 1.68 billion (in 2012) people. The population varies among countries, from 2.8 million in Mongolia to 1,408 million in China. Given varying geographical areas covered by the countries, population densities also differ, with the lowest being 2 persons/km² in Mongolia, and the highest being 489 persons/km² in South Korea.

Economic development measured in terms of GDP in 2012 reveals a distinct variation among the subregional countries; from 963 USD/capita (in Tajikistan) to 46,680 USD/capita (in Japan). The agricultural sector contributes only 6.1% to GDP in the subregion, but this figure varies from 1.2% in Japan and 2.5% in South Korea, to 26.6% in Tajikistan (FAO, 2015).

Ever-growing population, urbanization, industrialization, and climate variation/change are creating a range of environmental issues that are increasing the vulnerability of water resources and having subsequent impacts on the socioeconomic development of the subregion. Major water and environmental issues in the subregion associated with population, are a rising demand for water and other natural resources, water and soil pollution, depletion of groundwater resources, and aquifer contamination, among others.

15.3　WATER AVAILABILITY AND WITHDRAWAL

Major rivers in this subregion include the Yangtze, Yellow, Amu Darya, and Syr Darya, and notable lakes or water bodies are the Aral Sea, Balkhash Lake, and Issyk-Kul Lake. There are large variations in average annual precipitation and renewable water resources between the subregional countries (Table 15.2). Total precipitation of 8511 BCM/y is distributed unevenly among the subregional countries; from 92.2 BCM/y in Uzbekistan to 6192 BCM/y in China. In terms of the volume of total renewable water resources, China has the highest and Tajikistan has the lowest (Table 15.2). In terms of per capita water resources available, Mongolia is the richest country with 12,258 m^3/y of water resources and South Korea is poorest with only 1415 m^3/y. Groundwater contributes some 951 BCM/y or more than one fourth of total renewable water resources in the entire subregion (Table 15.2) and varies between 0.4 BCM/y in Turkmenistan and 828.8 BCM/y in Indonesia (Table 15.2).

Total freshwater withdrawal in the subregion is 803.4 BCM/y, which varies between 0.6 BCM/y in Mongolia and 554.1 BCM/y in China. The figure in per capita terms shows the highest withdrawal of 5753 m^3/person/y in Turkmenistan, and the lowest of 197 m^3/person/y in Mongolia (Table 15.1). The withdrawal as a percentage of total renewable water resources varies from 1.6% in Mongolia to 36.5% in South Korea. Water withdrawal for agricultural use is 67.7% of the total annual water withdrawal, which itself varies from 43.9% in Mongolia to over 90% in Turkmenistan, Tajikistan, and Uzbekistan (Table 15.1). Out of the total freshwater withdrawal, groundwater contributes 8%, the highest among countries whose data are available, in Mongolia. In all the other countries, the contribution of groundwater to the total withdrawal figure is less than 2%. However, in most of the urban centers or cities, groundwater constitutes a major source of water supply, necessary to meet both domestic and industrial demands (e.g., Chapters 16–19 in this book), and therefore, groundwater is an integral part of human health and socioeconomic development.

15.4　CASE STUDY CITIES

Groundwater is a key resource in the process of socioeconomic development, as it constitutes above one fourth of the total renewable water resources in the entire subregion. Due to a lack of understanding and previous poor governance, environmental issues relating to groundwater such as depletion of groundwater resources and well yield (e.g., Bishkek and Seoul), aquifer degradation, and land subsidence (e.g., in Seoul and Tokyo) are experienced in many aquifers that supply water to cities. This part of the book analyzes the status of the groundwater environment in four cities selected at random, based on the availability of contributors, but with the intention of encompassing as much diversity as possible. The cities in alphabetical order are Beijing (China), Bishkek (Kyrgyzstan), Seoul (South Korea), and Tokyo (Japan). They are detailed in Chapters 16–19. General characteristics of the cities are provided in Table 1.1.

REFERENCES

FAO, 2015. AQUASTAT database. Food and Agriculture Organization of the United Nations (FAO). Available from: http://www.fao.org/nr/water/aquastat/data/query/index.html?lang=en (accessed 19.06.2015.).

WB, 2015. The World Development Indicators. World Bank (WB). Available from: http://data.worldbank.org/indicator (accessed 14.04.2015.).

CHAPTER 16

Groundwater Environment in Beijing, China

Jingli Shao*, Jiurong Liu**, Qiulan Zhang†, Rong Wang‡, Zhiping Li**, Liya Wang**, and Qing Yang**
*School of Water Resources and Environment, China University of Geosciences, China
**Beijing Institute of Hydrogeology and Engineering Geology, China
†School of Water Resources and Environment, China University of Geosciences (Beijing)
‡Land Subsidence Research Institute of Beijing Institute of Hydrogeology and Engineering Geology, China

16.1 INTRODUCTION

As its capital, Beijing City with an area of 16,410.54 km², is the center of politics, culture, and economics of the People's Republic of China. The total volume of available water resources in the city is approximately 3800 million cubic meters (MCM/y). Groundwater constitutes 68% of the total water resources in the area. Water demand in the city has redoubled in nearly a half century, as a result of the rapidly increasing population and high-speed economic growth. In addition, Beijing is located in a semiarid and semihumid area, which is not rich in water resources. The mean annual available water resources per capita in Beijing has decreased recently to less than 200 m³/y (Beijing Water Authority, 2003), and this has been influenced by factors such as decades of successive aridness, among others.

Beijing has become one of the most water-scarce cities in the world, and water resources have become a major bottleneck to the sustainable development of its society and economy. Moreover, it is one of the cities in the world where groundwater accounts for over 70% of its total water supply. Learning from the continuous aridness and severe deficiency of surface water in the past, the city established a number of emergency water wells in 1999 in order to supplement the municipal water supply as and when required. This arrangement has helped Beijing to address the water scarcity crisis several times during the period from 1999 to 2014.

Since the 1970s, continuing overexploitation of groundwater and a lack of oversight in its protection have resulted in a series of groundwater environmental issues such as the depletion of the groundwater table, continuous expansion of depression cones of groundwater, and dwindling storage of

Groundwater Environment in Asian Cities
http://dx.doi.org/10.1016/B978-0-12-803166-7.00016-7
Copyright © 2016 Elsevier Inc.
All rights reserved.

groundwater year by year. The deficiency of water resources and related environmental issues, together with groundwater overexploitation and associated impacts, has constrained rapid socioeconomic development. Since the 1950s, various works such as groundwater surveys and monitoring, artificial recharge, utilisation of rainfall and floodwater, water diversion from extra watersheds, and the exploitation of unconventional water resources, have been extensively carried out to understand the groundwater resources and aquifer conditions of Beijing, and ultimately to achieve the sustainable utilization of groundwater resources. Based on these works, groundwater management systems and regulations have been established to fulfill the goal of a water supply for societal development.

This chapter follows the driver–pressure–state–impact–response (DPSIR) framework approach to analyze the status of Beijing's groundwater environment. It provides comprehensive documentation of aquifer conditions, groundwater resources, historical perspectives, and the current situation of groundwater development and management in the city. Table 16.1 provides a summary of DPSIR indicators for different time periods.

16.2 ABOUT THE CITY

16.2.1 Physiography and Climate

With the topography high in the northwest and low in the southeast, Beijing City is located in the transitional zone of the Taihang and Yanshan Mountains. The plain of the city lies within 115°25′–117°30′E and 39°28′–41°05′N. Taihang Mountain is to the west, with Yanshan Mountain to the north. The city lies in the alluvial plain with a southeast (SE) dip, covering a total area of 16,410 km² (Figure 16.1). The mountains cover an area of 9882 km², whereas the plain area covers 6528 km². Of the total plain area, the Beijing Plain and the Yanqing Basin cover 6400 km² and 128 km², respectively.

The climate in Beijing can be characterized as typically warm temperate, semihumid, and semiarid continental monsoon. The interannual precipitation varies greatly (Figure 16.2); the maximum recorded is 1116 mm, observed in 1956, and the minimum 267 mm in 1997. Precipitation in normal years varies from 300 mm to 800 mm. The long-term average precipitation was 622 mm between 1841 and 2005. There has been less precipitation in recent decades as evidenced by an average decrease to 533 mm for the period from 1990 to 2005. Some 80% of annual precipitation in the city is received between June and September.

Table 16.1 Available groundwater resources in Beijing (MCM/y). Mi-Huai-Shun is indicated by Miyun county, Huairou District, and Tongzhou District

Area	Town	Tongzhou	Daxing	Changping	Fangshan	Mi-Huai-Shun	Pinggu	Yanqing	Total
Plain	605	210	260	220	290	580	200	90	2455
Mountain Area	10	0	0	15	52	61	30	10	178
Total	614	210	260	235	342	641	230	100	2633

MCM, million cubic meters.

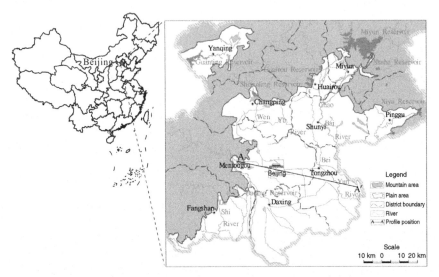

Figure 16.1 *Geographical Location and Distribution of Rivers in Beijing, China.*

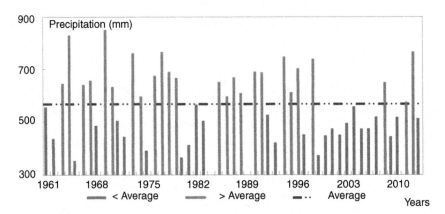

Figure 16.2 *Annual Precipitation Trends in Beijing from 1961 to 2013.*

Beijing belongs to the Haihe watershed, where the primary rivers (Figure 16.1) are the Chaobaihe River, Yongdinghe River, Jumahe River, Wenyuhe River, Juhe River, and Cuohe River. Upstream of the rivers, there are 85 reservoirs of varying capacity with a total reservoir capacity of 9200 MCM, and other relatively larger reservoirs such as Miyun, Guanting, and Huairou.

16.2.2 Hydrogeology

Groundwater in Beijing City is mainly contained in the porous aquifers of the plain and alluvial–diluvial fan, and partly in karst and bedrock fissures in the mountain area. Additionally, there is concealed karst water distributed sporadically in the plain. Based on the flowing and recycling characteristics of regional aquifer systems, the groundwater can be classified into three groundwater systems (Figure 16.3): (i) the Chaobaihe–Jiyun–Wenyuhe groundwater system; (ii) the Yongdinghe groundwater system; and (iii) the

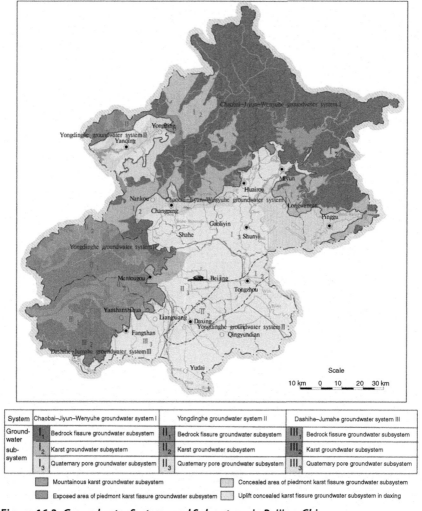

System	Chaobai–Jiyun–Wenyuhe groundwater system I		Yongdinghe groundwater system II		Dashihe–Jumahe groundwater system III	
Ground-water sub-system	I₁	Bedrock fissure groundwater subsystem	II₁	Bedrock fissure groundwater subsystem	III₁	Bedrock fissure groundwater subsystem
	I₂	Karst groundwater subsystem	II₂	Karst groundwater subsystem	III₂	Karst groundwater subsystem
	I₃	Quatemary pore groundwater subsystem	II₃	Quatemary pore groundwater subsystem	III₃	Quatemary pore groundwater subsystem

▓	Mountainous karst groundwater subsystem	▒	Concealed area of piedmont karst fissure groundwater subsystem
▓	Exposed area of piedmont karst fissure groundwater subsystem	☐	Uplift concealed karst fissure groundwater subsystem in daxing

Figure 16.3 *Groundwater Systems and Subsystems in Beijing, China.*

Dashihe–Jumahe groundwater system. Considering the type of aqueous medium, the systems above can be further classified into three subsystems, namely, fissure water, karst water, and pore water.

Porous aquifers of an unconfined nature can further be divided into six relatively independent hydrogeological units (Figure 16.4). The single unconfined aquifer is formed by sand and gravel in the upper part of the alluvial fan. Although the buried depth of groundwater is deep, the aquifer is rich in water. The sediments gradually become finer in the middle and lower parts of the alluvial–diluvial fan, where the layers of sand are

Secondary system	System code		Secondary system	System code	
Chaobai river groundwater subsystem	I_{3-1-1}	I_{3-1-2}	Yongding river groundwater subsystem	II_{3-1-2}	II_{3-1-2}
Wenyu river alluvial fan groundwater subsystem	I_{3-2}		Yongding river upstream intermontane basin groundwater subsystem	II_{3-2}	
Jiyun river(Ju River, Cuo River) alluvial fan groundwater subsystem	I_{3-3}		Dashi river–Juma river alluvial fan groundwater subsystem	II_{3}	

Figure 16.4 *Hydrogeological Units in the Uppermost Layer of an Unconsolidated Porous Groundwater Aquifer in the Plain of Beijing, China.*

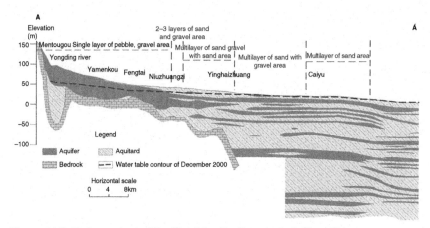

Figure 16.5 *Hydrogeological Profile of Aquifer Systems in Beijing, China.*

interbedded with clay, thus forming multiaquifers. The plain depth of the unconfined water is shallow and there are one or more layers of confined water below (Figure 16.5).

Under natural conditions, the main recharge modes of groundwater in the plain are infiltration by rainfall, rivers, and lateral infiltration of the mountains. The main discharge methods are evaporation and flowing out to the river and downstream. However, human activities can change the recycling conditions of groundwater, and exploitation of groundwater has become the main discharge method.

16.2.3 Groundwater Resources

In accordance with the newspaper article "Evaluation of Groundwater Resource and Environment Survey in the Capital Area", the quantity of multiannual average groundwater recharge in Beijing City was 3598 MCM/y from 1980 to 2000; allowable pumping is 2633 MCM/y (Table 16.1) and the status of groundwater pumping or abstraction in the city is 3027 MCM/y.

16.3 DRIVERS

The groundwater deficiency of Beijing City and a series of issues on the groundwater environment are the result of various factors, among which the major drivers are population growth, economic development, and urbanization. Table 16.2 provides the status of DPSIR indicators in Beijing, China.

Table 16.2 Summary of DPSIR indicators in Beijing, China

	Indicators	1980	2000	2013
Drivers	Population growth (within the district; area: 6528 km²)	Population: 9.12 million Population density: 1397 (persons/km²) Annual growth rate	13.64 million 2,089 2.01% (2000–1981)	21.15 million 3240 3.37% (2013–2000)
	Urbanization	Urban area (in 1990): 1470 km² or 8.92% Urban population (in 1949): <2 million	(in 2000): 2246 km² or 13.63% >10.23 million	(in 2012): 3440 km² or 20.88% >18.25 million
	Industrialization and economic development:	No. of registered industries: N/A	1399	3007
Pressures	Depletion in surface water resources	Ravi inflows (in 1984): 1076.70 MCM	In 2000: 1006.28 MCM	In 2005: 465.00 MCM
		Surface water allocated since inception of irrigation system to agricultural lands in periurban areas is no longer available. Due to land use change to urbanization, there is consistent reduction in recharge to groundwater. Additionally, surface water supply in Beijing mainly depends on water sources from Miyun Reservoir, Guanting Reservoir, and the river way. The water from Miyun Reservoir is the main source for municipal usage, while the water from Guanting Reservoir is utilized in industry and agriculture because of its low quality. The water storage of the two reservoirs both declined, due to the yearly increasing excessive volume of their water supply and the too much controlled runoff by the upstream.		
	Decline in groundwater recharge: land cover change	Cultivated area (in 1990): 5852 km²	(in 2000): 4915 km²	(in 2010): 3579 km²

State	Groundwater overexploitation	Groundwater abstraction from Beijing's aquifers is 3027 MCM/y, which is higher than estimated recharge of 2766 MCM/y.
	Well statistics	Numbered 17 before 1950, reached 51,700 in 2000, and no recent data are available.
	Groundwater recharge	Groundwater recharge is decreasing because of decline in river runoff due to water intake, damming, and storage in upstream reservoirs. Between 1989 and 2000, the annual groundwater recharge in the city's aquifer was estimated at 2766 MCM, which is 10% lower than groundwater abstraction.
	Groundwater abstraction	Abstraction (in 1950s): 400 MCM/y	In 2000: 2704 MCM/y	After 2007: 2400 MCM/y
	Groundwater level (mbgl)	In 1960s: average 5.0	In 1900s: average 16	In 2014: 24.5
	Groundwater quality	The quality of fracture and karst water in the mountain area of Beijing is ordinary, with total dissolved solids (TDS) of below 500 mg/L and hydrochemical type of HCO_3–Ca. It is nearly the same as that in the middle and upper part of different alluvial and diluvial fans, and the hydrochemical type of unconfined and confined groundwater gradually turns into HCO_3–Na·Ca and the TDS increases from 500 mg/L to 1000 mg/L at the edge of alluvial and diluvial fans and the alluvial plain. At present, due to increasing human activities, the hardness and TDS of the unconfined aquifer are beyond the third grade of groundwater quality standard, suffering from the infiltration of domestic refuse, leakage of sewage and contamination from point and nonpoint pollution.

(Continued)

Table 16.2 Summary of DPSIR indicators in Beijing, China *(cont.)*

	Indicators	1980	2000	2013
Impacts	Decline in groundwater level	Before 1980s: 0.3 m/y	1980–1990: 0.5 m/y	2000–2011: 2.0 m/y
	Depletion in groundwater storage	During 1961–2013, cumulative loss in groundwater storage from Quaternary aquifers was 10,177 MCM and the average loss was 192 MCM/y. However, after the late 1990s (or from 1999 to 2013), due to rapid urbanization and continuing aridity, groundwater was extremely overexploited, which further depleted groundwater storage.		
	Deterioration in groundwater quality	Deterioration of groundwater quality is being constantly faced; the exploited areas of groundwater gradually expand toward the suburbs. The upper layer of groundwater is being rapidly polluted with disposal of seriously contaminated surface water as well as leakage from the sewerage system. Comparison of the total hardness and nitrate–nitrogen in groundwater samples with previous studies shows that the quality of groundwater is deteriorating over time; and TDS increases as well after 1980.		
	Land subsidence and formation of ground fissures	Land subsidence in the Plain of Beijing is induced primarily by groundwater overexploitation. It can be divided into four stages: (i) forming period (1955–1973); (ii) developing period (1974–1983); (iii) expanding or spreading period (1984–1988); (iv) rapid development period (1999–2015). The total area with land subsidence was over 3900 km² in 2013 and the cumulative land subsidence was above 100 mm–3.2 times deeper than that in 1999.		
	Degradation of wetlands	Over 2000 km² (or 15% of the city's area) of wetland since the historical time reduced to around 800 km² in the early 1980s and 500 km² (or less than 3% of the city's area) in 2013. Groundwater overexploitation and subsequent enlargement in depression cone and desertification of underground are some of the reasons behind degradation of the wetlands.		

Responses	Groundwater explorations and studies	Several studies have been carried out since the 1950s to explore possibilities of groundwater development as well as ways of protecting and wisely utilizing the resource from the Beijing's aquifers.
	GW monitoring	Groundwater monitoring of Beijing aquifers started in the middle of 1950s and strengthened gradually over the time. In case of Beijing's aquifers, it includes monitoring of groundwater table, groundwater quality, land subsidence, and ground fissures. Total monitoring wells until end of 2012 were 635, among which 335 are equipped with automatic instruments. Monitoring wells are distributed in all the aquifer layers.
	Legal and regulatory re-forms/ interventions	Various laws have provisions for the development, sustainable utilization, and conservation of groundwater resources and environment. A series of policies, ordinances, and guidelines are also issued by relevant ministries and Beijing City authorities for the purpose of protecting groundwater resources (see Section 16.7.3).
	Initiating artificial recharge	Numbers of artificial recharge tests using various methods have been conducted in the alluvial fans since the need was felt in the 1960s. They include the tests in the early 1970s, 1978, 1979, 1983, 1995, and 1996. They have helped recover groundwater tables. For example, recharge tests in the early 1980s through the bed of Yongding River groundwater table along the bank of the river rose by more than 3 m and quality of groundwater were observed as improved.
	Increasing surface water availability	A variety of measures have been taken to arrange the surface, such as the utilization of rainfall flood, the SNWDP to divert water from adjoining areas. Until 2009, Beijing has built 51 rainfall flood utilization projects using different comprehensive operation modes. The middle line of the SNWDP is a significant strategic measure at national level to relieve the shortage of water resource in northern droughty region, especially in Beijing.

DPSIR, driver–pressure–state–impact–response; GW, groundwater; mbgl, meters below ground level; MCM, million cubic meters; SNWDP, South–North Water Transfer Project; TDS, total dissolved solids.

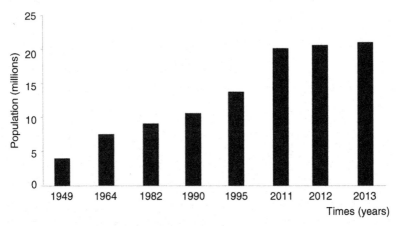

Figure 16.6 *Temporal Trend of the Population in Beijing, China.*

16.3.1 Population Growth

The population of Beijing City has been increasing since the 1950s (Figure 16.6).The resident population has increased from nearly 13.64 million in 2000 to 21.15 million in 2013; more than five times that of the 1950s when the population remained at approximately 4 million. The response to population growth is an overexploitation of groundwater. In recent years, groundwater has become almost the sole source of water supply for cities and towns (Yang et al., 2013). Districts such as Miyun, Huairou, Pinggu, Shunyi, Changping, and Fangshan in Beijing supply water, not only for local areas, but also for central water demand. All the aforementioned factors are fuelling the overexploitation of groundwater resources as a result of population growth. The native water sources, in areas supplied primarily by local groundwater, have been depleted. It puts the issue of water adequacy in the city at the forefront, and leads to the continuous depletion of groundwater.

16.3.2 Industrialization and Economic Development

With rapid economic development due to industrialization, the requirements for water resources are increasing year by year. Groundwater has remained the only source of water supply to support economic development, due to the shortage of surface water. Thus, rapid socioeconomic growth is one of the drivers putting stress on groundwater resources.Water consumption by agriculture, industry, and the municipality has manifestly increased with the expeditious growth of GDP in Beijing (Figure 16.7). Moreover, industrial development is a constant drain on water resources.Agriculture consumes the largest amount of water.Water-intensive service industries are also extensively

GDP (× 10⁹RMB)

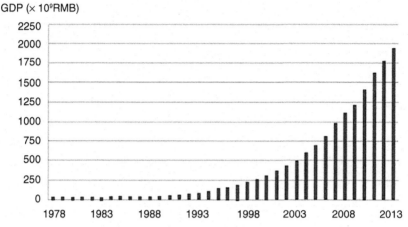

Figure 16.7 *Growth of GDP in Beijing from 1978 to 2013.*

developing. These include entertainment sites high in water consumption such as bathing centres, golf courses, ski slopes, and vehicle cleaning points. The rapid growth of various industries contributing to the national economy is aggravating the pressure on Beijing's groundwater environment.

16.3.3 Urbanization

Beijing City has over 3000 years of history. However, until 1949 there was only 40% urbanization. After 1949, as the capital of China, the city urbanised rapidly. By 2000, the rate of urbanization increased to over 75%; at least three quarters of the population of Beijing has primary jobs in areas other than agriculture. Urbanization in Beijing is affecting groundwater in the following ways:

- Increased industrial activities demand high quantities of water.
- The urban lifestyle is more water intensive than its rural counterpart. Per capita water demand in rural areas varies from 30 L/d to 70 L/d, which is far less than that of urban demand, which ranges from 90 L/d to 150 L/d. As an indication of the water–intensive urban lifestyle, Figure 16.8 depicts an increase in the annual volume of tap water sales and water supplies in Beijing.
- Urbanization changes agricultural areas, with the potential to recharge groundwater, into built-up areas as shown in Table 16.3 and therefore reduces the rate of recharge to groundwater.
- The accumulation of waste and discharge of wastewater also accelerates, infiltrating down into the aquifer, thus contaminating groundwater.

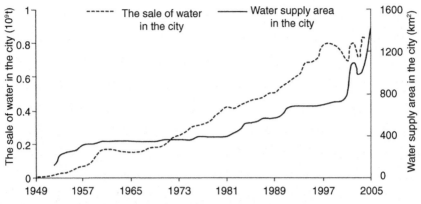

Figure 16.8 *Amounts and Areas of Tap Water Supply in Beijing, China.*

Table 16.3 Land cover change in Beijing from 1990 to 2012

Land cover type	1990 km²	%	2000 km²	%	2012 km²	%
Agriculture	5852.70	35.52	4915.21	29.82	3579.73	21.72
Forest	7278.95	44.17	7400.02	44.89	7396.33	44.88
Lawn	1361.41	8.26	1293.03	7.84	854.91	5.19
Territorial waters	399.60	2.43	513.84	3.12	790.88	4.80
Urban area	1470.00	8.92	2246.74	13.63	3440.86	20.88
Unused land	115.00	0.70	115.00	0.70	417.27	2.53
Total	16,477.66	100.00	16,483.84	100.00	16,479.98	100.00

16.4 PRESSURES

Pressure on the groundwater environment in Beijing City is exerted by depletion of surface water resources, groundwater overexploitation, and a decline in groundwater recharge.

16.4.1 Depletion in Surface Water Resources

There are 85 reservoirs upstream of the rivers in Beijing City. Most surface runoff has been controlled upstream by damming those reservoirs for water supply, and therefore runoff in the rivers passing through Beijing has obviously decreased (Figure 16.9). Rivers in the Beijing Plain are now almost dry, except in the high flow season.

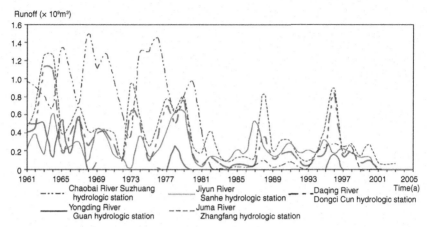

Figure 16.9 *Flow Process Lines of Exit Hydrological Stations at the Main Rivers of Beijing from 1961 to 2003.*

Surface water supply in Beijing mainly depends on water sources from the Miyun Reservoir, Guanting Reservoir, and the river way, where the amount received from the Miyun and Guanting Reservoirs accounts for 90%. The water from the Miyun Reservoir is the main source for municipal usage, while the water from the Guanting Reservoir is utilized in industry and agriculture because of its low quality. In 2000, the volume of surface water supply was 1340 MCM. From 1999 to 2005, the water inflow to the Miyun and Guanting Reservoirs was 258 MCM/y and 90 MCM/y, respectively. The volume of water storage in the Miyun Reservoir declined from 1540 MCM/y at the beginning of 2001 to 1036 MCM/y at the end of 2005. During the same time, water storage in the Guanting Reservoir declined from 420 MCM/y to 163 MCM/y. As shown in Table 16.4, the surface water volume supplied by the Miyun and Guanting Reservoirs in Beijing City declined after 2001 because the following years to 2010 were relatively dry due to relatively lower rainfall.

16.4.2 Groundwater Overexploitation

In the late 1950s, the amount of groundwater abstraction in Beijing was 400 MCM/y. Abstraction increased gradually in the 1960s with the fast development of suburbs, and reached 520 MCM in 1961 and 1211 MCM in 1965. The average abstraction between 1961 and 1970 was 1079 MCM. In the 1970s, abstraction increased more than twofold mainly because of urbanization and economic development activities. In the 1980s, abstraction increased further, exceeding the allowable amount of 2455 MCM/y

Table 16.4 Time trend of water supply from the Guanting and Minyun reservoirs (MCM)

Year	Minyun Reservoir	Guanting Reservoir	Total	Year	Minyun Reservoir	Guanting Reservoir	Total
1984	586.98	489.72	1076.70	1996	573.18	405.39	978.57
1985	315.78	358.46	674.24	1997	614.85	479.61	1094.46
1986	466.96	284.38	751.34	1998	645.67	328.52	974.19
1987	359.34	315.26	674.60	1999	820.31	335.04	1155.35
1988	408.35	299.86	708.21	2000	710.38	295.90	1006.28
1989	316.29	331.11	647.40	2001	496.50	227.10	723.60
1990	628.04	216.91	844.95	2002	542.20	144.60	686.80
1991	465.48	323.06	788.54	2003	501.50	176.20	677.70
1992	630.40	337.02	967.42	2004	289.80	154.90	444.70
1993	625.13	448.38	1073.51	2005	271.00	194.00	465.00
1994	673.87	231.47	905.34	Total	11,578.54	6643.77	18,222.31
1995	636.53	266.88	903.41	Average	526.30	301.99	828.29

MCM, million cubic meters.
Source: Ren et al. (2007).

Table 16.5 Decadal accounts of groundwater abstraction in Beijing

Years	1950s	1960s	1971	1978	1980s–1990s	2000
Abstraction volume (MCM/y)	400	1,079	1,380	2,559	2,600–2,800	2,704

(Beijing Groundwater, 2008), and concentrated in the plain area of Beijing. The decadal account for groundwater abstraction in Table 16.5 clearly shows the overexploitation of Beijing's aquifer since 1978.

To cater for growing water demand in the context of sharply depleting surface sources and subsequent deficiency in supply, five emergency groundwater fields were built between 2003 and 2006 to increase the total water supply capacity of 295.05 MCM/y. This further aggravated groundwater overexploitation in Beijing. Since 2007, abstraction has been gradually controlled to keep it at approximately 2400 MCM/y.

16.4.3 Decline in Groundwater Recharge

With urbanization and subsequent land cover changes, natural recharge areas are largely affected. This is also true for Beijing. The major reasons for declining recharge into Beijing's aquifers can be listed as follows:

• Due to the damming/storing of reservoirs and increasing water consumption in upstream areas, runoff in the rivers passing through Beijing has decreased (Figure 16.9), and this has led to a sharp decline in water infiltration from rivers.

- There are more built-up areas due to urbanization, increasing from 1470 km² in 1990 to 3440 km² in 2012. The construction of roads and buildings has led to a large decline in the rate of infiltration from precipitation. Assuming an annual rainfall of 550 mm and an infiltration coefficient of 0.25, the rate of infiltration from precipitation due to expansion of urban areas in Beijing from 1990 to 2012 has been reduced by 271 MCM/y compared with natural conditions.

16.5 STATE

16.5.1 Well Statistics

There were only 17 (WASA) wells before the 1950s, which increased to 51,700 in 2000. Recent statistics are not available, however, the current number is expected to be much higher than in 2000.

16.5.2 Groundwater Recharge

Groundwater aquifers in Beijing are recharged mainly by infiltration from precipitation and irrigation, leakage of surface water, and artificial injection. Annual groundwater recharge of the city's aquifers is estimated at 3598 MCM, of which 2766 MCM is recharged from the plain area, 1517 MCM from the mountains, and the remaining 685 MCM is due to the repeat volume of the plain being recharged by the lateral runoff from the mountains. The long-term average recharge components (Table 16.6) show that infiltration from precipitation represents a major share at 47.9% of the total recharge. The recharge volume was expected to decrease in recent years for many reasons, including urbanization, decreasing precipitation, and

Table 16.6 Recharge of groundwater in the plain of Beijing, China (MCM; calculated as an average between 1989 and 2000)

Components of recharge	Infiltration of precipitation	Lateral runoff of the mountains	Infiltration from rivers and channels	Infiltration from irrigation	Artificial recharge	Total
Recharge volume	1326	685	400	348	7	2766

MCM, million cubic meters.
Source: Beijing Groundwater (2008).

climate change. The same is true for the volume of recharge from surface water leakage because of a decline in river runoff due to water intake, damming, and storing in upstream reservoirs (Figure 16.9).

16.5.3 Groundwater Abstraction

Groundwater in Beijing is abstracted primarily by the centralized water supply source fields for tap water, self-reserved wells (wells owned by factories or industries), and wells installed for agricultural usage.

Before the 1950s, there was only one centralized water supply source field in Beijing, with a water supply capacity of 90×10^3 m^3/d. After the establishment of another five emergency centralized water supply source fields in 1999, capacity increased to 1100×10^3 m^3/d. Those five fields were constructed to alleviate the pressure of water shortage caused by the continuous aridity in the Beijing area since 1999.

In Beijing, the self-reserved wells include those in factories, institutions, government agencies, and schools. Based on available statistics, only 17 self-reserved wells existed in the city of Beijing before 1950, and this number increased to 2655 in 2000, with a water supply capacity of 726×10^3 m^3/d. Since the 1980s, the number of self-reserved wells has started to dwindle gradually (Figure 16.10) due to an increase in water supply coverage.

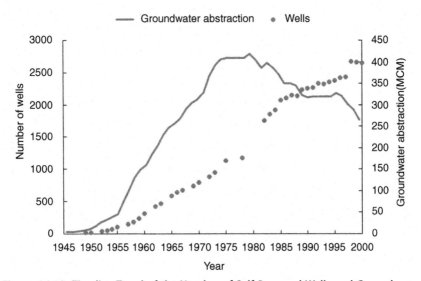

Figure 16.10 *Timeline Trend of the Number of Self-Reserved Wells and Groundwater Abstraction in Beijing, China.*

Irrigation wells, which were not installed until the 1950s, reached 3000 in 1961 and 43,000 in 1989. Total abstraction from those wells was 106 MCM/y in 1961 rising to 1615 MCM/y in 1989.

In 2000, the total number of groundwater abstraction wells in Beijing for various uses was 51.7×10^3, including 732 concertized wells, 7320 self-reserved wells, and 43.9×10^3 agricultural wells. Those wells abstracted 2708 MCM/y of groundwater, of which industrial, domestic, and agricultural sectors used 411,600, and 1641 MCM/y, respectively. Unknown amounts were lost through leakage of water pipes and storage tanks.

16.5.4 Groundwater Level

Knowledge of aquifer systems was poor prior to the 1970s and only the contour of the groundwater table in the focused aquifer could be depicted (Figure 16.11). The groundwater condition at that time could be considered

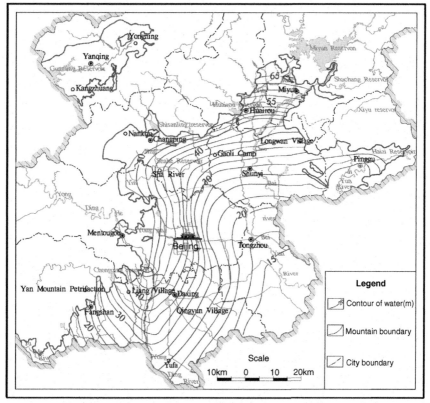

Figure 16.11 *Spatial Distribution of the Groundwater Level in the Plain of Beijing in 1965.*

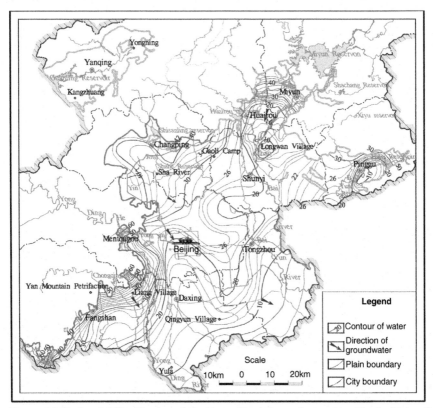

Figure 16.12 *Spatial Distribution of the Unconfined Groundwater Level in the Plain of Beijing in 2005.*

as in a state of natural flow. The general direction of groundwater flow was from north to south and from west to east, finally flowing toward the southeast of the plain. Following the increased abstraction in the 1970s, the flow field was altered. The impacts of significant abstraction can be seen in the groundwater table contours of 2005 (Figures 16.12 and 16.13). The figures show that multicones of depression in the two aquifers occurred and the direction of dominant groundwater flow changed toward the center of those cones.

16.5.5 Groundwater Quality

The quality of fracture and karst water in the mountain area of Beijing is ordinary, with total dissolved solids (TDS) of below 500 mg/L and a hydrochemical type of HCO_3–Ca. It is nearly the same as that in the middle and upper part of different alluvial and diluvial fans, and the hydrochemical

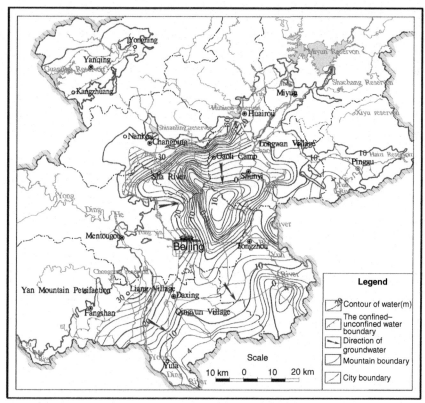

Figure 16.13 *Spatial Distribution of the Confined Groundwater Level in the Plain of Beijing in 2005.*

type of unconfined and confined groundwater gradually turns into HCO_3^- Na·Ca, and the TDS increase from 500 mg/L to 1000 mg/L at the edge of alluvial and diluvial fans and the alluvial plain. In the old city and the downstream area (namely, downtown and southeast Beijing), the hydro-chemical type for groundwater is $HCO_3 \cdot Cl \cdot SO_4$–Ca·Mg, etc., as a result of contamination into the unconfined aquifer from the refuse and sewage generated by long-term human habitation.

At present, due to increasing human activities, the hardness and TDS of the unconfined aquifer are beyond the third grade of groundwater quality standards, suffering from infiltration of domestic refuse, leakage of sewage, and contamination from point and nonpoint pollutants (Ministry of Land, 2002). Nitrate–nitrogen concentrations also exceeded the guideline value in the old city and in the 200 km² area south of Beijing. Additionally, the concentrations of Fe, Mn, and F in unconfined groundwater were higher than that for the third-grade groundwater quality standard in the southeast of Beijing, and the area was about 1000 km². However, these compositions were just scattered beyond the standard in the groundwater of the deep confined aquifer. The high concentrations of Fe, Mn, and F are natural, and not caused by human pollution.

16.6 IMPACTS

The overexploitation of groundwater, in order to cater for the ever-increasing water demands of Beijing City, has had several impacts including a decline in groundwater levels, eventually forming into large-scale depression cones, depletion of groundwater storage, deterioration of groundwater quality, land subsidence, ground fissures, and degradation of wetlands.

16.6.1 Decline in Groundwater Level

The steady increase in groundwater abstraction during the last four decades has resulted in a decline in the groundwater level of Beijing's aquifer. In the 1960s, the shallow groundwater level in the plain of Beijing was at a depth of 5 meters below ground level (mbgl) or less. The average annual rate of depletion, which was 0.3 m before the 1980s, reached 0.5 m in the 1990s and 2 m during the first decade of the twenty-first century. The groundwater level up to the end of June 2014 reached 24.5 mbgl in the plain of Beijing, and the shallow aquifer has been drained out in the west, in places such as the Fengtai District and adjoining areas. In some locations, groundwater tables are as deep as 40 mbgl. A sharp decline in the groundwater table for some wells (Figure 16.14) is due to the construction of an emergency supply for water source fields at Huairou in 2003, and other abstraction by water plants in various districts.

Figure 16.15 shows the timeline trend of groundwater levels at a 50 m deep monitoring well (17-D) in an unconfined aquifer at Beijing Normal University (BNU) in Haidian District, and another deep well (22-C) in a confined aquifer at Peking University (PKU). From the figure, the historical change in groundwater level can be divided into several stages:

- From 1958 to 1974, the groundwater level declined slowly and almost uniformly throughout the aquifer by 3 m in 17 years.
- From 1975 to 1981, it declined rapidly; almost 8 m in 7 years, at the rate of 1.14 m/y.
- From 1982 to 1995, the groundwater level became more or less stable.

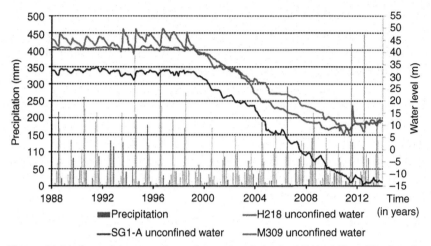

Figure 16.14 *Annual Average Groundwater Levels in the Mi-Huai-Shun Districts. See Figure 16.16 for the location of observation wells.*

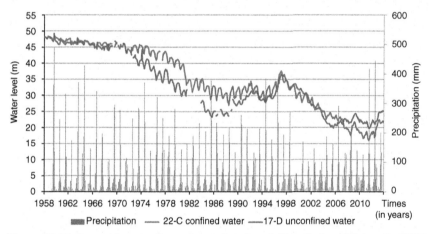

Figure 16.15 *Annual Average Groundwater Levels at Beijing National University (BNU) and Peking University (PKU). See Figure 16.16 for the location of observation wells.*

- From 1996 to 1997, the groundwater level had a tendency to rise as a result of partial wet flow in 1996.
- From 1997 to 2010, the groundwater level started declining again by 20 m in 13 years, at the rate of 1.54 m/y, resulting from successive dry years and the large volume of abstraction by emergency groundwater source fields.

The decline in groundwater levels has led to the formation of large depression cones. In the 1970s, reduced recharge and increased groundwater abstraction caused overexploitation of groundwater in suburban areas and a regional decline in groundwater levels. In the early 1980s, low rainfall in successive years reduced the recharge rate. However, groundwater abstraction continued at a gradually increasing rate. With the cumulative effect of a gradually increasing imbalance between recharge and abstraction, groundwater levels declined, depression cones formed, and those cones severely expanded. Based on historical data, the depression cones in Beijing were initially formed in 1971, when the center of the depression cone was located in the northeast suburbs of the city, which only covered a small area. With the increasing volume of groundwater abstraction since the 1970s, the rate of decline in groundwater levels became greater and large-scale depression cones formed in the east suburb, where the amount of abstraction was relatively large. With further exploitation of groundwater, depression cones occurred in many places, including Niulanshan and Tianzhu. Due to continuous aridity since 1999 and abstraction of groundwater for emergency uses, the depression cones have been continuously extending. The total extent of areas with a depression cone reached 1900 km^2 in 2013 (Figures 16.16–16.18).

16.6.2 Deplet ion in Groundwater Storage

Before the 1970s, groundwater abstraction in the city was more or less balanced. By the end of the 1970s, there was overexploitation in the suburban areas. In the early 1980s, as a result of successive dry years and the cutoff at the Yongdinghe and Chaobaihe Rivers, groundwater recharge decreased but groundwater abstraction continued to increase. These events collectively aggravated the deficiency of groundwater, enlarging the volume of accumulative storage year by year.

From 1961 to 2013, the accumulative loss in groundwater storage from Quaternary aquifers was 10,177 MCM (Table 16.7), and the average loss was 192 MCM/y. However, from 1999 to 2013, due to rapid urbanization and continuing aridity, groundwater was extremely overexploited, and groundwater storage was depleted further (The Hydrological Geological Brigade of Beijing, 2013a).

16.6.3 Deterioration in Groundwater Quality

Before the 1950s, groundwater quality was basically good with a very simple type of hydrochemistry (i.e., mainly HCO_3–Ca·Mg) and 450 mg/L of TDS (i.e., beyond the third standard of groundwater according to the *"Quality Standard for Groundwater."* From the 1960s to the 1970s, rapid urbanization

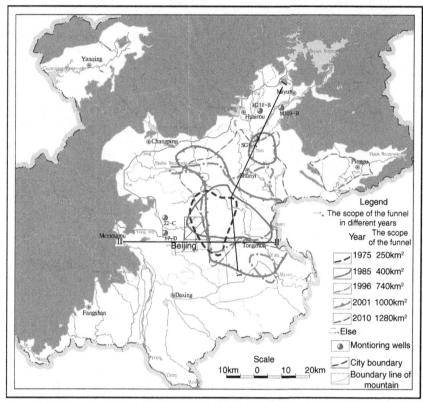

Figure 16.16 *Shift in Depression Cones at Exploited Aquifers in Beijing, China. See Figures 16.17 and 16.18 for profiles along I-I and II-II.*

and subsequent socioeconomic development of the city increased municipal water requirements and the volume of sewage discharge. Disposal of sewage in surface water bodies led to the serious contamination of surface water and degraded suburban groundwater quality year by year. During this period the major contaminant indexes were total hardness and nitrate–nitrogen. From

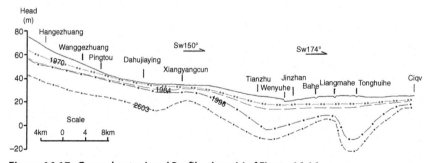

Figure 16.17 *Groundwater Level Profile along I-I of Figure 16.16.*

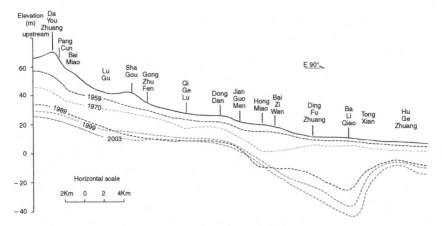

Figure 16.18 *Groundwater Level Profile along II-II of Figure 16.16.*

1980 to 2000 Beijing started to develop rapidly, industrialization intensified, and therefore water demand, as well as the generation and disposal of waste water, increased accordingly. The volume of sewage discharge in 1990 in the suburbs reached 782 MCM/y, 32 times more than for the early 1950s. The aforementioned statistics indicate that groundwater quality started to

Table 16.7 Variation of groundwater storage in the Beijing Plain (MCM)

Years	Storage variation (recharge– discharge)	Cumulative storage variation	Years	Storage variation	Cumulative storage variation
1961–1979		−1331	1997	−791	−3907
1980	−771	−2102	1998	311	−3596
1981	−1079	−3181	1999	−1209	−4805
1982	143	−3038	2000	−940	−5745
1983	−443	−3481	2001	−656	−6401
1984	−516	−3997	2002	−675	−7076
1985	326	−3671	2003	−687	−7763
1986	−242	−3913	2004	−417	−8180
1987	29	−3884	2005	−580	−8760
1988	175	−3709	2006	−414	−9174
1989	−569	−4278	2007	−429	−9603
1990	336	−3942	2008	254	−9349
1991	127	−3815	2009	−501	−9850
1992	−565	−4380	2010	−465	−10,315
1993	−677	−5057	2011	−121	−10,436
1994	872	−4185	2012	601	−9835
1995	229	−3956	2013	−342	−10,177
1996	840	−3116			

Table 16.8 Area (km²) of Beijing with total hardness exceeding the China Water Quality Standard of 450 mg/L

	Year						
Area	1937	1965	1975	1980	1990	2000	2005
Town	13	87.4	178	205	274	327	340
Entire area	13	87.4	178	205	507	820	840

Source: Beijing Groundwater (2008).

degrade after the early 1980s and this decline continued thereafter. Apart from total hardness and nitrate–nitrogen, TDS became one of the major contaminant indices for groundwater quality assessment after 1980. From the early twenty-first century, the contaminant indices in groundwater have been stable or shown improvement (The Hydrological Geological Brigade of Beijing, 2014). This has been due primarily to increasing awareness of the need for environmental protection, and positive responses from industries by treating their sewage before disposal.

From Table 16.8, it is evident that the areas with excessive total hardness were widely distributed before 1980. However, those areas subsequently started to expand toward the suburbs. By 2005, out of the entire area of 840 km² with excessive hardness, only 340 km² was in the urban area, with the rest in the suburbs. In the case of nitrate–nitrogen, areas with excessive concentrations above the standard were increasing until 1990, becoming stable in an area of approximately 185 km² after 2000 (Table 16.9).

16.6.4 Land Subsidence and Formation of Ground Fissures

There is evidence of land subsidence and ground fissures in the plain of Beijing, induced primarily by groundwater overexploitation (Yang et al., 2013). This has implications for development activities and the overall economy of the city. The fissures are formed in complicated geological environments such as in Shunyi and Changping (Wang et al., 2014).

Table 16.9 Area (km²) of Beijing with nitrate–nitrogen exceeding the China Water Quality Standard of 20 mg/L

	Year			
Area	1980	1990	2000	2005
Town	66	137	170	171
Entire area	66	150	184	185

Land subsidence in Beijing can be divided into several stages (Table 16.10):
- Formation period from 1955 to 1973
- Developing period from 1974 to 1983
- Expansion or spreading period from 1984 to 1988
- Rapid development period from 1999 to date

The cumulative total amount of land subsidence during these periods is reported in Table 16.10. Even recently, two new land subsidence areas have been formed in Beijing's plain (Figure 16.19), covering several districts in the north (such as Chaoyang, Haidian, Shunyi, Tongzhou, and Changping) and major parts of Daxing in the south. Monitoring data shows that in areas of cumulative land above 100 mm, subsidence accounts for more than half of the plain's total area: 3.2 times larger than that in 1999 (Beijing Institute of Hydrogeology and Engineering Geology, 2013b). The study indicates that the total cumulative amount and rate of land subsidence are closely comparable with groundwater levels. The majority of land subsidence occurs in areas with excessive groundwater abstraction (Tian et al., 2012).

Table 16.10 Stages of land subsidence in Beijing, China

Period	Main subsidence area (cumulative settlement, mm)	Subsidence area (km²)
1955–1966	Dongbali Zhuang (58), Jiuxian Qiao (30)	No statistics
1965–1973	Dongbali Zhuang–Dajiaoting (230), Laiguangying (126)	No statistics
1974–1983	Dongbali Zhuang–Dajiaoting (590), Laiguangying (307)	No statistics
1984–1987	Dongbali Zhuang–Dajiaoting (652), Laiguangying (367), Changping Shahe–Baxian Zhuang (303), Daxing Lixian–Yuzhan (298)	860
1988–1999	Dongbali Zhuang–Dajiaoting (722), Laiguangying (565), Changping Shahe–Baxian Zhuang (688), Daxing Lixian–Yuzhan (661), Shunyi Pinggezhuang (250)	1800
2000–2005	Dongbali Zhuang–Dajiaoting (750), Laiguangying (677), Changping Shahe–Baxian Zhuang (1086), Daxing Lixian–Yuzhan (831), Shuanyi Pinggezhuang (420), Tongzhou Liyuan–Taihu (265), Shuanyi Yangfang and Changping Yandan (200)	2851
2006–2013		>3900

Source: Beijing Groundwater (2008).

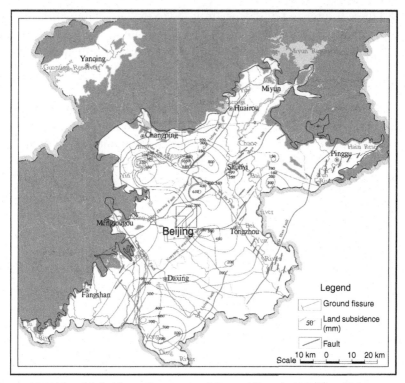

Figure 16.19 *Land Subsidence Contours and Ground Fissures in Beijing, China.*

Groundwater overexploitation is also thought to be responsible for the development of groundwater fissures in the area (Jia et al., 2011; Liu et al., 2013). Since the 1980s, groundwater fissures have been actively developing in Beijing (Figure 16.19). The fissures at Shunyi, Tugou–Gaoliying, and Yangfang–Miaojuan areas are the most frequently active and affect domestic and economic activities in the city. For example, the Tugou–Gaoliying Fissure, which was first discovered in the 1990s, starts from the zone of Baxian Zhuang Cottage in the Changping District, passes northeast through to the Wenyuhe River and spreads to the north of Xiwanglu village in the Shunyi District, along the development zone. These fissures have created serious damage to buildings and other infrastructure (Jia et al., 2011). The Gaoliying Fissure is located at the point where the Huangzhuang–Gaoliying active fault converges with the Changping–Baxianhe depression cone, and the Shunyi Fissure occurs where the Shunyi fault converges with the Pinggezhuang depression cone (Figure 16.19). These facts support the theory that groundwater overexploitation is partly responsible for the formation and/or development and/or expansion of groundwater fissures.

All these fissures and subsidence may add to the instability of urban flood control facilities and uplifting/inclining/riving/ruining of deep well pipes, which in turn creates obstacles for traffic and increases maintenance costs.

16.6.5 Degradation of Wetlands

Groundwater is also a significant component of the ecological environment. Caused by successive declines in the groundwater table and enlarged depression cones year by year, local aquifers were drained out and the underground area desertified. Wetland areas in Beijing have measured over 2000 km² since historical times, occupying 15% of the city's total area. This area had reduced to around 800 km² by the early 1980s. The areas are degrading and declining year by year for many reasons, including groundwater overexploitation. According to the latest statistics, the residual area of wetland in Beijing is just 500 km², which is less than 3% of the total area. The natural wetland areas are only around 350 km². As an ancient capital, some of Beijing's previous charm is fading since five of the proud waterscapes ("Peking's Eight Great Sights") have lost their former beauty. The Yuquan-shan Spring disappeared more than four decades ago; it was named as the first spring by the Emperor Qian Long (Yang et al., 2008).

16.7 RESPONSES

The central and municipal governments in Beijing have taken a variety of measures to meet the demand for water resources and improve and control groundwater environmental problems by the rational development and utilization of groundwater resources. These include groundwater exploration and study, groundwater monitoring, legal and regulatory reforms/interventions, initiating artificial recharge, setting up emergency groundwater sources, utilization of rainfall flood, the south–north water diversion project, and the prohibition and limiting of groundwater exploitation.

16.7.1 Groundwater Exploration and Studies

Several studies have been carried out since the 1950s to explore possibilities for groundwater development, and to investigate ways of protecting and utilizing the resources from Beijing's aquifers wisely. Major efforts made in previous decades are summarized as follows:

- In the 1950s: regional hydrogeological surveys were carried out on the city's water supply.

- In the 1960s: an investigation focusing on the water supply to farmland; hydrogeology mapping (scale 1:50,000) of the entire Beijing City.
- In the 1970s: various forms of exploration for hydrogeology were carried out.
- In the 1980s: a water supply survey for cities and towns was undertaken, enterprises and institutions were set up, an evaluation of groundwater resources was conducted.
- In the 1990s: groundwater exploration was carried out for emergency supplies to ensure water security for the city in an emergency.
- After 2000: the extensive development and utilization of groundwater created geological problems for the environment, such as a decline in the groundwater table and deterioration of water quality. Groundwater exploration shifted gradually from single water supply to a combined comprehensive survey and evaluation of hydrogeology, environmental geology, and engineering geology. Comprehensive projects continue with various titles such as the "Evaluation of groundwater resources and environmental survey for the capital area" and "Evaluation of groundwater for sustainable utilization in the North China Plain (Beijing)." Meanwhile, research activities have been initiated and are continuing, focusing on the control of groundwater pollution, combined regulation of surface and groundwater, artificial recharge and utilization of rainfall flood, as well as issues relating to the geological environment.

16.7.2 Groundwater Monitoring

Groundwater monitoring of the Beijing aquifers began in the middle of the 1950s and has strengthened gradually over time. The monitoring includes the groundwater table, groundwater quality, land subsidence, and ground fissures.

Monitoring of Groundwater Level

Monitoring of the groundwater level began in 1956. The total number of monitoring wells up to December 2012 was 635, of which 335 were equipped with automatic instruments. The wells monitor all the aquifers: 293 in the unconfined aquifer, 222 in the first confined aquifer, 72 in the second confined aquifer, 42 in the third confined aquifer, and 16 in the bedrock. They measure the "dynamic water table on a monthly basis" (Figure 16.20).

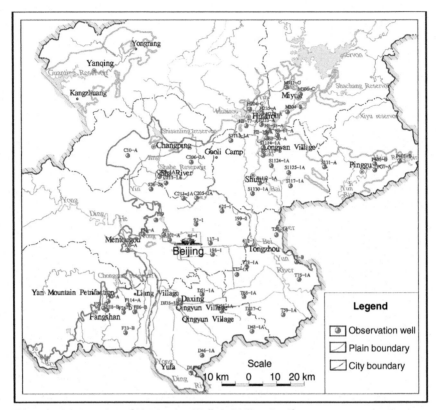

Figure 16.20 *Location of Monitoring Wells in Beijing Aquifers.*

Monitoring of Groundwater Quality

The monitoring of groundwater quality in Beijing started at the same time as water table monitoring. Until 1972, monitoring focused on conventional water quality parameters at a few typical wells only. After the early 1970s, the monitoring program was comprised of a network of 212 wells and covered 1100 km² of the urban and suburban areas. Monitoring frequency was twice a year: once in the dry season and once in the wet season. More parameters – total hardness, ammonia, nitrogen, and the "five evils" index (phenol, cyanide, arsenic, mercury, chromium) – were monitored in addition to the conventional indicators. After 1982, the monitoring network was expanded to include the entire Beijing Plain (area = 6540 km²) with 564 observation wells.

At the end of 2009, the Beijing Plain groundwater environment-monitoring network was designed and implemented. This network of 1182-observation sites/wells aims to monitor the local groundwater environment as well as key pollution sources (Zhou et al., 2008). The local groundwater environment-monitoring network consists of 822 monitoring wells, which can monitor four aquifers in the plain area. It monitors 32 inorganic parameters and 26 organic parameters at the frequency of twice a year: once in the dry season and once in the wet season. The pollution source-monitoring network consists of 360 monitoring wells covering industrial development zones, industrial enterprises, large-scale livestock and poultry farms, landfill, golf courses, and reclaimed water irrigation zones. The monitoring depth is located in the unconfined aquifer. It monitors 36 inorganic and 39 organic parameters at a frequency of four times a year (Beijing, 2014b).

Monitoring of Land Subsidence and Ground Fissures
A project for a land subsidence monitoring, forecasting and prediction system was initiated in 2001 with the support of the Beijing Municipal Commission for Development and Reform (BMCDR). The network so far comprises seven land subsidence stations (e.g., Tianzhu, Wangjing, Wangsiying, Zhang Jiawa, etc.), 114 GPS monitoring sites, 300 special monitoring sites, and one ground fissure monitoring site at Xiwanglu Street. The network has formed a combination of points, lines, and surface monitoring, and has achieved land subsidence monitoring under the ground, on the ground, and at multiple angles in the air. The multiple methods used for monitoring include ground deformation at stations, groundwater hydrodynamics, and real-time monitoring of subsidence at different depths. It provides basic data and documents for the study of land subsidence and its prevention.

16.7.3 Legal and Regulatory Reforms/Interventions
China has formulated laws such as the Law of the People's Republic of China on Prevention (Water Law of the Peoples Republic of China, 2002) and Control of Water Pollution 1984 (Anon, 1984) and Water Act of the People's Republic of China 2002. These laws contain provisions for the development, sustainable utilization, and conservation of groundwater resources and the environment. A series of policies, ordinances, and guidelines are also issued by relevant ministries and Beijing City authorities with the purpose of protecting groundwater resources, and include:
- Implementation Methods for the Water Abstraction Permission Policy 1993
- Announcement from the People's Government of Beijing City for the Bounding of Groundwater in a Serious Overexploited Area, Overexploited

Area and with no Overexploitation in the Plain Zone of Beijing and for Further Enhancement of the Groundwater Resources Management 1995

• Regulation of Beijing City on Collecting Conservation Funds for Groundwater 2001

• Regulations on the Management of Water Abstraction Permission and Collection of Water Resources Fee 2006

• Technical Guidelines for Delineating Source Water Protection Areas 2007 (Anon., 2007)

• Regulation on Groundwater Resources of Water Works in Beijing 2007

• Ordinance of Beijing City on Protection and Control of Water Pollution 2010

• Regulation of Beijing City on Drainage and Reclaimed Water 2010

• Regulation on Pollution Prevention of Source Water Protection Areas 2010

• Technical Guidelines for Environmental Impact Assessment on the Groundwater Environment 2011

• Guidelines for Water Resources Assessment of Construction Projects 2013

In addition, monthly accounting of groundwater withdrawals from municipal pumping wells (around 20% of the total) and industrial-pumping wells (around 14%) has started and the process continues. Monitoring of groundwater abstraction for irrigation is being promoted with an increasing number of instruments installed for its monitoring.

The aforementioned legal and regulatory responses/interventions over time have yielded some positive results, reducing pressure on the groundwater environment. As an indicator of this, groundwater withdrawal in the area has reduced from 2800 MCM in 1995 to 2100 MCM in 2014.

16.7.4 Initiating Artificial Recharge

The necessity to augment groundwater storage by artificial means to cater for the water demand of future populations was felt as early as the 1960s. Since then, Beijing has conducted a number of artificial recharge tests using various methods in the upstream area of the Yongdinghe and Chaobaihe Rivers' alluvial fan. In the early 1970s, the Xihuangcun artificial recharge test field of gravel pits was built in the western urban area of Shijiang Shan. The recharge tests were carried out in 1978, 1979, 1983, 1995, and 1996, respectively. In the early 1980s, groundwater artificial recharge tests were conducted through the bed of Yongdinghe River. After those recharge tests, the groundwater table along the bank of the Yongdinghe River rose by more than 3 m and an improved water quality was observed. In 1965, recharge through pumping wells was initiated with the slogan: "recharge in winter and pump in summer," at the third weaving factory in the eastern suburbs. In 1980, recharge tests through large open wells were chosen at Shougang in the upstream area of the Yongdinghe fan, with wells dug 25 m deep and 8 m in diameter. The groundwater table in the area was over 30 mbgl. Recharge rates during the tests were controlled at 0.5, 0.3, and 0.1 m^3/s, respectively. The monitoring results showed that initial recharge can reach a maximum of 0.5 m^3/s, reducing gradually with blocking of the wells, and then remain stable at 0.3 m^3/s (Beijing, 2007). Earlier studies have also suggested the Chaobaihe River's alluvial fan as a potential site for artificial recharge.

16.7.4 Increasing Surface Water Availability

A number of measures have been taken to increase the availability of surface water resources in the city. They include the South–North Water Diversion Project (SNWDP) and utilization of rainfall flood.

South–North Water Diversion Project

The middle line of the SNWDP is a significant strategic measure at the national level to relieve the shortage of water resources in the northern drought-affected region, especially in Beijing. The project diverts water to Beijing from the Danjiangkou Reservoir located at the border of Shanxi, Hubei, and Henan Provinces. The project aims to divert 1000 MCM/y of water to Beijing, 500 MCM of which is for reducing excessive groundwater abstraction and 200 MCM is for increasing water resources by means such as recharge. The project successfully transferred water to Beijing for the first time on December 27, 2014. The project is expected to reduce pressure on groundwater resources by reducing abstraction and increasing recharge.

Utilization of Rainfall Flood

With the rapid development of cities, impermeable areas increase in size, leading to increases in surface runoff and a greater demand on resources for flood control. If the large volume of precious rainfall flood is stored and utilized appropriately, it will contribute to efficient water use, reduce pressure on flood control and drainage, recharge groundwater, and improve the ecological environment. Beijing City has tapped into this opportunity by designing and implementing "urban rainfall flood utilization projects." Up to 2009, Beijing built 51 rainfall flood utilization projects, using different comprehensive operation modes, which are able to collect over 9 MCM of rainfall a year (Pan et al., 2009). Beijing has also encouraged each department to construct rainfall flood utilization facilities to suit their own circumstances. Furthermore, Beijing municipal government has set suburban rainfall flood utilization projects as the 50th practical work. The main details of projects include dredging and clearing existing dam gates, swag, sinkage, and wasteland such as sand pits, to construct rainfall utilization facilities and retain rainfall. They promised to build 150 rainfall utilization projects and store 9 MCM/y. The efforts to store and utilize rainfall flood can reduce groundwater overexploitation, improve the local water environment, create green landscapes, and finally help increase recharge to groundwater aquifers.

16.8 SUMMARY

Groundwater is the main source of water supply in Beijing. The shortages of water resources and groundwater environmental issues have become the bottleneck restricting sustainable socioeconomic development of the city. The groundwater environment in Beijing is facing a serious challenge. DPSIR analysis shows that the groundwater in the city is under pressure from drivers such as population growth, socioeconomic development, and urbanization. They have not only resulted in the excessive use of groundwater, but have also limited its recharge. Growing water consumption in upstream areas, and climate variability and change have resulted in the reduction of surface water supplies. All these factors have impacted the groundwater environment and can be observed in the form of depletion of groundwater levels and groundwater storage, degradation of groundwater quality, land subsidence, formation of ground fissures, and the degradation of wetlands. To address the problem in Beijing, the central and municipal governments have taken a variety of measures. These include groundwater exploration and study, groundwater monitoring, legal and regulatory reforms/interventions, initiating artificial recharge, setting up emergency groundwater sources, utilization of rainfall flood, the SNWDP, and the prohibition and limit of groundwater exploitation. Despite these efforts, sustainable use of the resource is still a major concern. For the sustained use of resources and conservation of the groundwater environment, implementation of regulatory measures on groundwater abstraction, control on groundwater pollution and the artificial recharge of aquifers is recommended.

ACKNOWLEDGMENTS

The authors acknowledge the Beijing Institute of Hydrogeology and Engineering Geology for providing valuable data as well as useful maps and figures.

REFERENCES

Beijing, 2008. Beijing groundwater. Bureau of Geology and Mineral Resources Exploration, Beijing Institute of Hydrogeology & Engineering Geology.

Beijing, 2007. Study on regulation and joint operation of groundwater and south-to-north water. Beijing Institute of Geological & Prospecting Engineering.

Beijing, 2014. Protection and utilization of groundwater resource in Beijing City. Beijing Institute of Hydrogeology & Engineering Geology.

Beijing, 2013a. The evaluating report of groundwater resources exploitation in the plain area of Beijing in 2013. Beijing Institute of Hydrogeology & Engineering Geology.

Beijing, 2013b. The monitoring report of land subsidence in Beijing. Beijing Institute of Hydrogeology & Engineering Geology.

Beijing, 2014b. The running of monitoring networks for groundwater environment in the plain area of Beijing. Beijing Institute of Hydrogeology & Engineering Geology.

Beijing Water Authority, 2003. Beijing Water Statistical Yearbook, 2003.

Jia, S., Wang, H., Ye, C., et al., 2011. Investigation and survey methods appropriate for ground fissures in Beijing. J. Eng. Geol. 19 (Suppl), 104–111.

Anon., 1984. Law of the People's Republic of China on Prevention and Control of Water Pollution 1984.

Liu, M., Wang, R., Jia, S., 2013. The risk assessment methods of the ground fissures in Beijing area. City Geol. 8 (4), 29–34.

Ministry of Land, 2002. Quality Standard for Ground Water 2002. Ministry of Land and Resources of the People's Republic of China.

Pan, A., Zhang, Sh., Meng, Q., et al., 2009. Initial concept of stormwater and flood management in Beijing City. China Water Wastewater 25 (22), 9–12.

Ren, X., Lu, Z., Cao, Y., 2007. Water Resources Appraisement of Haihe Basin. China Water Power Press, Beijing, 160–161.

Anon., 2007. Technical Guideline for Delineating Source Water Protection Areas 2007.

Tian, F., Guo, M., Luo, Y., et al., 2012. The deformation behavior of soil mass in the subsidence area of Beijing. Geology in China 39 (1), 236–242.

Wang, R., Lium, M., Jia, S., 2014. Research based on the high-speed railway affect the dynamic quantitative relationship between groundwater and land subsidence. Chin. J. Geol. Hazard Control (2), .

Water Law of the Peoples Republic of China, 2002.

Yang, T., Fu, Y., 2008. The reason and treating measures of spring water change in Beijing Haidian. Groundwater 30 (5), 55–57.

Yang, Y., Zheng, F., Liu, L., Dou, Y., Jia, S., 2013. Susceptibility zoning and control measures on land subsidence caused by groundwater exploitation. Geology in China 40 (2), 653–658.

Zhou, L., Wang, Y., Lin, J., et al., 2008. Optical design of monitoring network of groundwater quality in the Beijing Plain. Hydrogeol. Eng. Geol. 35 (2), 1–9.

CHAPTER 17

Groundwater Environment in Bishkek, Kyrgyzstan

Rafael G. Litvak*, Ekaterina I. Nemaltseva**, and Gennady M. Tolstikhin***
*Ground Water Modelling Laboratory, Kyrgyz Research Institute of Irrigation
**Laboratory rational ground water use, Institute of Water Problems and Hydropower, National Academy of Science, Kyrgyz Republic
***Kyrgyz Hydrogeological Survey, State Agency on Geology and Mineral Resources of the Kyrgyz Republic

17.1 INTRODUCTION

The city is 100% aquifer dependent for potable, domestic, commercial and industrial water supplies, which are provided by both intraurban and periurban wellfields. A highly productive but much localized periurban valley fill wellfield located only 8 km south of the city center provides about half of the city's water demand, the balance coming from boreholes of various depths distributed throughout the city. Surface water is used only for irrigation in the city. The majority of abstraction boreholes are operated by the municipal water supply utility Bishkekvodokanal, which provides water for both domestic and industrial purposes. There are two separate reticulation systems for domestic water: cold water for potable use, and hot water for nonpotable use and for district heating use, the last being a closed system (within each block of flats) which operates only during the winter. All come under the description of "public water supply." Private urban water use is much less important both numerically and volumetrically; a small number of factories have private wells for potable or nonsensitive supply, and there are also a few private domestic and municipal irrigation wells in seasonal use. Owner-operated boreholes for commercial premises, hospitals, and large state administrative buildings appear to be insignificant.

Although there is a very extensive piped water infrastructure (pressurized drinking water and hot-water mains, piped sewerage), widespread on-site sanitation is practised in single/two-story residential areas. The main environmental problems connected with the groundwater of Bishkek are: groundwater flood in northern part of the city and the threat of groundwater pollution. Special attention in this chapter is devoted to these issues.

Groundwater Environment in Asian Cities
http://dx.doi.org/10.1016/B978-0-12-803166-7.00017-9

Copyright © 2016 Elsevier Inc.
All rights reserved.

17.2 ABOUT THE CITY

Bishkek, the capital city of Kyrgyzstan with a population of approximately one million, lies on the northern flanks of the Alatau range of the Tien Shan Mountains, in the northern part of the republic (Figure 17.1). It is the most economically developed and densely populated city of the Kyrgyz Republic. Altitude in the city ranges between 550 m and 900 m. The city lies at 42°50′N and 74°35′E. Bishkek City was founded in 1878 and was originally called Pishpek. The city is the greenest in Central Asia with more trees per head of population than any other country in the region.

Bishkek is a manufacturing center, as its factories produce about half of Kyrgyzstan's output, specializing in textiles, footwear, and heavy engineering (a particular legacy of the Second World War, when a number of factories were transferred from the European part of the Soviet Union). Currently, the majority of large companies (which were built in Soviet times) is practically not working.

17.2.1 Climate and hydrology

The city has an arid climate. Based on analysis of weather data from Bishkek Station (altitude = 756 m, and the only weather station in the city territory), the coldest months are December, January, and February, where the monthly average temperature drops to −10°C (in 2008). The hottest month is July, where the absolute maximum monthly average temperature has been known to reach +25.8°C (in 1991). Within the period of observations

Figure 17.1 *Location of Bishkek, capital of Kyrgyzstan, Central Asia.*

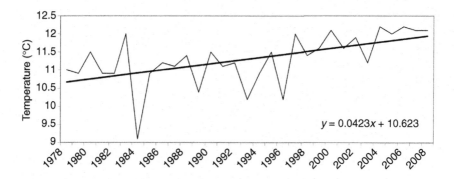

Figure 17.2 *Average Annual Air temperatures Based on Data at Bishkek Station.*

(1978–2008) there was an upward trend in air temperature observed, and the yearly average temperatures of air were above the long-time average annual values by 6–14% (Ljanov and Mamrenko, 1995). The trends of long-time average annual temperatures have positive values, indicating a general warming trend, see Figure 17.2.

Average annual precipitation in the Bishkek area during 1978–2008 was 462 mm/y. Figure 17.3 depicts changes in annual precipitation and Table 17.1 shows the distribution of precipitation on a monthly basis. This table contains the data from the years with minimal (1995) and maximal (2002) precipitations, as well as multiyear averages. The maximum amount of precipitation Bishkek received was from April through May.

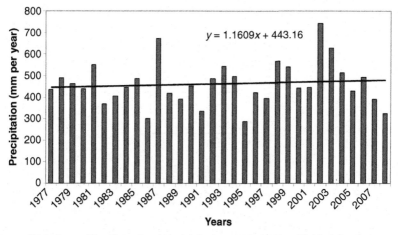

Figure 17.3 *Annual Precipitation (mm/y) during 1978–2008 at Bishkek Station.*

Table 17.1 Precipitation (mm/month) at Bishkek Station

Months	1	2	3	4	5	6	7	8	9	10	11	12	Total (mm/y)
In 2002 (maximum precipitation)	51.4	31.3	122	137.7	147.7	86.3	23.9	12.7	16.5	42.5	20.5	49.0	589
In 1995 (minimum precipitation)	13.1	33.3	23.0	11.7	31.6	26.6	22.0	9.5	13.6	55.3	24.5	23.8	193
Multi year average	28.3	34.0	52.4	75.0	65.0	35.7	20.6	13.9	17.4	40.2	44.5	35.5	462

The winter period is characterized by a rather stable snow cover, which is most often formed in the middle of December and is preserved until the end of February. On average, there are 75–90 days with a snow mantle (Ljanov and Mamrenko, 1995). The evaporation in the city territory is approximately 1000 mm/y.

The main rivers crossing the city are the Alamedin and the Ala-Archa (flowing down from the northern slopes of the Kyrgyz ridge). Table 17.2 shows the average annual monthly values of the river flows for the period from 1928 till 2008. In addition, data on the wettest year and the driest year are presented. Average annual water flow in the rivers for this 80-year period shows an increasing trend (Figure 17.4). Meltwaters of the seasonal snow cover, glaciers, and permanent snow cover are main sources of water in the Ala-Archa and Alamedin rivers. The contribution of rainwater to the river flows is of secondary importance and does not exceed 1–10 % (Ljanov and Mamrenko, 1995).

At the locations where rivers exit from ravines into the valley, the river water is generally broken up into canals for irrigation and to feed groundwater. In all the described areas, the artificial water system is highly developed (i.e., the hydrological regime of the rivers is completely changed by water resource utilization activities).

17.2.2 Hydrogeological Setting

The geology underlying Bishkek is comprised of an alluvial sequence fining outward from the flanks of the Tien Shan Mountains northward toward the plain. In the southern part of the city, the geology consists of a relatively homogeneous sequence of coarse-grained clastic alluvial fan cobbles, gravels, and sands. To the north, the geology comprises a multilayered system, with numerous low permeability silt and clay-rich horizons interspersed with higher permeability sands and fine gravels. As shown in Figure 17.5, more complex semi-confined aquifer conditions occur in the flatter northern part of the city, where thin but extensive surficial silty clay is present. Based on a combination of pumping test results and drilling returns, the aquifer system beneath the city proper has historically been divided into an upper, middle, and lower aquifer (Morris et al., 2006).

Scope therefore exists for significant pumping-induced vertical leakage of potentially contaminated urban recharge to depths ostensibly remote from the land surface. This is especially likely in the southern parts of Bishkek, where the coarseness of the piedmont clastics and relative paucity of intervening low-permeability horizons favour strong vertical connectivity.

Table 17.2 Discharge into Ala-Archa and Alamedin rivers

Months	Ala-Archa River (m³/s)			Alamedin River flow (m³/s)		
	1997	1951	Average long-term values	1920	1938	Average long-term values
1	1.3	1.4	1.5	1.5	2.2	1.8
2	12	1.3	1.4	1.7	1.9	1.6
3	1.1	1.2	1.4	1.4	1.4	1.5
4	1.7	1.1	1.5	1.6	1.3	1.6
5	4.3	3.2	3	11.6	4.1	4.1
6	11.5	5.2	7.8	33.4	5.8	11.2
7	19.2	7.6	12.9	20.9	14	18.9
8	17	9.3	12.6	17.7	12.2	18.6
9	8.9	4.4	5.9	9.5	4.9	9.2
10	4.4	2.6	2.8	5.8	2.9	4.5
11	2.5	1.7	2	3.9	2.1	2.8
12	2.2	1.4	1.7	3	1.5	2.2
Average	6.3	3.4	4.5	9.3	4.5	6.52

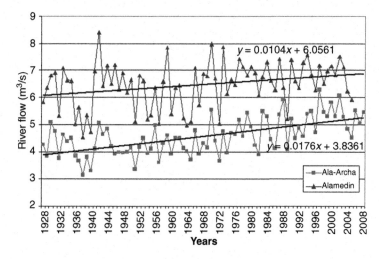

Figure 17.4 *Average Annual Discharge in the Ala-Archa and Alamedin Rivers between 1928 and 2008.*

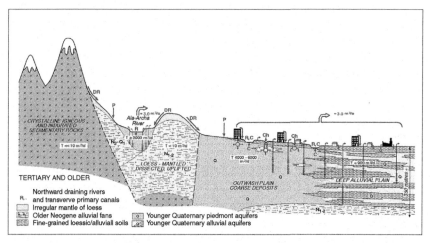

Figure 17.5 *Groundwater Systems of Bishkek, Kyrgyzstan.* Source: *Morris et al. (2006).*

The unconsolidated alluvial and fluvioglacial deposits comprising this intergranular flow aquifer system have high transmissivity (900–6000 m²/d) and vertical permeability, although unmeasured, is likely to be large.

The city's hydrogeological setting is complex, with a laterally heterogeneous intergranular fluvioglacial/alluvial multiaquifer system of Quaternary age, in excess of 350 m thick in northern districts of the city. There is a large lateral and

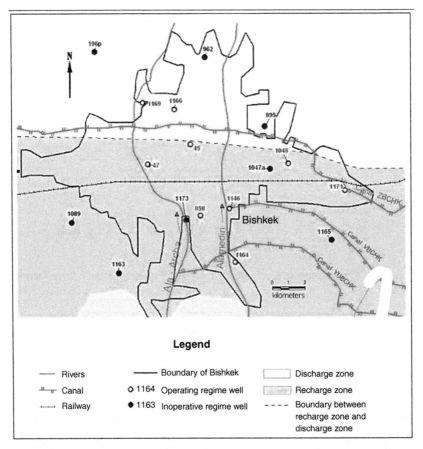

Figure 17.6 *Schematic Map of the Bishkek City Area.*

vertical variability, but as a first approximation the system fines laterally northwards, over a distance of less than 10 km, from coarse clastic deposits composed of coalesced piedmont fans fronting the foothills into more stratified deep alluvial precipitations forming an extensive plain to the north (Figure 17.6).

Bishkek is situated within the area of the Ala–Archa groundwater deposits. In the south it is confined by low piedmonts of the Kyrgyz mountain range, and in the north by a regional drain of groundwater of the Chu River. The total area of the deposit is 960 km² (Tolstihin, 2005).

Useful groundwater resources in the deposit in 1995 were estimated as 9.5 m³/s (Tolstihin, 2005; Tolstihina and Ponomareva, 2008). This deposit is used as a source of supply to domestic and industrial uses.

A highly productive but much localized periurban valley fill wellfield located only 8 km south of the city center provides about half of the city's water demand, the balance coming from boreholes of various depths distributed throughout the city. The latter are screened extensively in the middle aquifer (typically> 120 m intake depth), but the lower part of the upper aquifer (40–120 m) is also widely tapped.

The area of interest (according to its hydrogeological regionalization) is included into the Chu-Talas hydrogeological region. The Chu artesian basin has three-storied structure in the vertical cut. (Hydrogeology of USSR, 1971; Grigorenko, 1979). The area is distinguished by the lithological composition of the water-bearing rocks, their water-transmitting capabilities and the peculiarities of groundwater distribution. The upper floor of the groundwater is of great practical interest as it is the main source of water for utility and drinking, as well as a partial source of irrigation water. In the vicinity of Bishkek there are two large hydrogeologically distinct areas, or hydrogeological zones (see Figure 17.6): (i) zone of groundwater flow formation (recharge zone), and (ii) discharge and groundwater evaporation zone (discharge zone).

- *Recharge zone*: On the abruptly sloping plain of an alluvial bench, a hydrogeological zone of groundwater flow formation is developed. It is distinguished by a specific structure of water balance, horizontal movement of free-flowing waters, and by significant transmissivity from 1000 m^2/d to 6000 m^2/d. Within soft detrital sediments of Quaternary age, there lies a single aquifer from 300- to 500-m thick containing groundwater at depths up to 150–200 m. The wells can yield at a rate of 20–40 L/s at a drawdown of 0.3–1.6 m.

- *Discharge zone*: To the south of the Big Chu Canal, the alluvial bench transforms into discharge hydrogeology zone. The boundary between recharge and discharge zones is characterized by flow from the permeable rocks of the alluvial bench entering a row of aquifers dissected by semi-permeable layers. All wet intercalations of the water head complex and the ground source, are hydraulically interconnected and form a single system. The piezometric level of groundwater is established above the land surface, the transmissivity is 500–900 m^2/d. The ground water level slopes is 0.005–0.01. Capacities of wells make 10–20 L/s at decrease of the groundwater level 5–14 m. Along the area of aquifer distribution there is a vertical overflow from the bottom layers in the upper aquifers. The main aquifer feeding source is groundwater flow from the recharge zone.

17.3 DRIVERS

Major drivers exerting pressures on the Bishkek groundwater environment are population growth and urbanization (Table 17.3).

Table 17.3 Status of DPSIR indicators in Bishkek, Kyrgyzstan

Component	Indicator	1999	2009	Data sources
Drivers	Population growth			
	Population (million people)	0.76	0.84	*Census of Population and Housing Censuses of the Kyrgyz Republic*, Vol. 3 (2010), Department of Statistics, Bishkek, pp. 190.
	Population density (persons/km²)	–	4,953	
	Annual growth rate	2% (1989–1999)	1% (1999–2009)	
	Urbanization			
	Urban area (km²)	127.3	169.6	General Plan of Bishkek Development
	Urban area in 2025 (km²)	278.8		
Pressures	Groundwater flood	The northern part of the city (40% of territory) is in the state of groundwater flood.		Data of Kyrgyz Research Institute of Irrigation
	Land cover change	Construction of buildings, services in parks, and protected areas		
State	Well statistics	No. of abstraction wells = 220 (in 1998) Observation wells: 16 (in 1990) and 7 (in 2009)		Data of Kyrgyz Hydrogeology Survey
	Groundwater abstraction	Total volume of abstraction is 109 MMC/y (or 3.46 m³/s)		Data of Kyrgyz Hydrogeology Survey
	Groundwater level	Groundwater intakes (located in the city) have reduced groundwater levels in the southern part of the city by 8–12 m, although this has stabilized over the last 30 years.		Data of Kyrgyz Research Institute of Irrigation
	Groundwater quality	Characterized by high concentration of nitrate, hexavalent chromium, and total hardness due to industrial activities and lack of a centralized sewerage system in some areas of the city. Salinity: 0.3–1.0 mg/L; copper: 0.02–0.04 mg/L; manganese: 0.03–0.04 mg/L; etc. More details are provided in Section 17.5.4.		Data of Kyrgyz Hydrogeology Survey
	Recharge	Groundwater recharge fully covers withdrawal		

Impacts	Depletion in groundwater level in south	Groundwater in southern Bishkek declined by 8–12 m between 1965 and 1980 due to commissioning of groundwater intakes.	Data of Kyrgyz Hydrogeology Survey and Kyrgyz Research Institute of Irrigation
	Rise in groundwater level in north	Almost 40% of land in northern Bishkek is under groundwater flood as a result of groundwater level rise due to the destruction of natural and man-made water collectors and drains.	
	Degradation in groundwater quality	Groundwater is polluted with nitrate, hexavalent chromium, and total hardness mainly due to anthropogenic activities.	
Responses	Groundwater monitoring	The first observation well was drilled in Bishkek in 1933. The wells numbered 16 in 1990 but reduced to 7 in 2009. More investment is required to expand and strengthen the monitoring infrastructure.	
	Scientific studies for drainage system installation	To design and construct a workable and efficient drainage system in northern Bishkek and to eradicate the groundwater-flooding problem, scientific studies were carried out with the aim of lowering the groundwater level to at least 2 m below ground surface.	
	Policy and legal reforms	Several state agencies are responsible for determining environmental standard guidelines related to groundwater and several laws and standards for the protection and sustainable utilization of groundwater have been put in place.	

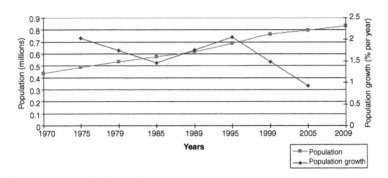

Figure 17.7 *Population Increase and Population Growth Rates in Bishkek city, Kyrgyzstan.*

The population of the city has grown continuously since 1970. Total population from 1999 to 2009 increased from 0.76 to 0.84 million (Table 17.3). The annual population growth rate since 1970 has remained at 1% with the only exception being the period 1989–1999, when it increased to 2.06% (Figure 17.7) due to the migration of people from other areas to the capital, following Kyrgyzstan's gaining of independence in 1991.

The city area has increased both in a planned manner, and as a result of the revolutions of 2005 and 2010 (these revolutions were accompanied by illicit land seizures for housing construction, although the process is now partially legalized). According to official data, the area of the city in 1999 was 127.3 km², which increased to 169.6 km² in 2009 and is expected to reach 278.8 km² by 2025 as per the General Plan of Bishkek Development. It should be noted that the city boundary and area are determined by the municipality and other state organizations and are not directly linked to population growth.

17.4 PRESSURES

17.4.1 Groundwater Flood

As stated earlier, the groundwater regimes in the north and south of the city are essentially different as they belong to different hydrogeological zones. Observation wells at depths over 50 m in northern part of Bishkek, show water heads above the land surface. Typical fluctuations of groundwater depths in shallow wells (less than 10 m) are described in (Figure 17.8). Observed groundwater level data show a high standing of groundwater levels. Almost the entire area of northern Bishkek is situated

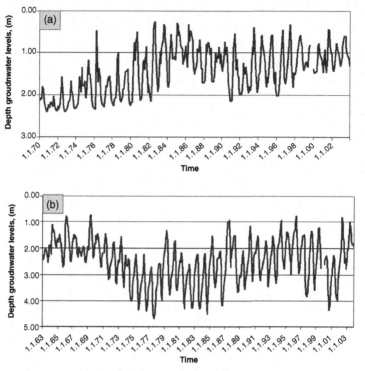

Figure 17.8 *Groundwater Levels at Shallow Depths in the Northern Part of Bishkek (mbgl).* (a) at Well No. 196 with a depth of 5.2 m; (b) at Well No. 962 with a depth of 10 m.

within the groundwater flood zone. It is especially dangerous when the depths of the groundwater level are less than one meter. It threatens not only buildings, but also the health of communities through water-related diseases (e.g., malaria).

17.4.2 Land Cover Change

Bishkek was one of the greenest cities in the former Soviet Union and during the last decade, various laws and instructions dealing with the protection and rational use of water and land resources have been enacted. Nevertheless, against this background we observe a steady redevelopment of the groundwater in highly vulnerable areas with elite private residences, automobile repair shops and car washes, cafes and restaurants. It can be clearly seen in the area adjoining the river Alamedin. There was a specially planted woodland belt some 10–15 years ago, but now almost the whole area is built up. Due to site development, the area of Bishkek's recreational forests has gradually shrunk. Natural and man-made collecting drains are being built on in the city area, thus hastening the flooding processes. An example of this process is the former large natural outfall drain situated to the north of the city's bus terminal. It has been backfilled and had residential houses built on it, which resulted in the flooding of the underground communication system of the bus station.

17.5 STATE

17.5.1 Well Statistics

Changes in groundwater level in the city area are traced by observation wells. The number of observation wells in Bishkek decreased from 16 to 7 within 20 years (1990–2009) due to a lack of financial resources as well as a decrease in the level of responsibility for implementation of governmental programs. The chemical composition of groundwater is monitored by water samples taken from production wells.

Regarding production or abstraction wells, there were as many as 220 wells in Bishkek in 1998, and this increased significantly in subsequent years owing to private water consumers. According to the Kyrgyz Hydrogeological Survey, the number of production wells has now reached 500. According to indirect evidence, such as the regime of groundwater levels and water consumption rates, it is apparent that newly constructed producing wells basically are switched off.

17.5.2 Groundwater Abstraction

The total volume of groundwater abstraction through some 500-abstraction wells is estimated as 109 MCM/y. There are 30 groundwater abstraction intakes in Bishkek, with each consisting of 1–20 production wells. The largest is the Orto-Alysh groundwater intake, which is separated from Bishkek by poorly permeable Neogene-Quaternary sediments and its influence is not considered here.

17.5.3 Groundwater Level

Information on groundwater levels is provided through a network of observation wells of the Kyrgyz Hydrogeological Survey. A list of observation wells in the southern part of the city is provided in Table 17.4. Groundwater levels are essentially different in southern Bishkek (i.e., recharge zone) and northern Bishkek (i.e., discharge zone). Figure 17.9 shows depths of the average annual groundwater level at below ground level in the southern part of Bishkek at six observation wells established before 1970. Two (S_1047a and S_1048) out of these six wells are still

Table 17.4 Details of observation wells in the southern part of Bishkek

No. of well	Altitude of wellhead (m)	Start of observation (year)	Depth of well (m)	Position of screen (m)		Year of closing (year)	Observation period (years)
				From	To		
S_45	732.08	1933	15	7	13	2000	67
S_47	739.85	1933	22.3	18	20.7	2000	67
S_858	812.2	1954	92	75	90	2000	46
S_1047a	761.26	1964	80	69.02	80		
S_1048	757.53	1964	150	69.18	92	2000	36
S_1089	773.85	1966	65	60	65		
S_1146	809.38	1973	300	125.91	294	2000	27
S_1163	850.25	1975	155	123	150		
S_1164	878.82	1976	200	158.6	189.45	1999	23
S_1171	747.72	1976	88	70.3	83	1995	19
S_1173	805.1	1976	402	119	395		

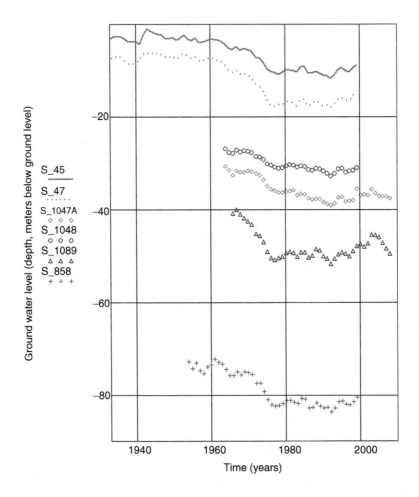

Figure 17.9 Depths of Average Annual Groundwater Levels below Ground Surface in Southern Bishkek at Observation Wells of Kyrgyz Hydrogeology Survey (wells established before 1970 are also shown).

in use for groundwater monitoring. It is clear that there was a considerable decrease in groundwater levels within the period 1960–1980 in all the wells that existed during the period in the area. Groundwater levels in the southern part are of the greatest interest as they are considered as recharge areas for the city's aquifers. A decline in groundwater levels of up to 10 m (approx.) was observed during the period when the construction of groundwater intakes was carried out.

To check whether the municipal groundwater withdrawals are the reasons for the decline in groundwater levels in the area during 1960–1980, a modeling study of the mentioned groundwater intakes was carried out. The groundwater model of the central part of the Chu valley developed by the authors of this chapter was used for the purpose. The modeling results confirmed that the recession in groundwater levels can be attributed to massive groundwater abstraction through those groundwater intakes.

17.5.4 Groundwater Quality

Regular studies of the groundwater condition carried out by the Hydrogeological Service of the Kyrgyz Republic in accordance with the program of groundwater monitoring in the area of the Ala-Archa deposit, have discovered zones of pollution with nitrate and hexavalent chromium in the vicinity of Bishkek. The depth of the pollution zone is generally less than 100 m.

In addition, the chemical composition of groundwater aquifers deeper than 100 m was investigated (Morris et al., 2006). The outcome for the Bishkek area at 18 randomly selected sites (see Figure 17.10 for their locations

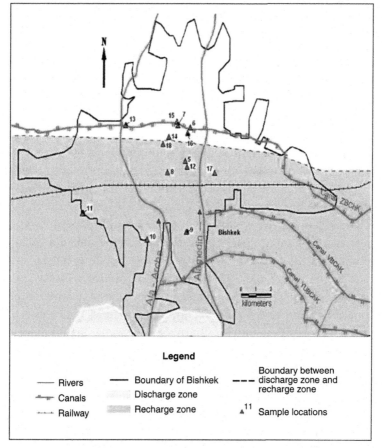

Figure 17.10 *Location of Sampling Sites for Groundwater Quality Testing in Bishkek.*

Table 17.5 Description of groundwater-sampling sites in terms of representation of various hydrogeological settings

Setting	Unsaturated zone (m)	Significance	Site No.
Unconfined piedmont fans	200–20	Zone of periurban coalescing coarse alluvial fans with deep unsaturated zone shallowing northward into southern suburbs and city center	5, 8, 9, 10, 11, 12, 17
Unconfined alluvial outwash plain	20–0	Urbanized zone with industrial, residential, and commercial land use; relatively vulnerable due to shallow water table and absence of low-permeability surficial silty clay layer	14, 16, 18
Confined alluvial outwash plain	Artesian	As previous zone but much lower aquifer vulnerability due to artesian conditions, surficial silty clay layer, and silt/clay bed interstratification	6, 7, 13, 15

Source: After Morris et al. (2006).

and Table 17.5 for details of the sites) is shown in Table 17.6. Results show the dominance of high-quality groundwater in Bishkek's aquifers. Water at all the sampling sites is suitable for potable use. Within the zone of groundwater formation, salinity ranges from 0.1–0.3 g/L to 0.4–1.0 g/L. The dominant type of mineralization is calcium hydrogencarbonate and magnesium. Chlorides and sulfates are also found near to the groundwater discharge zone.

Anthropogenic activities are contributing to groundwater pollution in some areas. The greatest anthropogenic pressure on the city is observed near the railway line in the west and the central parts of the Ala-Archa groundwater deposits. Wells situated in the western industrial zone of the city near to the railway, characterize the aquifer both in the plane and on the depth in the range of 30–250 m. Stable nitrate pollution surfaced since the beginning of the 1980s (Baeva and Tolstihina, 2001) and now covers a large area in the southwest and western industrial zones of the city, and along the railway line. Although chromium pollution also covers a large area, its level is small. Pollution by salts of total hardness is found in local sites at groundwater intake areas mainly in the western part of the city.

Nitrate Pollution

Nitrate pollution is widely observed in the groundwater of the Ala-Archa deposit. The length and width of the nitrate pollution halo, taken along the line of maximum permissible concentration (MPC = 45 mg/L) is 7–8 km.

Table 17.6 Selected minor inorganic substances in Bishkek groundwater (Morris et al., 2006). Please refer Figure 17.10 for the location of sampling sites

Site No.	Ca	Mg	Na	K	Cl	SO$_4$	HCO	NO$_3$, N	SiO$_2$	B	Sr	Fe	Zn	T	pH
	(mg/L)									(µg/L)				(°C)	
5	88	8.11	21	3.3	14	46.1	244	8	13.9	143	510	85	13	16.7	7.33
6	95	8.78	30.2	2.9	15	52.7	253	7.9	16.7	158	608	32	13	18.3	7.2
7	55	6.62	15	1.6	11	34.8	162	4.2	13.9	70	420	52	12	12.6	7.72
8	51	5.56	13.2	1.3	7.3	32.2	152	3.2	12.6	46	341	53	10	12.5	7.83
9	79	7.05	17.4	2.5	12	37.1	218	6.4	12.6	129	450	46	12	14.7	7.52
10	54	5.65	15	1.4	6.4	32.8	162	3.3	12.1	52	347	52	11	12.4	7.85
11	101	12.8	26.4	2.8	11	46.9	279	12.7	17.2	163	672	60	14	13.5	7.31
12	90	7.67	22.2	3.2	14	44.8	235	8.3	15.3	152	471	70	16	18.5	7.36
13	112	11.2	34	2.4	37	65.3	246	16	14.5	286	684	30	14	13	7.6
14	164	18.7	50.8	3.5	52	158	314	27.1	19.4	47	572	38	18	13.9	6.78
15	91	9.65	26.1	2.2	15	54.8	238	8.3	14.7	155	635	33	13	12.8	7.42
16	215	21.9	101	4.8	94	169	344	66.8	18.8	45	618	63	17	18	6.75
17	94	9.58	30.3	.3	26	41	267	8.3	16.1	123	664	46	13	13.2	7.28
18	96	9.64	28.4	2.5	19	55.8	256	10.8	15.1	157	646	59	14	11	7.36

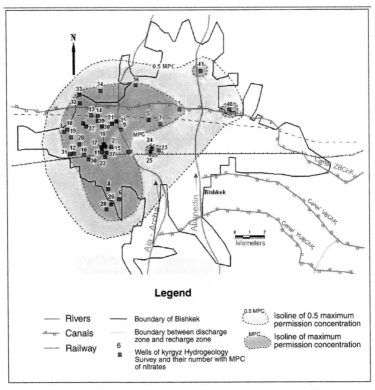

Figure 17.11 *Nitrate Pollution in the Groundwater of the Ala-Archa Deposit in Bishkek*
Source: Schematic map drawn according to data of Kyrgyz Hydrogeology Survey.

As shown in Figure 17.11, nitrate concentration in some wells exceeds the MPC. The pollution halo's elliptical shape extends along the groundwater flow in the southwest part of the city. The area of intensive nitrate pollution extends to 40–50 km². The boundary of the groundwater pollution halo runs along the groundwater intakes that capture the upper part of the producing aquifer where maximum pollution is observed. The boundary of the transitive area of groundwater pollution with nitrate concentrations of above 0.5 MPC covers almost the entire area of the city. Of 300 tested wells, only 41 had nitrate pollution (i.e., NO_3: 45–111 mg/L) and 20 more had nitrate concentrations exceeding 40 mg/L (i.e., 0.9 MPC) (Table 17.7).

The dynamics of nitrate pollution with time and aquifer depth are shown in Figure 17.12. Along the depth, the upper 70–110 m of aquifer contains the highest nitrate concentration (exceeding MPC) and then decreases gradually and does not exceed MPC in the depth interval of 150–230 m. In the northwest part of the city, the highest level of nitrate pollution, 111 mg/L at depths 41–75 m (Well No. 26), is discovered. Nitrate

Table 17.7 Nitrate concentrations (NO₃) in wells exceeding MPC of 45 mg/L

Well No.	NO$_3$ (mg/L)	Well No.	NO$_3$ (mg/L)	Well No.	NO$_3$ (mg/L)
1	50	15	73	29	62
2	49	16	66	30	59
3	53	17	64	31	50
4	56	18	80	32	53
5	57	19	83	33	52
6	60	20	50	34	53
7	56	21	61	35	95
8	70	22	47	36	48
9	53	23	46	37	74
10	60	24	45	38	52
11	84	25	48	39	64
12	45	26	111	40	50
13	64	27	75	41	54
14	58	28	59		

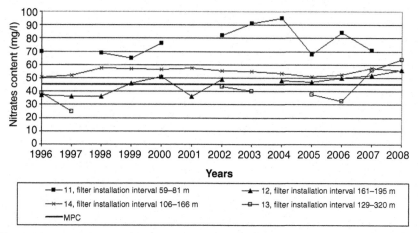

Figure 17.12 *Dynamics of Nitrate Pollution in the Groundwater of Selected Wells in Bishkek. Source: Drawn by the authors with data from Kyrgyz Hydrogeology Survey.*

concentrations at other wells are in the range of 50–93 mg/L (i.e., consistently exceeding MPC).

The worst situation is seen in the territory of the former Agricultural Machinery Plant (Wells No. 15–17). The former powerful production facility was demolished, and now a large number of various small industrial businesses (e.g., foundry production, leather industry, papermaking, construction operations, food production, etc.) are in operation on the site.

Severe groundwater pollution, with a nitrate concentration of $2 \times$ MPC (or 90 mg/L), was observed during 1999–2000. In subsequent years, it decreased slightly to 70 mg/L, and then to 50–57 mg/L by the end of 2008. However, the nitrate concentration still remains high enough to exceed MPC.

A condensed summary of nitrate pollution in Bishkek's groundwater is provided below;

• The Bishkek aquifer is generally polluted to a depth of 70–110 m. With increasing depth, nitrate pollution decreases and comes into line with the standards of a potable water supply.

• The western part of the city, along the railway line, where the Agricultural Machinery Plant was situated, has more nitrate pollution than other parts.

• No obvious trend in nitrate pollution has been observed for the last 10–15 years.

Hexavalent Chromium Pollution

The areal chemical pollution of groundwater with hexavalent chromium was discovered in the 1990s in the western part of the city in the land adjoining the railway line. Subsequent studies have revealed increased concentrations in other parts of the city (Figure 17.13). The results have led to the conclusion that several sources are responsible for the pollution. Ubiquitous testing of all city water intake areas revealed that 57 out of 300 wells have a chromium presence. Chromium concentrations in 14 wells have exceeded MPC (i.e., 0.05 mg/L). The chromium pollution map (Figure 17.13) developed from 2007 test results discovered three major zones and a number of local sites of pollution (Tolstihina and Ponomareva, 2008).

• *Zone 1:* This zone has a length, width, and area of 4–5 km, 1.5–2.0 km, and 6–8 km², respectively. The pollution extends to a depth of 30–300 m; however, there is no distinct trend as evident with nitrate. Boundaries with potential for transport of this contaminant are significantly large, with a length and width of 5–6 km each. The major source of pollution is the technogenic impact of large industrial enterprises. The unsatisfactory condition of industrial sites, electroplating plants, leakages of chromic effluents from old sewer systems, and inordinate storage of chromic reactants has led to pollution of soils, as well as to the subsequent pollution of the groundwater system. Spatial and temporal variation in chromium concentration in Bishkek's groundwater is provided in Table 17.8. For example, maximum chromium concentration of 0.142 mg/L within Zone 1 is observed at Well No. 11 located in the Youth Park (Soroka Lake) in 2007, exceeding MPC 2.8 times.

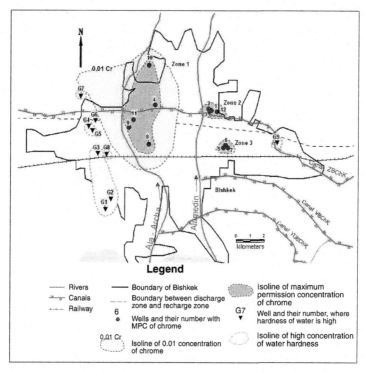

Figure 17.13 *Hexavalent Chromium Pollution in Groundwater of the Ala-Archa Deposits in Bishkek.* Source: *Schematic map drawn according to data of Kyrgyz Hydrogeology Survey.*

Table 17.8 Dynamics of hexavalent chromium pollution in selected wells in Bishkek

Well No.	Zone No.	Concentration of hexavalent chromium (mg/L) (MPC = 0.05)						
		2002	2003	2004	2005	2006	2007	2008
1	2	0.90	0.624	0.520	0.13			0.64
2	2	0.008	0.18	0.064	0.008			0.016
3	2		0.005	0.014		0.04	0.006	0.06
4	1				0.044	0.016	0.05	0.046
5	3			0.026		0.16	0.023	0.014
6	3					0.072	0.012	
7	3		0.176	0.048	0.11			
8	1	0.104	0.02		0.01	0.0792	0.056	0.08
9	1				0.012	0.0544	0.022	
10	1						0.053	
11	1						0.142	
12	2							0.07

• *Zone 2:* The pollution halo of this zone is about 1 km wide. The actual length is unknown due to the absence of observation wells. Chromium concentration at Well No. 1 in this zone in 2002 was 0.90 mg/L, which in 2008 was still excessively high at 0.6–0.64 mg/L or 12–12.5 MPC, at depths of 87–133 m. Chromium concentration in excess of MPC was observed in 4 out of 12 groundwater wells. The reasons for the pollution are the same as in Zone 1.

• *Zone 3:* The chromium pollution in this zone was first identified in 2003 in Wells 5, 6, and 7. The concentration was in a range of 0.01–0.176 mg/L (up to 3.5 times MPC). The concentration in water intake areas was in excess of MPC until 2006. During 2007–2008, it dropped and did not exceed 0.5 MPC.

Total Hardness

The groundwater in Bishkek is characterized by a higher concentration of total hardness, in a range of 3.0–5.0 meq/L. Higher concentrations of calcium and magnesium salts are present in water samples from all the groundwater intakes. Total hardness in areas with higher nitrate concentrations is observed up to 8–10.2 meq/L. Total hardness decreases with depth, with the minimum value observed at depths of 250–300 m. The halos of total hardness pollution form ellipse-like patches in the western part of Bishkek (Figure 17.13).

In addition, heavy metal contents are also observed in the groundwater samples. The manganese in groundwater is in a range of 0.03–0.04 mg/L (MPC = 0.1), copper in 0.02–0.04 mg/L (MPC = 1.0), zinc in 0.1–0.5 mg/L (MPC = 5.0), and aluminium, lead, and molybdenum at the level of natural background values or responsiveness of measuring instruments.

17.5.5 Recharge

The main source of aquifer recharge in Bishkek is groundwater flow from the southern side through the regional recharge zone. In the first decade of operation of urban water withdrawals (1965–1975), lateral inflow and other recharge sources did not fully compensate for the withdrawal of groundwater. In those years, there was a steady decrease in groundwater level, which according to observation at some wells, reached 12 m. However, since 1980, groundwater recharge began to cover withdrawals, and the steady reduction in groundwater level was halted.

17.6 IMPACTS

Major impacts on the groundwater environment of Bishkek due to various anthropogenic activities are: (i) depletion of groundwater level in the south; (ii) increase in groundwater level in the north; (iii) groundwater pollution.

17.6.1 Depletion of Groundwater Level in the South

As discussed in Section 17.5.3, groundwater level in the southern part of the city, the recharge zone, is depleting due to significant abstraction from groundwater intake sites. Figure 17.9 reveals a considerable decrease of 10–12 m in groundwater level during 1960–1980 in the wells located in the southern part of the city. A decline in groundwater level of up to 10 m was observed in the period of construction of groundwater intakes. The groundwater level after the 1980s, however, has more or less stabilized. It is estimated that volumes of groundwater intake in Bishkek could be increased by 20–30% without any negative impacts on the environment.

17.6.2 Rise in Groundwater Level in the North

Groundwater level increase and subsequent flooding has been a key issue in the northern part of the city, the discharge area, for a long time. This is a result of natural causes (the geological structure of the Chui Valley) and human activity causing the destruction of natural and man-made water collectors and drainage ways. For example, before the construction of the city bus terminal, the territory was a natural water collector, which could reduce groundwater level. The area was gradually filled up with earth and the city bus terminal was built on it. As a result, the underground communication system of the bus terminal is now flooded. Almost 40% of the territory in the northern part of the city is under groundwater flooding. It impacts negatively on city infrastructure and threatens human health through water-related diseases (e.g., malaria, tuberculosis). It is especially dangerous when the depth of the groundwater level is less than 1 m.

17.6.3 Degradation in Groundwater Quality

As discussed in Section 17.5.3, nitrate pollution, hexavalent chromium pollution, and total hardness pollution are observed in the groundwater of Bishkek. Nitrate and hexavalent chromium pollution are mainly due to anthropogenic activities such as discharge from industrial enterprises and sewage from the part of city lacking a centralized sewerage system. Zones of groundwater pollution with the aforementioned pollutants are shown in Figures 17.11 and 17.13.

17.7 RESPONSES

17.7.1 Groundwater Monitoring

The first monitoring well was drilled in Bishkek in 1933. Currently, monitoring is carried out by the Kyrgyz Hydrogeological Survey, Bishkekvodokanal, and the Department of Sanitary and Epidemiological Control. The number of observation wells in Bishkek has decreased from 16 to 7 over 20 years (1990 to 2009) due to a lack of financial resources as well as a decrease in the level of responsibility for implementation of governmental programs. Chemical composition of groundwater is determined by water samples taken from production wells. The current monitoring infrastructure seems inadequate to allow an understanding of the changes in groundwater level and quality over an adequate temporal and spatial scale. More investment is required to expand and strengthen the groundwater-monitoring infrastructure. Improved monitoring and tighter controls over private owners of production wells are also required. Their operation must comply strictly to existing licenses.

17.7.2 Scientific Studies for Drainage System Installation

To design and construct a workable and efficient drainage system in northern Bishkek to eradicate the groundwater-flooding problem, scientific studies were carried out. The aim was to lower the groundwater level to at least 2 m below ground surface (mbgl), the admissible depth in that area. Drains should be capable of draining water from the following four sources: (i) additional lateral inflow (caused by a decrease in groundwater level); (ii) additional inflow from the underlying layers; (iii) inversion of evaporation; and (iv) inversions of the existing drainage effluent. Estimates from those scientific studies are provided below.

Additional Lateral Inflow (Q_1)

The territory in the northern part of Bishkek can be represented in the form of a circular zone of 28.7 km^2, with a radius (R) of 3.02 km. Solution of the mathematical problem of radial filtration in a bedded water-bearing formation gives an estimate of Q_1. This filtration problem without evaporation inversion is solved in Averianov (1978). Considering evaporation inversion as the major factor for drainage water formation, the authors of this chapter modified the solution to the following equation (17.1) to estimate Q_1.

$$Q_1 = 2 \cdot \pi \cdot R \cdot \frac{T}{P_2} \cdot S_R \cdot \frac{K_1\left(\dfrac{R}{P_2}\right)}{K_0\left(\dfrac{R}{P_2}\right)} \qquad \text{(Equation 17.1)}$$

where T is transmissivity of the first aquifer (counted from the surface downward) and has the value of 500 m^2/d (from hydrogeological investigation); P_2 is a factor dependent upon T, k_2, a, and m_2 and is given by Equation (17.2):

$$P_2 = \sqrt{\frac{T \cdot m_2}{k_2 + am_2}} \qquad \text{(Equation 17.2)}$$

where a is a coefficient in the linear relationship between the receding water table and the evaporation inversion and given as $a = 0.0012$ 1/d (at the water table lowering by 1.5 m, the evaporation inversion will be 0.7 m/y, taking into account building on the area of 50%); k_2 and m_2 are the filtration coefficient and thickness of the lower semi-permeable layer, respectively ($k_2 = 0.01$ m/d and $m_2 = 10$ m); $K_0(x)$, $K_1(x)$ are the zero and first-order modified Bessel Function of the second kind; S_R is the water table lowered at the border of the studied area, and is taken as equal to 1.5 m (if considering the initial depth levels of groundwater as 0.5 m, and the forecasted depths on the border as 2 m).

The side inflow estimate based on Equation (17.1) is 370 L/s or 0.13 L/s per hectare.

Additional Inflow from the Underlying Layers (Q_2)
This estimate of this value, based on Equation (17.3), is 100 L/s or 0.35 L/s per hectare:

$$Q_2 = \frac{k_2 \cdot S_R \cdot \pi \cdot R_2}{m_2} \qquad \text{(Equation 17.3)}$$

Inversion of Evaporation (Q_3)
This value is estimated as 570 L/s or 0.2 L/s per hectare.

Inversion of the Existing Drainage Effluent (Q_4)
The existing discharge to the available drainages is 600 L/s. It is supposed that the inversion will make 50% of the existing discharge or 0.1 L/s per hectare.

Total discharge to be drained by the drainage system, estimated as a sum of the above four components, is estimated as 0.78 L/s per hectare. It means the average drainage module should not exceed 0.78 L/s per hectare.

17.7.3 Policy and Legal Reforms

The State Agency on Geology and Mineral Resources of Kyrgyzstan, the State Agency on Environment Protection and Forestry, and the Ministry of Health are the ministries responsible for determining environmental standard guidelines concerning groundwater. The main laws and standards for the protection and sustainable use of groundwater in the Kyrgyz Republic are listed below.

- Water Code of the Kyrgyz Republic (with alterations and amendments as of October 26, 2013)
- Law of Kyrgyz Republic "About Drinking Water" (last modified October 10, 2012)
- Law of Kyrgyz Republic "About Subsurface Resources" (last modified May 24, 2014)
- Law of Kyrgyz Republic, "Technical Regulations on the Safety of Drinking Water" (May 30, 2011)
- Act of the Kyrgyz Republic "Provision on the Protection of Groundwater in the Kyrgyz Republic" (March 2, 2015)
- Sanitary–Epidemiological Rules and Norms, No. 2.1.4.018-07, Ministry of Health of Kyrgyz Republic (November 1, 2007)

17.8 CONCLUSIONS AND RECOMMENDATIONS

The hydrogeological setting and general conditions in the Bishkek zone described in this chapter provide an understanding of the groundwater systems in the city. In addition, analysis by the government, as well as the interrelationships of key factors such as drivers, pressures, state, impacts, and responses, helped to understand the current situation of the groundwater environment, which would be the starting point for any future programs or interventions aimed at groundwater conservation and management.

Based on the groundwater level regime in the southern part of Bishkek, the following measures to improve the effectiveness of groundwater monitoring are proposed:

- In the available observation wells it is quite sufficient to measure the groundwater level at a monthly frequency, even though the currently established standard makes provision for a frequency of 5 times a month.

- In the four closed observation wells with rather longer observation records (i.e., Wells 45, 47, 858, and 1048) it is recommended to restore the missing data by means of calculations (Litvak, R.G. et al., 2013). This can be accomplished to a high degree of accuracy with the help of multivariate regression. It is recommended, wherever practical, to restore field measurement in the closed wells at least 2–4 times a year for the purpose of cross-check analysis.
- In abnormal situations (in cases of essential deviation from the observed regimes from the long-standing regularities) it is recommended to use groundwater modeling and additional observations to identify causes of observed phenomena.

Based on scientific studies on a drainage system design to lower groundwater level by at least 2 mbgl in the northern part of the city, which is affected by the groundwater-flooding issue, it was found that the capacity of an average drainage modulus should exceed 0.78 L/s per hectare.

Based on groundwater pollution studies, the following measures on rational groundwater use and the prevention of further pollution are proposed:

- Timely reconstruction of available sewage conduits, and construction of new ones in accordance with the growing requirements of the city.
- All large industrial enterprises of the city should inspect their own primary treatment facilities, effluent neutralization stations, sewage ponds, and sewage collectors in order to provide for elimination of polluted flows seepage to the aquifer.
- It is suggested that pumping wells (depth 60–80 m) be put into operation for irrigation of green planting and other technical needs (especially in places of unavoidable pollution). Contaminated water pumping from the top part of the aquifer contributes to the improvement of the groundwater conditions.
- To ensure regular governmental control, first control the activities of, and take action on, already revealed and potential groundwater polluters, which are the businesses such as leather making, fur production, dye and galvanic shops, producers of detergents, and fuel stations.
- For the future development of Bishkek city, when deciding on the location of various enterprises, it is necessary to take into consideration the groundwater vulnerability map of the city. One version of the mentioned map has been improved on by the authors of this paper and submitted for consideration to the Bishkek Mayor's Office.

ACKNOWLEDGMENTS

The authors acknowledge the Kyrgyz Hydrogeology Survey for making available hydrogeology information; the British Geology Survey, especially Brian L. Morris, for sampling and chemical analysis; and Jyldyz Aidakeeva for help in preparing this chapter.

REFERENCES

Averianov, S.F., 1978. Salinization Control of the Irrigated Lands, Kolos, Moscow, pp. 288.

Baeva, N.B., Tolstihina, G.G., 2001. The Environmental and Hydrogeological Conditions of Kyrgyzstan. Collection Water and Sustainable Development of Central Asia, Soros Kyrgyzstan Foundation, Bishkek, pp. 126–130.

Grigorenko, P.G., 1979. Groundwater of the River Chu Basin and Prospects of Their Use. Ilim, Frunze, pp. 187.

Hydrogeology of USSR, 1971. Hydrogeology of the USSR (Kirghizia), vol. 40. Nedra, Moscow, pp. 487.

Litvak, R.G., Nemaltseva, E.I., Prilepskaya, S.V., et al., 2013. Scientific Rationale and Development of the Chu Valley Groundwater Monitoring Plan on the Groundwater Modelling Basis. Scientific Report. Institute of Water Problems and Water-Power Engineering, Academy of Sciences of Kyrgyzstan, Bishkek, 55.

Ljanov, T.D., Mamrenko, A.V., 1995. Report on the Results of the Detailed Assessment of Fresh Groundwater of the Existing Ala-Archa Groundwater Basin in the Central Part of the Chu Valley. Kyrgyz Hydrogeology Survey, Bishkek, pp. 355.

Morris, B.L., Darling, W.G., Gooddy, D.C., Litvak, R.G., Neumann, I., Nemaltseva, E.I., Poddubnaya, I.V., 2006. Assessing the extent of induced leakage to an urban aquifer using environmental tracers: an example from Bishkek, capital of Kyrgyzstan, Central Asia. Hydrogeol. J. 14, 225–243.

Tolstihina, G.G., Ponomareva, E.Z., 2008. Groundwater Monitoring in Northern Areas of the Kyrgyz Republic, Kyrgyz Hydrogeology Survey, Bishkek, pp. 306.

Tolstihin, G.M., 2005. Scientific basis of potable groundwater resource management in the Kyrgyz Republic. Candidate of Geological and Mineralogical Sciences Thesis, Bishkek, pp. 112.

CHAPTER 18

Groundwater Environment in Seoul, Republic of Korea

Heejung Kim*, Jin-Yong Lee**, Woo-Hyun Jeon**, and Kang-Kun Lee*
*School of Earth and Environmental Sciences, Seoul National University, Seoul, Republic of Korea
**Department of Geology, Kangwon National University, Chuncheon, Republic of Korea

18.1 INTRODUCTION

The Republic of Korea (hereafter referred to as Korea) is located in Northeast Asia (Figure 18.1a), on the southern half of the Korean Peninsula, jutting out from the far east of the Asian landmass. Seoul, the capital city of Korea, is located at $37°25'-37°41'$N and $126°45'-127° 11'$E in the midwest of the Korean Peninsula (Figure 18.1b), covering a total area of 605.5 km^2 (Kim et al., 2001, 2011). The Han River flows through the city from east to west, dividing it into north and south (SMG, 2015). The city consists of 25 districts (Figure 18.1c). The population of the city has grown rapidly since the Korean War to about 10 million – representing around 20% of the total population of Korea (KOSIS, 2014). Seoul is well known for its high population density and its metropolitan area is one of the most densely populated cities in Asia (OECD, 2012). Economic growth and industrialization have stimulated population growth in Seoul, and this growth has been profitable to the city as it has become a center of governance, education, culture, commerce, and production.

Groundwater represents about 10% of the water supply in the city (Choi et al., 2005). The use of groundwater for drinking is increasing continuously. However, in 2000, the total use of groundwater in Seoul was estimated to be only about 0.1 m^3/y from about 14,921 wells (Table 18.1), which is considered to be about 23% of the estimated sustainable groundwater yield (Yun et al., 2000; Choi et al., 2005). However, the city has also experienced subsurface environmental problems such as a degradation of groundwater quality, water level decline, increasing groundwater temperature, and frequent occurrences of ground subsidence. The subsurface environment, especially groundwater, plays an important role in urban areas (Jago-on et al., 2009). The city has undergone various changes and has been seriously disrupted by certain human activities, such as the increasing coverage of the land surface,

Groundwater Environment in Asian Cities
http://dx.doi.org/10.1016/B978-0-12-803166-7.00018-0

Copyright © 2016 Elsevier Inc.
All rights reserved.

413

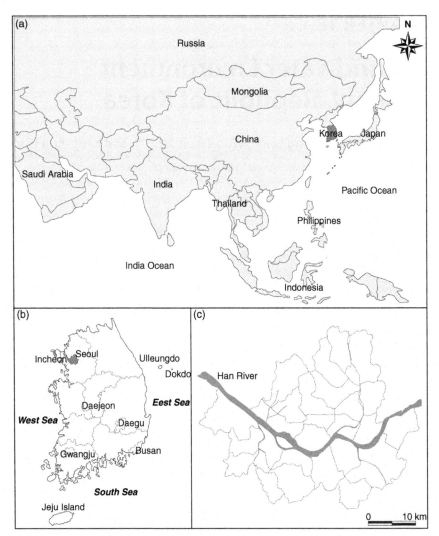

Figure 18.1 *Location of the Metropolitan City of Seoul, Korea.* (a) Northeast Asia, (b) Southern part of the Korean Peninsula, and (c) Seoul city.

construction of many large buildings, and expanding subway lines – all of which has made the urban groundwater system vulnerable to depletion and pollution. Due to various pollution sources and climate change (i.e., warming and the heat island effect) in urban areas, the quality and quantity of groundwater have become important issues for the urban groundwater environment (Collin and Melloul, 2003). The air temperature of the metropolitan city is higher than in rural regions due to artificial heat, air pollution,

Table 18.1 Status of DPSIR indicators in Seoul, Korea

	Indicators	1970	1990	2010
Drivers	Population growth (SMG, 2015)	Population: 5.43×10^6 (Figure 18.5)	10.6×10^6	10.5×10^6
		Population density: 8,863 persons/km²	17,532	7,473
	Urbanizations	Urban area: 36.0 km², in 1913	597.9 km² (1963)	605.5 km²
		Urban population: 5.43×10^6	10.6×10^6	10.5×10^6
	Climate change	Air temperature: 11.4°C (Figure 18.6)	12.8°C	12.1°C
Pressures	Land cover change	Buildings and roads area: 193.76 km²	267.02 km²	292.45 km²
		Agricultural area: 112.01 km² (Figure 18.7)	54.00 km²	28.45 km²
	Water (SMG, 2015) and groundwater use (NGIC, 2015)	100% of city residents are supplied with water by the piped public water supply system; 1334 MCM in 2003 and 1194 MCM in 2010 Groundwater use: 40.9 MCM in 2000 and 23.8 MCM in 2010 (Figure 18.8)		
State	Well statistics (NGIC, 2015)	Total production wells: 14,921 wells in 2000 and 10,166 wells in 2010		
		Total monitoring wells: 118 wells in 2005 and 223 wells in 2014		
	GW abstraction (NGIC, 2015)	40.8 MCM/y in 2000 and 24.0 MCM/y in 2010 (Figure 18.8)		
	GW level	GW level decrease in the central building and commercial areas (Figure 18.10a, b)		
	GW quality	GW qualities deteriorate in the building area (Table 18.4; Figure 18.11a, b)		
	GW temperature	11.6–18.3°C (average 15.2°C) during 2000–2005; 13.4–17.6°C (average 15.5°C) during 2010–2014		
		Recharge: 450 MCM/y by natural precipitation in 2001; 670 MCM/y by leakage from water supply and sewerage systems		

(Continued)

Table 18.1 Status of DPSIR indicators in Seoul, Korea *(cont.)*

	Indicators	1970	1990	2010
Impacts	Depletion in GW level (depth to water table, m)	0.51–27.61 m (average 8.25 m; n = 118) during 2000–2005		
		0.80–38.90 m (average 8.55 m; n = 83) during 2010–2014		
	Land subsidence	Accidents due to land subsidence increase year on year (Table 18.5)		
Responses	Establishment of an organization in charge of GW management	Groundwater team in the city and district in charge of groundwater affairs		
	GW monitoring (Kim, 2000)	Installation and operation of a local groundwater-monitoring network since 1997		
	Soil and GW environment information system	The system was launched in 2012, but its use is limited to official use only and is yet to open to the public		
	Reuse of GW seepage	Only 11.3% of GW seepage is reused, and much of the seeped groundwater is still directly discharged to ditches		
	Efforts for mitigating land subsidence	Replacement of superannuated sewer lines, investigation for potential subsidence areas, and a cooperation agreement with Tokyo, Japan		

DPSIR, driver–pressure–state–impact–response framework; GW, groundwater; MCM, million cubic meters.

and diminished green tracts, which can also cause groundwater temperatures to increase (Lee, 2006). The possibility of groundwater contamination is also higher in urban areas (Klimas, 1995; King, 2003). This groundwater contamination is closely linked to land use.

Therefore, contaminated groundwater is frequently observed in regions where industrial facilities are densely distributed (Nazari et al., 1993; Cox et al., 1996). Nitrate (NO_3), trichloroethylene (TCE), and tetrachloroethylene (PCE) are well-known contaminants in urban groundwater (Lee and Lee, 2004; Park et al., 2005). In particular, nitrate contaminants in urban groundwater mainly originate from leaking sewerage (Eiswirth et al., 2004). Subway tunnel seepage also affects groundwater quality (Chae et al., 2008).

18.2 ABOUT THE CITY

The geology of Seoul comprises Precambrian banded biotite gneiss, age unknown porphyritic gneiss, Jurassic granite, and Quaternary alluvium (Figure 18.2). The banded gneiss is mainly outcropped in the central and southern regions of Seoul. During the Jurassic period, granite intruded into

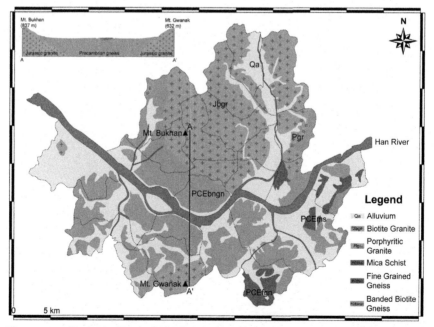

Figure 18.2 *Geological Map of Seoul Showing a Geological Section A-A. Source: Geology data are from the Korea Institute of Geoscience and Mineral Resources; www.kigam.re.kr.*

metamorphic rocks. In addition, Cretaceous felsites and dikes intruded into the bedrocks. Geological structures such as joints, folds, and faults were elaborately developed, and the inclines occurring on the schist are dominant in the southeast or northwest. The Seoul batholiths are known as biotite granite (Kwon et al., 1994). The granites belong to subsolvus biotite monzogranite (Hong, 1984). The area of Quaternary alluvium deposits adjacent to the Han River is topographically gently undulating with varying thicknesses (Kim et al., 1998; Lee et al., 2003).

Aquifers are mainly developed in the unconsolidated alluvium of the Han River and its tributaries, and fractured crystalline bedrocks such as gneiss, granite, and schist (Kim et al., 2001). The alluvium is mainly composed of coarse to fine-grained sediments with variable permeability. High hydraulic conductivity (10^{-2}–10^{-1} cm/s) is associated with the alluvial aquifers up to 20 m depth (Kim et al., 1998; Figure 18.3a), while conductivities decrease with greater depth. The hydraulic conductivity values for fractured bedrocks range from 2.3×10^{-5} to 7.4×10^{-3} cm/s (Kim, 2000). Groundwater flows from the surrounding mountainous regions toward the Han River. The alluvial aquifer system adjacent to the Han River is characterized by fluctuating water tables reflecting the tidal variations in river levels affected by the tide of the Yellow Sea in the west (Kim, 2000). The aquifer system and groundwater levels are largely affected by the operation and maintenance of subway lines where groundwater seepage into the tunnels is pumped and mostly discharged into ditches (Kim and Lee, 1999; Figure 18.3b).

The climate of the city shows four distinct seasons with the obvious seasonal variations in air temperature being largely influenced by the North

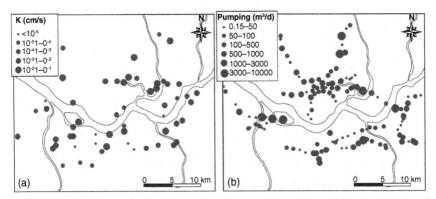

Figure 18.3 *Distribution of (a) Hydraulic Conductivity (K, cm/s) and (b) Removal Rates of Groundwater Seepage Caused by the Operation of Subway Stations (Kim and Lee, 1999).*

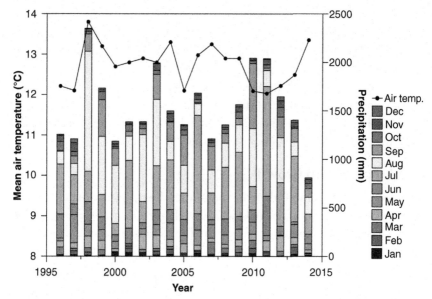

Figure 18.4 *Annual Mean Air Temperature and Precipitation from 1996 to 2014.* Source: www.kma.go.kr.

Pacific high-pressure system (Kim, 2000). From 1996 to 2014, mean annual air temperature and precipitation were 12.7°C and 1532.5 mm, respectively (Figure 18.4). Precipitation is mostly rainfall with a lesser amount of snowfall. The dry and wet seasons are generally October–May and June–September, respectively (Lee and Lee, 2000). In particular, over 70% of annual precipitation is concentrated in the wet season of June–September due to the monsoon, which is characteristic of the climate in the Korean Peninsula (Lee and Lee, 2000; Choi et al., 2005). Maximum temperatures of 38.4°C for summer, and a minimum of −16.4°C for winter were recorded over the past 20 years (KMA, 2015).

18.3 DRIVERS

We evaluated the groundwater environment in the urban area of Seoul, through collected groundwater data from the groundwater level, electrical conductivity (EC), water temperature, ground subsidence, and concentrations of NO_3, TCE, and PCE for evaluation of the groundwater environment using the DPSIR approach. Groundwater levels, EC, and water temperature data were obtained from the Seoul metropolitan groundwater-monitoring wells that have been in operation since 1997. However, monitoring data is only available from January 2000 to December 2005 and from January 2010 to December 2014. Daily average values were used to evaluate the variation and distribution of the three parameters. Nitrate concentration data for 2005 from 1998 wells was also obtained. In addition, groundwater quality data, including TCE and PCE, were gathered from 206 to 286 wells per year for the period 2004–2010 from the Korean Ministry of Environment (KME, 2015). Ground subsidence data for 2010–2014 were collected from Seoul Metropolitan Government (SMG, 2015).

18.3.1 Population Growth

The city is home to more than 10 million residents, approximately 20% of the nation's total population (KOSIS, 2014; Figure 18.5). Due to the massive population migration from many rural areas to Seoul in search of better jobs, industrial workers accounted for 52% of the total city population in

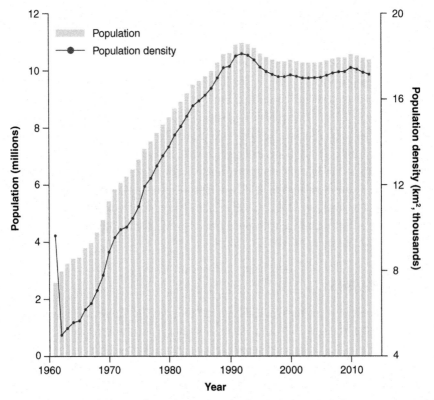

Figure 18.5 *Population and Population Density from 1960 to 2013. Source: www.seoul. go.kr.*

1970 (Kim et al., 2001; McGee, 2001). The city's early stage of urbanization from the 1970s to the 1990s shows an explosion in the urban population (Table 18.1).

In effect, the population of the Seoul metropolitan city area grew rapidly following the Korean War and peaked in the 1990s (Figure 18.5). The city has subsequently maintained a rather stable population of around 10 million (population density of 17,532 persons/km^2) over the past two decades. Therefore, Seoul is one of the biggest cities and the sixth most densely populated area in the world (SMG, 2015). According to the Korea Research Institute for Human Settlements (KRIHS, 2009), the population density of Seoul is the highest among large cities in advanced nations belonging to the Organisation for Economic Co-operation and Development (OECD, 2012). Even though there has been no increase in residents of the city itself since 2010, populations in the surrounding (satellite) cities have skyrocketed and the number of regular commuters into the city has also increased substantially.

18.3.2 Urbanization

The city currently covers an area of 605.5 km^2 (2015). It has undergone significant expansion since 1913 (36 km^2), as shown in Table 18.1. The area increased massively from 36 km^2 in 1913 to 597.9 km^2 in 1963. Furthermore, the population also increased enormously from 5.43 million in 1970 to 10.6 million in 1990. In 2010, the population density amounted to 17,473 persons/km^2 (Table 18.1). Urbanization of this city is characterized by the continuous construction of large commercial buildings and extension of subway lines.

18.3.3 Climate Change

Urban areas, with their large volume of impermeable pavements, and building surfaces that have low reflectance and increased thermal storage compared with agricultural areas, are subject to an urban heat island effect (Feyisa et al., 2014; Zhu et al., 2015). Furthermore, urban areas have a large energy demand for the cooling and heating systems of various buildings and facilities (Jo et al., 2009). The urban heat island effect is reflected in the shallow subsurface and hence groundwater temperature (Taniguchi et al., 2007; Zhu et al., 2015). Over the last five decades (1970–2014), the annual mean air temperature of the city showed a marked increasing trend of +0.03°C/y, which is closely related to the urban heat island effect and global climate change (warming) (Figure 18.6).

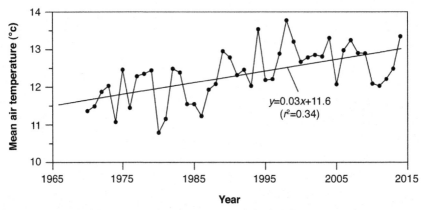

Figure 18.6 *Annual Averages of Air Temperatures of Seoul for 1970–2014. Source: Data are from the Korea Meteorological Administration; www.kma.go.kr.*

18.4 PRESSURES

18.4.1 Land Cover Change

Land cover/use data for 1970, 1990, and 2010 from the Seoul Metropolitan Government (SMG, 2015) are shown in Figure 18.7. The fast pace of urban growth between 1970 and 1990 led to significant changes in land use (Table 18.1). The main changes were reflected in the large reduction of agricultural fields and forest, and sharp increase in buildings and roads. There was a significant increase in buildings and roads between 1970 and 1990, from 25.7% to 33.8% and from 6.3% to 10.3%, respectively. However, agricultural fields and forests largely decreased from 18.5% to 9.0%, and 31.7% to 27.1%, respectively, during the same period. However, factories occupied only a small proportion of land, and remained almost unchanged.

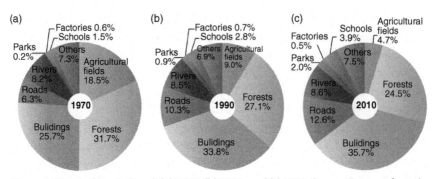

Figure 18.7 *Land Use in Seoul: (a) 1970, (b) 1990, and (c) 2010. Source: Data are from the SMG; www.seoul.go.kr.*

The statistics for 2010 show further land use change, with the reduction of agricultural fields and forests, and the expansion of buildings and roads. For 2015, of the total 605.5 km^2 area in the city, green tracts and open spaces account for 241 km^2 (39.8%), buildings 243 km^2 (40.1%), and transportation facilities and roads 77 km^2 (12.7%). Buildings are scattered all over the city. However, green tracts and open spaces are mostly distributed near the Han River, and the north and south city boundary areas.

18.4.2 Water and Groundwater Use

All the residents of Seoul have been supplied with water from the public municipal piped waterworks system since 2004 (Choi et al., 2005; Lee et al., 2005a). The piped water supply system predominantly depends on surface water from the Han River. The amount of supplied water was 1,334 million cubic meters (MCM) in 2003, decreasing by 10.5% to 1194 MCM in 2010. This is largely attributable to a decrease in the amount lost through leakage, thus enhancing the usage rate. Daily water use in the city, which was 284 liters per capita per day (lpcd) in 2010, increased to 286 lpcd in 2014, which is approximately six times greater than Washington DC, and three times greater than New York (Money Today, 2014). Meanwhile, the use of groundwater from wells has also dropped by one half since 1996 (Figure 18.8). The large decrease in groundwater use can be mostly attributed to the full implementation of the municipal waterworks and the relocation of a large number of factories. Even though groundwater use from wells in the city has decreased substantially, groundwater seepage (most of which is not recycled, but is instead discharged directly into ditches) has greatly increased. This is due to the construction of many tall buildings with deep foundations, and the expansion of subway lines, depleting about three times as much water than is being pumped from groundwater wells (Inews24, 2014).

18.5 STATE

18.5.1 Well Statistics

In the city, there were 14,593 production wells in 1996, 14,921 wells in 2000, and 10,166 wells in 2010 (SMG, 2006; NGIC, 2015). Annual production per well for each of those years was 3242, 2743, and 2346 m^3/y, respectively. In 2010, 6508 (64.0%), 3270 (32.2%), 317 (3.1%), and 71 (0.7%) wells were developed for living, agricultural, industrial, and other purposes, respectively (see Figure 18.8). Many of the wells are shallow (<20 m, 47.4%) but

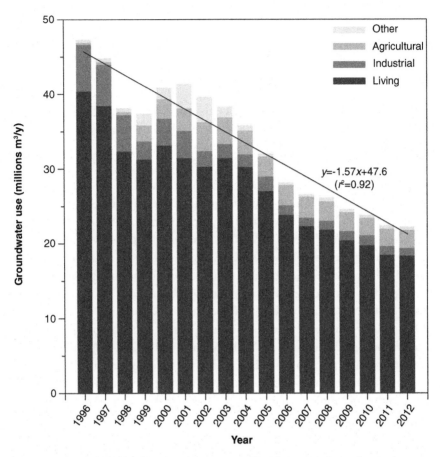

Figure 18.8 *Groundwater Use for 1996–2012 in Seoul. Source: Groundwater data are from the National Groundwater Information Center; www.gims.go.kr.*

some deep wells over 100 m (10.6%) were installed, mostly in the fractured rock aquifers (SMG, 2006). Meanwhile, Figure 18.9 shows the location of the metropolitan groundwater-monitoring wells (a total of 118), operating since 1997 (Lee et al., 2005b). The black and gray dots mean old and new monitoring wells, respectively. The monitoring wells are evenly distributed throughout the city, and an average of 4.7 groundwater-monitoring wells were installed for each district (Kim et al., 2011).

18.5.2 Groundwater Abstraction

In Seoul, 2–10% of total water use originates from groundwater (Choi et al., 2005). Domestic use predominates, followed by agriculture, which is

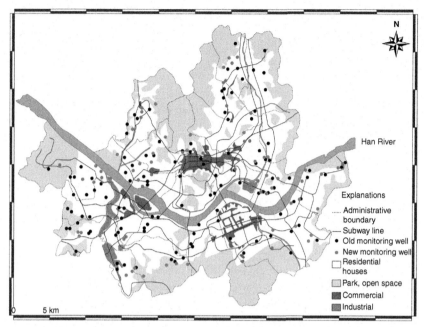

Figure 18.9 *Distribution of Metropolitan Groundwater-Monitoring Wells, Subway Lines, and Land Use of the Study Area.*

only practiced in the outermost parts of the city (Yang et al., 1997). However, the volume of abstraction has been steadily dropping from 1996 to 2012: from 47 MCM in 1996 to 22 MCM in 2012 – a decrease of 47% (see Figure 18.8). Even though groundwater abstraction from wells has continually decreased, groundwater removal (mostly directly discharged to ditches) for seepage control, maintenance of building construction sites, and expanding subway tunnels, is largely increasing.

18.5.3 Groundwater Level

Contour maps of mean groundwater levels for 2000–2005 and 2010–2014 are shown in Figure 18.10a,b. The mean groundwater levels for 2000–2005 were relatively high compared with those from 2010–2014. The groundwater levels of the central and southwestern regions were much lower than those of other regions. Generally, the decline in water tables is likely to be caused by overabstraction or pumping in urban areas for various purposes (Yang et al., 1997). However, the central and southwestern regions are heavy commercial and industrial districts, respectively (see Figure 18.9), and most of the residential and industrial water is supplied from the waterworks in these regions. In addition, the use of groundwater for residential and

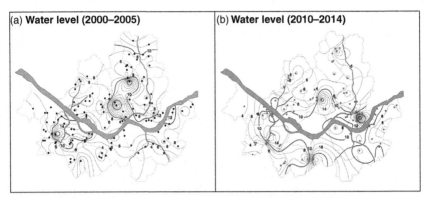

Figure 18.10 *Distribution of Mean Groundwater Level (Depth to Water below Ground Surface) for (a) 2000–2005 and (b) 2010–2014, respectively. Source: www.seoul.go.kr.*

industrial purposes has gradually decreased in the city (see Figure 18.8). Therefore, it is difficult to explain the decrease in groundwater levels by overabstraction alone. The low groundwater levels in these regions is due to the increase of impermeable pavements, causing a reduction in groundwater recharge, and the pumping (to direct discharge) of groundwater for the operation and maintenance of subway lines, super high-rise building sites, and numerous underground parking lots.

Urbanization and the attendant increasing population causes an increase in impermeable surfaces and a decrease in green tracts and open spaces, by the expansion of large-scale residential land development and urban infrastructure (Ku et al., 1992; Kim et al., 2001; Slonecker et al., 2001; Lerner, 2002). In addition, while most precipitation permeated sufficiently well underground before urbanization, well-organized drain systems and impermeable lined streams have brought about a large decrease in groundwater recharge and an increase in hydraulic velocity after urbanization (Otto et al., 2002; Lee et al., 2006). This has caused an additional decline in the groundwater levels of metropolitan cities (Kim et al., 1998; Shah et al., 2001, 2003; Lee et al., 2003; Morris et al., 2003; Lee et al., 2005b; Vázquez-Suñé et al., 2005; Chae et al., 2010; Chung, 2010; Park et al., 2011; Su et al., 2011). Taking into consideration future plans for the construction of numerous skyscrapers (sometimes over 100 stories), further declines in groundwater levels are expected, especially in commercial areas.

18.5.4 Groundwater Quality

In 2005 there were 1998 groundwater wells investigated for groundwater quality in the city and these can be classified by three different land uses

Table 18.2 Concentrations of nitrate (as NO$_3$-N) in groundwater wells for land use in Seoul

Land cover/use	Area (km²)	Number of wells	Concentration of NO$_3$-N (mg/L)		No. of wells with NO$_3$-N above 10 mg/L
			Max	Min	
I*	237	1563	84	0.001	362
II**	70.1	68	20.8	0.001	10
III†	298.4	367	80	0.1	91
Total	605.5	1998	84	0.001	463

* I – Building area.
** II – Roads and others.
† III – Open space (i.e., forests, agricultural areas, rivers, and parks).

(Table 18.2). Concentrations of NO$_3$-N over 10 mg/L (Korean Drinking Water Standard) were observed at 463 wells (23.2%), indicating that almost one fourth of the city's groundwater wells were contaminated with nitrate. Approximately 78.2% of the wells contaminated by NO$_3$-N (>10 mg/L) were distributed in building areas including residential, commercial, and industrial land. There are more complicated networks of sewage lines for Land Use I than in Land Use III (forests, green tracts, and agricultural area).

However, high concentrations of NO$_3$-N (>10 mg/L) were also found in Land Use III regions. This means that the use of fertilizer has a greater effect on the rise in NO$_3$-N concentrations than leakage from the sewerage system. The nitrate contaminant is directly penetrated underground in an open space, and categorized as Land Use III rather than II. In the case of Land Use II, high pavement coverage hinders the pollutant influx to the groundwater system. The concentration of NO$_3$-N in resident areas (Land Use I) ranged from 0.001 mg/L to 84.0 mg/L with 23.2% wells over 10 mg/L. The maximum concentration of NO$_3$-N in Land Use II was only 20.8 mg/L and 2.2% of wells were over 10 mg/L, which is the lowest percentage of NO$_3$-N contaminated wells. There are 91 wells (19.7%) from a total of 367 in Land Use III, showing concentrations of NO$_3$-N over 10 mg/L. The widespread groundwater contamination by nitrate is obvious in this city and is caused by leaking sewer lines in the inner areas and agriculture in the outer areas (Won et al., 2004).

Generally, EC values of uncontaminated groundwater in Korea range from 200 µS/cm to 400 µS/cm (Lee and Lee, 2004). Figure 18.11a,b show the distribution of the average EC for 2000–2005 and 2010–2014 in Seoul. The average EC ranged widely from 144 µS/cm to −1645 µS/cm

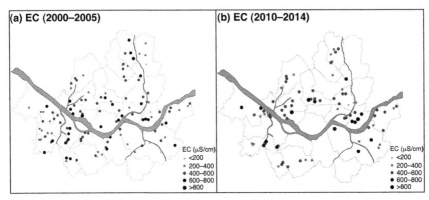

Figure 18.11 *Distribution of Mean EC (μS/cm) Values of Groundwater.* (a) 2000–2005 and (b) 2010–2014, respectively. *Source: www.seoul.go.kr.*

(mean = 478 μS/cm) and 116–2014 μS/cm (mean = 487 μS/cm), respectively. In Figure 18.11a, EC showed a range from 200 μS/cm to 400 μS/cm at 32 wells (27%), from 400 μS/cm to 600 μS/cm at 50 wells (42%), and from 600 μS/cm to −800 μS/cm at 6 wells (5%). In Figure 18.10b, EC ranged from 200 μS/cm to 400 μS/cm at 21 wells (25%), from 400 μS/cm to −600 μS/cm at 42 wells (50%), and from 600 μS/cm to 800 μS/cm at 4 wells (5%). Moreover, EC had values over 800 μS/cm at 2 (2%) and 4 wells (5%) during the two periods. The locations of the monitoring wells showing high values of EC (over 800 μS/cm) were in the Mapo and Nowon Districts of the city (Figure 18.11a). The Nanjido, Seoul's oldest official dumping site, constructed in 1978 (Kang et al., 2001), is located in the Mapo District. In the following 15 years, the Nanjido landfill site received urban waste and was transformed into a huge mountain of garbage, but the garbage on the landfill is no longer exposed (Kang et al., 2001) and it is now used for golf courses and camping sites for citizens (Kim et al., 2015).

The barrier walls around the landfill were constructed deep into the ground to prevent the seepage of contaminated water into the groundwater and Han River. Despite these efforts, the EC values of groundwater in the Mapo District are generally very high. The landfill is close to the Han River and leachage from the landfill was discharged into the river and groundwater. The leachage has had a negative effect on local groundwater quality (Kim et al., 2001). It seems that the Nanjido landfill is a main contamination source for groundwater in the district. Furthermore, the Nowon District, one of the most densely populated areas in Seoul, also showed high values of ECs in its groundwater. The groundwater quality in the most populated areas commonly fails to meet the Korean drinking water standards for nitrate, TCE, PCE, and bacteria (Sung et al., 1997). The superannuated leaking sewer systems in the city also have a profound effect on groundwater EC values. Figure 18.11b shows the high EC values in the Mapo, Jongro, Gangnam, and Songpa Districts – highly urbanized areas compared with other districts in the city. Consequently, it is noted that the groundwater resource of the city is endangered with respect to its quality. Groundwater contamination is not limited to certain districts, and is not just local, but widespread and regional.

For the evaluation of changing trends in ECs, linear regression analysis and Mann–Kendall analyses were conducted with respect to the daily mean value and the monthly median value (Tables 18.3 and 18.4). The linear regression results for daily mean values for the periods 2000–2005 and 2010–2014 indicated that 76 (64.4%) out of a total 118 monitoring wells and 43 (51.8%) out of 83 monitoring wells showed positive slopes, and the remainder (35.6% and 48.2%) showed negative slopes, respectively. The positive EC trends for over half of the monitoring wells in both periods are indicative of progressive groundwater contamination in the city (Park et al., 2011). The increasing ECs in the northern boundary are due to intensive horticulture (Figure 18.12a,b). The groundwater in this region is affected by applied fertilizers and pesticides. Mismanagement and inappropriate dumping of agricultural waste aggravates the groundwater quality. However, the administrative control of the city to prevent such illegal activity cannot reach this region. Groundwater contamination by uncontrolled agricultural waste is not limited to this city, and is widespread in the country.

The Mann–Kendall trend test also revealed similar increasing trends. Results of the trend analysis of the monthly median ECs for 2000–2005 and 2010–2014 showed that 34.7% and 22.9% of the total monitoring wells represented increasing trends, while 28.8% and 18.1% showed a decreasing trend at a confidence level of 95%. The remaining 36.5% and 59.0% showed no trends (Table 18.4). Additionally, the increasing trend is overcoming the decreasing trend, which again confirms progressive groundwater contamination (Park et al., 2011). Figure 18.13a,b show the spatial distribution of the changing trends of ECs at a confidence level of 95%. The increasing trends are scattered all over the city, indicating that progressive groundwater contamination is widespread, not localized.

Trichloroethylene (TCE) and tetrachloroethylene (PCE) are the most frequently occurring contaminants in the urban groundwater of Korea (Lee and Lee, 2004; Baek and Lee, 2011), because they have been widely used

Table 18.3 Linear regression analysis results for groundwater level, EC, and temperature in 2000–2005 and 2010–2014

| | 2000–2005 | | | 2010–2014 | | |
Slope	WL*	EC**	Temp†	WL	EC	Temp
Positive	44.1% (52‡/118)	64.4% (76/118)	47.4% (56/118)	38.5% (32/83)	51.8% (43/83)	59.0% (49/83)
Negative	55.9% (66§/118¶)	35.6% (42/118)	52.6% (62/118)	61.5% (51/83)	48.2% (40/83)	41.0% (34/83)

* Water level.
** Electrical conductivity.
† Groundwater temperature.
‡ Number of wells showing positive slope.
§ Number of wells showing negative slope.
¶ Number of wells analyzed.

Table 18.4 Results of Mann–Kendall trend analysis for monthly median values in the periods from 2000–2005 and 2010–2014

| | 2000–2005 | | | 2010–2014 | | |
Trend	WL	EC	Temp	WL	EC	Temp
Increasing	20.3% (24*/118)	34.7% (41/118)	58.5% (69/118)	24.1% (20/83)	22.9% (19/83)	33.7% (28/83)
Decreasing	39.0% (46**/118)	28.8% (34/118)	21.2% (25/118)	34.9% (29/83)	18.1% (15/83)	31.3% (26/83)
No trend	40.7% (48†/118‡)	36.5% (43/118)	20.3% (24/118)	41.0% (34/83)	59.0% (49/118)	35.0% (29/83)

EC, electrical conductivity; Temp, groundwater temperature; WL, water level.
* Number of wells showing increasing trend.
** Number of wells showing decreasing trend.
† Number of wells showing no trend.
‡ Number of wells analyzed.

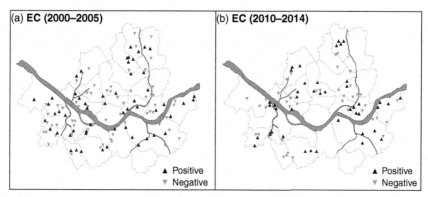

Figure 18.12 *Spatial Distribution of Linear Regression Analysis Results for EC.* (a) 2000–2005 and (b) 2010–2014.

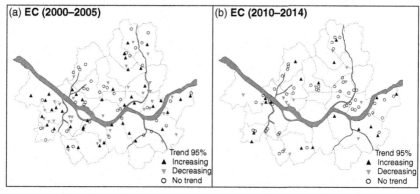

Figure 18.13 *Variation Trends of EC Determined Using the Mann–Kendall Test at a Confidence Level of 95%.* (a) 2000–2005 and (b) 2010–2014.

in dry cleaning, metal degreasing, and electrical and electronic industries (Wiedemeier et al., 1999; Kim et al., 2011). TCE is a probable carcinogen (Huff et al., 2004) and it has a low solubility and high density. Therefore, TCE and PCE as dense nonaqueous phase liquids (DNAPL) tend to migrate vertically in the aquifer, until they reach the impermeable layers at the bottom (Poulsen and Kueper, 1992; Baek and Lee, 2011). The remediation of polluted groundwater by TCE and PCE requires much more energy and higher costs, and is technically difficult to achieve, relative to other petroleum contaminants, and in fractured rocks its remediation is the most challenging, even in recent times (Lee and Lee, 2004; Yun et al., 2014). The maximum TCE contaminant level for drinking groundwater set by the Korean Ministry of Environment is 0.03 mg/L, and the PCE standard is 0.01 mg/L.

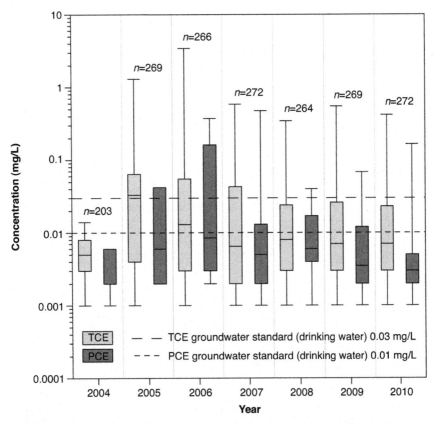

Figure 18.14 *TCE and PCE Concentrations in the Urban Groundwater of Seoul for 2004–2010. Only Detected Values are Plotted. Source: Data from the Korean Ministry of Environment; www.me.go.kr – the number of investigation wells varies by year.*

Figure 18.14 shows the detected TCE and PCE concentrations in the groundwater of Seoul (in total 206–286 samples each year, but they vary over the years). The TCE and PCE concentrations in the urban groundwater of the city for 2004–2010 show that they are detected (detection limit = 0.001 mg/L) in nearly all the groundwater wells investigated, even though their median values do not exceed Korean standards. However, at several groundwater wells, the TCE and PCE concentrations are 125 and 50 times greater than the standards, respectively. These high concentrations are mostly found in the industrial areas of the city (Figure 18.9; Park et al., 2005; Lee, 2011). However, the SMG (and also the Korean Government) has hesitated to clean up the contaminated groundwater, with the excuse that there are not many people drinking it. The remediation of contaminated groundwater is costing too much, and there is no effective practical technology in Korea to clean it up, especially in fractured aquifers (Lee and Lee, 2013). Thus, active remediation of contaminated groundwater is not on track in the city, nor in the whole of Korea (Lee and Lee, 2013; Lee, 2015). The difficulty in deciding which polluter is responsible for groundwater contamination is another obstacle to preventing remedial action due to possible financial implications, and their status in the country.

Interestingly, though currently unconfirmed, there seems a decreasing trend of TCE and PCE concentrations in urban groundwater (levels in recent years are shown in Figure 18.14) even though no remedial action for such contamination has been undertaken in the city (Lee, 2015). In addition, natural attenuation cannot be readily expected by recharged clean water mixing because groundwater recharge from rainfall is limited in Seoul due its widespread impermeable pavements. If contamination at any groundwater wells is found to exceed the standards, such wells are banned from further use by the city's administrative authority, and are generally closed. Therefore, contaminated wells are excluded from further investigation into groundwater quality (Lee and Lee, 2004), and thus the decreasing contamination levels may be misleading. In order to grasp the exact status and progress of groundwater contamination, the investigation wells should continue to be monitored even though they are contaminated.

Recently, groundwater contamination by petroleum hydrocarbons including BTEX (benzene, toluene, ethylbenzene, and xylenes) in and around US Forces Korea (USFK) military facilities stationed in the heart of the city, has surfaced as an environmental issue (Lee, 2011; Yang et al., 2014). However, due to military security and political reasons, the metropolitan government has not undertaken a comprehensive investigation into such contamination, and the military has not taken the necessary remedial action (Lee and Lee, 2013). Thus, groundwater contamination is still spreading. All these things aggravate the groundwater environment and citizens perceive groundwater to be an unusable water resource (Lee, 2015).

18.5.5 Groundwater Temperature

The spatial distribution of the average daily groundwater temperatures from 2000–2005 and 2010–2014 is shown for each monitoring well in Figure 18.15a,b, respectively. Groundwater temperatures of 14–15°C (93 and 39 wells – 78.8 and 46.9%, respectively, for each period) were most prevalent in the city. This groundwater temperature range is rather high in relation to agricultural areas and green spaces (Zhu et al., 2015), but it

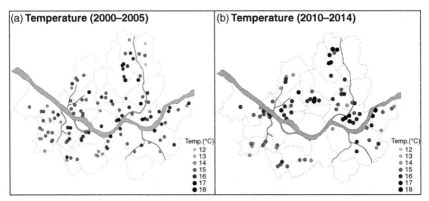

Figure 18.15 *Distribution of Mean Groundwater Temperatures (°C) for (a) 2000–2005 and (b) 2010–2014, Respectively (www.seoul.go.kr).*

is fairly similar to the mean groundwater temperature throughout Korea (14.5°C; Park et al., 2011). Well No. 4 and Well No. 1 (3.4 and 1.2%) had the range of 12–13°C and Well No. 21 and Well No. 43 (17.8 and 51.8%) showed range of 16–18°C. The lowest groundwater temperatures are generally observed in the outermost forest areas (Figure 18.15a), while the highest temperatures are seen in the central commercial areas (Figure 18.15b). The numerous high buildings in these areas are emitting a great deal of heat into the surrounding environment, resulting in an increase in shallow-groundwater temperatures (Taniguchi et al., 2007; Zhu et al., 2015). However, the very high groundwater temperatures at the northern boundary (see Figure 18.15b), especially in the intensive horticulture area, are of interest and the inference is that they are caused by multiple return flows of irrigation.

Linear regression results of groundwater temperatures from 2000–2005 for the daily mean values indicated that 56 (47.4%) from a total of 118 monitoring wells showed a positive slope, while the remaining 62 (52.6%) showed a negative slope. However, the regression results for 2010–2014 indicated that 59.0% showed a positive slope and 41.0% showed a negative slope (Table 18.3; Figure 18.16). The positive slope increased by 11.6%. Considering the obvious increasing trend of urban air temperatures (see Figure 18.6), this is due to a warming air temperature and the urban heat island effect (Menberg et al., 2013; Zhu et al., 2015).

Results of the Mann–Kendall trend tests for 2000–2005 and 2010–2014 (Table 18.4; Figure 18.17) showed a decreasing trend of 21.2% and 31.3% and an increasing trend of 58.5% and 33.7% at a 95% confidence level

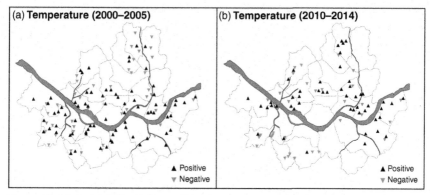

Figure 18.16 *Spatial Distribution of Linear Regression Analysis Results for Groundwater Temperatures.* (a) 2000–2005 and (b) 2010–2014.

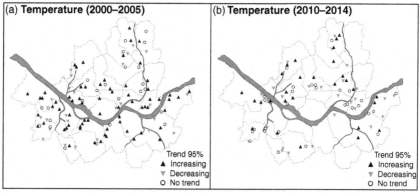

Figure 18.17 *Variation Trends of Groundwater Temperatures Determined Using the Mann–Kendall Tests at a Confidence Level of 95%.* (a) 2000–2005 and (b) 2010–2014.

for the two periods, respectively (Table 18.4). Thus, the increasing trend of groundwater temperatures in this city is predominant. The increasing groundwater temperature is not limited to this city, but it is widespread throughout the country at rates of 0.04–0.09°C/y (Park et al., 2011), which far exceeds the increase in air temperature rates (0.03°C/y). The increasing trend is not limited to the central commercial area but is general for the entire city (Figure 18.17a). However, in green tracts and open spaces, the increasing rates are relatively low.

18.5.6 Recharge

Groundwater recharge in this metropolitan city is somewhat complicated relative to rural areas, being controlled by anthropogenic factors, and to

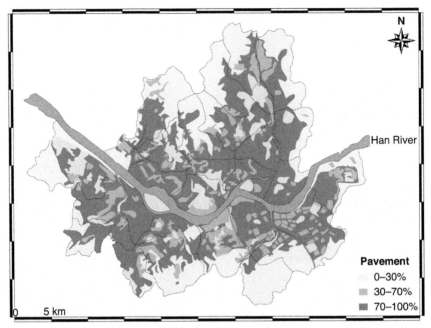

Figure 18.18 *Pavements in the City (as at 2005). Source: Data are from SMG; www.seoul. go.kr.*

a lesser extent, natural precipitation. Groundwater recharged by the leak-ing metropolitan water supply and sewerage systems was estimated to be 670 MCM/y in the city (Kim, 2000). The recharge quantity derived from both systems is approximately 15 times greater than that from natural pre-cipitation (mostly rainfall), which is estimated to be 45 MCM/y in the Seoul area (Kim et al., 2001). System leakage is the primary recharge factor for the city. Although the city generally receives a large amount of precipi-tation (averaging 1532 mm/y for 1996–2014; see Figure 18.4), the aquifer recharge is limited due to widely distributed impermeable materials. Over 70% of the land surface is covered with impermeable materials throughout the city (Figure 18.18). Major rechargeable areas, green tracts, and open spaces, are mostly located in the outermost parts of the city. To a certain ex-tent, artificial irrigation from river water in the parks and open spaces plays a role in groundwater recharge (Lee and Koo, 2007). In high-water seasons (especially in summer), groundwater is recharged to some extent by water from the Han River, but it is limited to an area proximal to the river, near to an alluvial aquifer (Kim, 2000).

18.6 IMPACTS

18.6.1 Depletion of Groundwater Level

Results from linear regression analysis of groundwater levels for 2000–2005 and 2010–2014 are shown in Table 18.3 and Figure 18.19a,b. For the period from 2000–2005, 55.9% ($n = 66$) of the total wells ($n = 118$) and from 2010–2014, 61.5% ($n = 51$) of the total wells ($n = 83$) showed a negative slope (decreasing groundwater levels), while 44.1% ($n = 52$) and 38.5% ($n = 32$) of those had a positive slope.

Figure 18.19a,b show the spatial distribution of changing slopes in groundwater levels. These decreasing slopes are obviously distinct in the central commercial area from 2000 to 2005, while they are less obvious in the data from 2010 to 2014. The sporadic increase of groundwater levels, especially in the outermost areas of the city, can be attributed to the reduction of substantial groundwater pumpage and, to some extent, artificial recharge due to leakage from waterworks (Kim, 2000; Lee and Koo, 2007).

However, the scattered distribution of the regression slopes in groundwater levels cannot be easily explained by a limited number of factors, because there are a variety of causes affecting groundwater levels in this complicated city. Often, local sources or sinks near monitoring wells (e.g., artificial surface water irrigation from rivers or pumping stations to mitigate urban flooding) may play a critical role in the behavior of groundwater levels. However, considering the recent general increasing trend in annual precipitation (except the 2014 drought) and decreased groundwater use, declining groundwater levels are the result of anthropogenic activities.

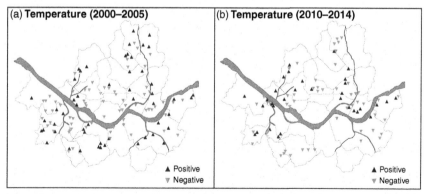

Figure 18.19 *Spatial Distribution of Linear Regression Analysis Results for Groundwater Levels.* (a) 2000–2005 and (b) 2010–2014.

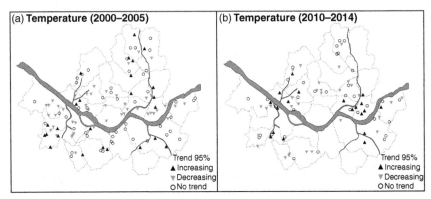

Figure 18.20 *Variation Trends of Water Levels Determined Using Mann–Kendall Trend Tests at a Confidence Level of 95%.* (a) 2000–2005 and (b) 2010–2014.

In order to give a statistical explanation of the variation, Mann–Kendall trend tests for groundwater level data sets from 2000–2005 and 2010–2014 were carried out using the median monthly value of groundwater levels (Table 18.4). The 39.0% and 34.9% decreasing trend among total wells, indicate a confidence level of 95%, while 20.3% and 24.1% of the total wells showed an increasing trend for each period.

Figure 18.20a,b show the spatial distribution of the changing trends in groundwater levels for 2000–2005 and 2010–2014, respectively. The determined trends are much similar to those determined by linear regression analysis. The decreasing trend in the central commercial area is reconfirmed by this analysis. There are many negative factors affecting groundwater levels in the city. In particular, the high coverage of impermeable pavement has a large impact on urban groundwater quantity. The majority of rainfall is concentrated in the rainy season and is often torrential, thus almost all rainfall goes through the urban sewage line, and does not recharge groundwater (Lee et al., 2014). Consequently, the groundwater resource of the city is endangered from a quantitative point of view. Since the city is in a semiarid area when it comes to precipitation and evaporation, increasing air temperature and the resulting increased evaporation with decreasingly efficient rainfall for recharge, will aggravate the physical condition of groundwater in the city.

18.6.2 Land Subsidence

The major cities of Asia, such as Tokyo in Japan (Hayashi et al., 2009), Taipei in Taiwan (Chen et al., 2007), Jakarta in Indonesia (Abidin et al., 2011), and Bangkok in Thailand (Piancharoen, 1976; Phien-Wej et al., 2006) have been built on geologically soft ground as sediments are piled up to several hundred meters or more. Therefore, from the 1960s until recently, these cities have experienced serious safety problems, originating from ground/land subsidence partly due to a decline in groundwater levels. Underground structures in urban areas have suffered from these side effects (Jago-on et al., 2009; Lee, 2015). To tackle these problems, cities have made efforts to maintain groundwater levels. However, the situation in Seoul is somewhat different from that of other cities. Rocks are distributed at several meters beneath the city of Seoul except in areas near to the Han River (SMG, 2015). The rocks, mostly granite and gneiss, are composed of mineral ingredients, which do not easily dissolve in flowing groundwater within a short time period (Kim, 2000). Therefore, in Seoul, there is scant opportunity for the formation of sinkholes in flowing groundwater as is the case for cities on karst terrains (Gao et al., 2005; Galve et al., 2009).

Trust toward vague safety from land subsidence such as settlement, sinking, sinkhole, and cavity problems is commonplace in the urban area of Seoul. Nevertheless, a large number of cavities and land subsidence have occurred in the main urban areas (see Figure 18.9; Figure 18.21). In particular, beneath the ground in Seockchon, in the Songpa District, an abrupt cavity

Figure 18.21 *Ground/Land Subsidence Occurring in Seoul.* Source: *Subsidence occurrence date and sources are noted in each photograph.*

Table 18.5 Causes of ground/land subsidence in Seoul occurring from 2010–2014

Causes	2010	2011	2012	2013	2014.1–7	Sum
Sewer pipe damage	409*	413	611	754	449	2636
Water supply pipe damage	5	11	13	8	15	52
Construction sites (subways, building foundations, etc.)	21	149	65	92	104	431
Total	435	573	689	854	568	3119

* Numbers represent cases of ground subsidence and cavities larger than 0.1 × 0.1 m in surface dimension.

developed measuring 5–8 m wide, 4–5 m deep, and 80 m long (see the photograph at the top left in Figure 18.21; SMG, 2015). Table 18.5 shows the causes of ground subsidence occurring in the city from 2010 to 2014. The number of cases (subsidence development) is increasing year by year.

The main causes of detected land subsidence or cavities are superannuated sewer pipes, the fraudulent construction of subways and deep-grounded buildings, and deficiencies in the management of groundwater (SMG, 2015). The old sewer lines often leak wastewater and ultimately break, resulting in land subsidence. However, this kind of land cavity is small, while groundwater-induced land subsidence may create larger sized holes (Lee, 2015). Recently, more serious ground subsidence has occurred at construction sites (Table 18.5). The construction of high-rise buildings and subway tunnels inevitably lowers groundwater levels through pumping. Maintenance of buildings and subway operations also requires the removal of groundwater seepage. Therefore, the lowering of hydraulic pressure by depleted groundwater triggers land subsidence (Sun et al., 1999). The surrounding area near subway tunnels and the heavily commercialized area in the central region show much lower groundwater levels in relation to development and are fairly susceptible to land subsidence, especially in the area with alluvial deposits near the Han River.

18.7 RESPONSES

The SMG has implemented various measures to tackle the groundwater problems mentioned earlier, including the establishment in 1997 of an organization in charge of groundwater management, installation, and operation of the local (metropolitan) groundwater-monitoring network (Lee et al., 2005b), and the launch of a soil and groundwater information system in 2012. The following sections detail some of the SMG's main efforts.

18.7.1 Establishment of an Organization in Charge of Groundwater Management

In 1995 the SMG implemented a groundwater team, consisting of three officials, complying with the Groundwater Law, which controlled all groundwater affairs under the Department of Water Management until 2002, when it became the Department of Water Quality. Since 2005 it has been known as the Department of Water Management Policy (soil and groundwater team) (Kim and Lee, 2001; SMG, 2006, 2015). There are similar teams in the 25 self-governing districts (county level). The groundwater team in the city is responsible for: (i) groundwater development, use and conservation management; (ii) designation of groundwater conservation areas; (iii) operation and management of the metropolitan groundwater-monitoring wells; and (iv) operation of the Groundwater Management Board (SMG, 2006). The groundwater team in each district takes charge of: (i) permission for groundwater development and use; (ii) issuance of orders for preventative groundwater pollution measures; and (iii) investigation of the status of groundwater development and use. The Metropolitan Groundwater Management Board is the top organization in the city, established by Metropolitan Groundwater Ordinance (No. 3683, Article 11; November 15, 1999), and has the final say on the following matters: (i) implementation and changes to the metropolitan groundwater management plan; (ii) designation of groundwater conservation areas; and (3) disputes regarding groundwater development and use (SMG, 2006, 2015). The board consists of 15 members including a chairperson (Director of the Division of Water Cycle Strategy).

The groundwater team in the city has conducted many groundwater works. The team completed a baseline survey of the groundwater management plan for Seoul in 1996, and established the SMG groundwater plan in 2006 (SMG, 2006, 2015). This plan was revised in 2014, in order to comply with climate change and changes in groundwater and city conditions. There have been increasing groundwater issues and civil petitions, and thus the present officials cannot deal appropriately with all those tasks. Enforcement by the groundwater team (at departmental level) with respect to budget and manpower is an essential requirement.

18.7.2 Groundwater Monitoring

The SMG has operated a local groundwater-monitoring network consisting of 118 monitoring wells (see Figure 18.9), complying with the Groundwater Law enforced in 1994. According to the law (Article 17), the mayor (local government) has to monitor groundwater levels within his territory and report any variation to central government. Some monitoring wells in Seoul are in use while others are exclusively for monitoring purposes. Almost all are evenly distributed throughout the city (see Figure 18.9). Each monitoring well has automated sensors to measure groundwater levels, water temperature, and EC. However, there has been no integrated groundwater data management system and thus monitoring is managed by each district. Consequently, monitored data are often missing according to local officials in charge. In addition, sometimes the city and district officials are reluctant to disclose groundwater data to the public (Lee, 2015). Groundwater monitoring has also been conducted regarding quality. The SMG conducts groundwater quality investigations twice a year (April–May and September–October) using 103–143 wells at a time. The 20 analyzed parameters include heavy metals (Pb, Cr^{6+}, Hg, Cd) and chlorinated compounds (TCE, PCE, 1.1.1-TCA). The SMG reports the investigation results to the central government (Ministry of Environment) and gives the public access to them (Yun et al., 2014). In the event that groundwater quality exceeds the designated groundwater standard, the SMG relays the results to the well owner and bans the use of contaminated groundwater. However, to our disappointment, the central and metropolitan governments do not take any remedial measures — they simply repeat the regular monitoring procedure.

18.7.3 Soil and Groundwater Environment Information System

In 2012, the SMG announced the implementation of an environmental information system for soil and groundwater, based on a geographic information system (GIS), to efficiently control soil and groundwater contamination. The SMG has conducted periodic investigations of both soil and groundwater contamination since 1994. However, the investigation data are not managed systematically and the two are dealt with separately. Thus, city officials have not used this valuable data in an efficient way. The SMG will place all the soil and groundwater data including contaminant concentrations in soil and groundwater, potential contamination source, and groundwater use on an electronic grid map of the city (Environment Daily, 2012). With this system, the officials (decision makers or policy makers) are expected to make an efficient decision on soil and groundwater-related administrative measures, but this system is for official use only, and is not open to the public.

18.7.4 Reuse of Groundwater Seepage

There is much groundwater seepage around deep underground buildings and facilities such as subway tunnels (Lee et al., 2005a). Even though the

Groundwater Law (Article 9.2) enforces the obligatory reuse (by the facility owner) of unwanted groundwater seeping into underground facilities, the seeped groundwater is mostly discharged to nearby sewer lines. Since there are numerous high-rise buildings and many long lines of subways (see Figure 18.8), the careless discharge of this kind of groundwater is one of the main reasons for the decline in groundwater levels in the city (Lee, 2015), and can be related to land subsidence. In the Seoul metropolitan city area, groundwater seepage into subsurface buildings and tunnels is estimated to be 178,599 m^3/d, three times greater than daily groundwater use from wells (61,470 m^3/d) in the area (Inews24, 2014). Only 11.3% of them are reused for road cleaning, public latrines, and park irrigation. As part of its reuse efforts, the SMG has used the groundwater seepage from 13 subway stations (22,000 m^3/d) to supply maintenance water (18% of the total water requirement) for an artificial stream, the Cheonggyecheon, which has flowed through the central part of northern Seoul since 2005 (Seoul Economy, 2005). However, much of the seeped groundwater is still discharged directly into the ditches.

18.7.5 Efforts for Mitigating Land Subsidence

The city has recently experienced an increasing number of land subsidence incidents or ground cavities (see Table 18.4). Hence public grievance regarding unexpected land subsidence is growing. To mitigate public concerns and reduce land subsidence itself, the SMG has started the replacement of superannuated sewer lines with the support of the central government. The city is also conducting a comprehensive investigation into potential land subsidence areas near subway lines and high-rise buildings using groundwater and geophysical surveys. For this purpose, the SMG entered into cooperation with Tokyo in Japan, where there are advanced mitigation technologies for land subsidence, in order to exchange experience, information, and technology (SBS, 2015). Taking the issue seriously, the central government (Ministry of Science, ICT and Future Planning, and National Research Council of Science and Technology) launched an integrated research group (funded by USD30 million for 3 years), consisting of four research institutions and 11 commercial companies, to develop an urban underground monitoring and management system based on an Internet of Things (IoT). The major participating institutions include KIGAM (Korea Institute of Geoscience and Mineral Resources) and ETRI (Electronics and Telecommunications Research Institute). The most important areas of expertise for the success of the research are considered to be hydrogeology and information technology.

18.8 CONCLUSIONS AND RECOMMENDATIONS

The situation of the groundwater environment in the city of Seoul is evaluated in this chapter based on collected groundwater data and available internet resources. The groundwater environment in the city is largely influenced by two comprehensive factors: a high population and land use change. The physicochemical groundwater conditions are affected by the 10 million residents and a variety of buildings and facilities. The groundwater usage of the city from wells has dropped continuously. Nevertheless, the groundwater level is declining. The decreasing water levels are mainly caused by groundwater removal for construction and seepage control, not for any use, and reduced recharge due to the progressive growth of impermeable pavements. Therefore, an appropriate strategy to reuse or recharge the seeped groundwater is essential requirement. The groundwater temperature is increasing due to warming air temperatures and the urban heat island effect. There are many pollution sources in the city such as gas stations, laundries, factories, wastewater treatment facilities, and leaking sewer lines, creating complicated groundwater contamination. The high and increasing EC levels are observed in heavy commercial and industrial areas, coincident with high concentrations of nitrate and chlorinated solvents. However, the widespread occurrence of these contaminants is indicative of the extensive groundwater contamination in this metropolitan city. Recently, frequent land subsidence or large ground cavities near large construction sites seem to be related to lower groundwater levels, even though these can also be attributed to superannuated sewer lines.

It is very difficult and costly to restore contaminated groundwater, but in order to control the spread of groundwater contamination, the SMG has to take prompt remedial action against contaminated groundwater. The groundwater is the source of last resort for water resources in this era of changing climate. Moreover, there is public unrest in densely populated regions concerning safety due to land subsidence caused by the decline in groundwater levels, and the SMG must devise measures to sustain these levels, such as the implementation of permeable blocks. As part of these efforts, a proportion of reuse from groundwater seepage into subways and high-rise buildings should be promoted by active administrative procedures. Most of all, the first step to protect groundwater hazards or mitigate groundwater-related problems, is to monitor the groundwater quantity and quality on a regular basis. The current quality of monitoring practice is not good enough to meet the need for accurate and reliable analysis.

The monitoring data should be stored and managed in an integrated way. The groundwater data quality should be improved and the locations of the monitoring wells should not be changed. The SMG should exert further efforts to raise the profile of the value of groundwater resources with citizens and officials.

ACKNOWLEDGMENT

The authors acknowledge support from the Basic Science Research Program through the National Research Foundation of Korea (NRF), funded by the Ministry of Education (NRF-2011-0007232).

REFERENCES

Abidin, H.Z., Andreas, H., Gumilar, I., Fukuda, Y., Pohan, Y.E., Deguchi, T., 2011. Land subsidence of Jakarta (Indonesia) and its relation with urban development. Nat. Hazards 59 (3), 1753–1771.

Baek, W., Lee, J.Y., 2011. Source apportionment of trichloroethylene in groundwater of the industrial complex in Wonju Korea: a 15-year dispute and perspective. Water Environ. J. 25 (3), 336–344.

Chae, G.T., Yun, S.T., Choi, B.Y., Yu, S.Y., Jo, H.Y., Mayer, B., Kim, Y.J., Lee, J.Y., 2008. Hydrochemistry of urban groundwater, Seoul, Korea: the impact of subway tunnels on groundwater quality. J. Contam. Hydrol. 101, 42–52.

Chae, G.T., Yun, S.T., Kim, D.S., Kim, K.H., Joo, Y., 2010. Time-series analysis of three years of groundwater level data (Seoul, South Korea) to characterize urban groundwater recharge. Q. J. Eng. Geol. Hydrogeol. 43, 117–127.

Chen, C.T., Hu, J.C., Lu, C.Y., Lee, J.C., Chan, Y.C., 2007. Thirty-year land elevation change from subsidence to uplift following the termination of groundwater pumping and its geological implications in the Metropolitan Taipei Basin Northern Taiwan. Eng. Geol. 95 (1), 30–47.

Choi, B.Y., Yun, S.T., Yu, S.Y., Lee, P.K., Park, S.S., Chae, G.T., Mayer, B., 2005. Hydrochemistry of urban groundwater in Seoul, South Korea: effects of land-use and pollutant recharge. Environ. Geol. 48, 979–990.

Chung, S.Y., 2010. Groundwater obstructions and countermeasures for groundwater discharge from subway in Seoul, Korea. J. Geol. Soc. Korea 46 (1), 61–72, (in Korean with English abstract).

Collin, M.L., Melloul, A.J., 2003. Assessing groundwater vulnerability to pollution to promote sustainable urban and rural development. J. Clean. Prod. 11 (7), 727–736.

Cox, M.E., Hillier, J., Foster, L., Ellis, R., 1996. Effects of a rapidly urbanising environment on groundwater, Brisbane, Queensland, Australia. Hydrogeol. J. 4, 30–47.

Eiswirth, M., Wolf, L., Hotzl, H., 2004. Balancing the contaminant input into urban water resources. Environ. Geol. 46, 246–256.

Environment Daily, 2012. Seoul city, smart management of soil and groundwater contamination. Available from: www.hkbs.co.kr (accessed 27.02. 2015.).

Feyisa, G.L., Dons, K., Meilby, H., 2014. Efficiency of parks in mitigating urban heat island effect: an example from Addis Ababa. Landscape Urban Plan. 123, 87–95.

Galve, J.P., Gutierrez, F., Lucha, P., Bonachea, J., Remondo, J., Cendrero, A., Gutierrez, M., Gimeno, M.J., Pardo, G., Sanchez, J.A., 2009. Sinkholes in the salt-bearing evaporite karst of the Ebro River valley upstream of Zaragoza city (NE Spain): geomorphological mapping and analysis as a basis for risk management. Geomorphology 108 (3–4), 145–158.

Gao, Y., Alexander, Jr., E.C., Barnes, R.J., 2005. Karst database implementation in Minnesota: analysis of sinkhole distribution. Environ. Geol. 47 (8), 1083–1098.

Hayashi, T., Tokunaga, T., Aichi, M., Shimada, J., Taniguchi, M., 2009. Effects of human activities and urbanization on groundwater environments: an example from the aquifer system of Tokyo and the surrounding area. Sci. Total Environ. 407 (9), 3165–3172.

Hong, Y.K., 1984. Petrology and geochemistry of Jurassic Seoul and Anyang Granites, Korea. J. Geol. Soc. Korea 20, 51–71.

Huff, J., Melnick, R., Tomatis, L., LaDou, J., Teitelbaum, D., 2004. Trichloroethylene and cancers in humans. Toxicology 197, 185–187.

Inews24, 2014. Abrupt groundwater level decline and rise: closely related to sinkhole. Available from: http://news.inews24.com (accessed 10.03.2015.).

Jago-on, K.A.B., Kaneko, S., Fujikura, R., Fujiwara, A., Imai, T., Matsumoto, T., Taniguchi, M., 2009. Urbanization and subsurface environmental issues: an attempt at DPSIR model application in Asian cities. Sci. Total Environ. 407 (9), 3089–3104.

Jo, J.H., Golden, J.S., Shin, S.W., 2009. Incorporating built environment factors into climate change mitigation strategies for Seoul, South Korea: a sustainable urban systems framework. Habitat Int. 33 (3), 267–275.

Kang, D.H., Cho, W.C., Lee, J.Y., 2001. The behavior of leachate on the transient condition in the Nanji waste landfill. J. KoSSGE 6 (2), 57–67, (in Korean with English abstract).

Kim, H., Jeon, W.H., Lee, J.Y., Lee, K.K., 2011. Urbanization and groundwater condition in a metropolitan city of Korea: implication for sustainable use. Res. Earth Resources 26, 1–24.

Kim, Y.J., Koo, J.Y., Lee, J.H., Lee, M.J., Kim, S.I., 2015. Study for Han River Park Nanji campsite revitalization – focus on cultural art space. Digital Design Res. 15 (1), 731–742, (in Korean with English abstract).

Kim, Y.J., Lee, S.M., 2001. The methods and principles of groundwater conservation area designation and its management. Seoul Development Institute, Seoul, Korea, 216 p. (in Korean).

Kim, Y.Y., 2000. Analysis of hydraulic properties of an urban groundwater system: groundwater system in Seoul area, Korea. PhD thesis, Seoul National University, Seoul, Korea.

Kim, Y.Y., Lee, K., 1999. GIS application to urban hydrogeological analysis of groundwater system in Seoul area. J. GIS Assoc. Korea 7 (1), 103–117, (in Korean with English abstract).

Kim, Y.Y., Lee, K.K., Sung, I.H., 1998. Groundwater systems in Seoul area: analysis of hydraulic properties. J. Eng. Geol. 8 (1), 55–73, (in Korean with English abstract).

Kim, Y.Y., Lee, K.K., Sung, I.H., 2001. Urbanization and the groundwater budget metropolitan Seoul area, Korea. Hydrogeol. J. 9, 401–412.

King, C., 2003. Urban groundwater system in Asia. UNU/IAS Working Paper No. 101, The United Nations University, Institute of Advanced Studies.

Klimas, A.A., 1995. Impacts of urbanisation and protection of water resources in the Vilnius district, Lithuania. Hydrogeol. J. 3, 24–35.

KMA (Korean Meteorological Administration), 2015. Annual climate report (in Korean).

KME (Korean Ministry of Environment), 2015. Homepage of Ministry of Environment. Available online at http://www.me.go.kr

KOSIS (Korea Statistical Information Service), 2014. Population census. Available from: www.kosis.kr (accessed 12.01.2015.).

KRIHS (Korea Research Institute for Human Settlements), 2009. Homepage of Research Institute for Human Settlements. Available from: http:// www.kris.re.kr

Ku, H.F.H., Hagelin, N.W., Buxton, H.T., 1992. Effects of urban storm-runoff control on ground-water recharge in Nassau County, New York. Groundwater 30 (4), 507–514.

Kwon, S.T., Cho, D.L., Lan, C.Y., Shin, K.B., Lee, T., Mertzman, S.A., 1994. Petrology and geochemistry of the Seoul granitic batholith. J. Petrolog. Soc. Korea 3 (2), 109–127.

Lee, B., Hamm, S.Y., Jang, S., Cheong, J.Y., Kim, G.B., 2014. Relationship between groundwater and climate change in South Korea. Geosci. J. 18 (2), 209–218.

Lee, J.H., Lee, Y.J., Lee, S.J., 2005a. A Study on Mitigation Plan for Environmental Impacts of Groundwater in Tunnel. Korea Environmental Institute, Seoul, 164 p.

Lee, J.Y., 2006. Characteristics of ground and groundwater temperatures in a metropolitan city, Korea: considerations for geothermal heat pumps. Geosci. J. 10 (2), 165–175.

Lee, J.Y., 2011. Environmental issues of groundwater in Korea: implications for sustainable use. Environ. Conserv. 38 (1), 64–74.

Lee, J.Y., 2015. Lessons from three groundwater disputes in Korea: lack of comprehensive and integrated investigation. Int. J. Water (accepted).

Lee, J.Y., Choi, M.J., Kim, Y.Y., Lee, K.K., 2005b. Evaluation of hydrologic data obtained from a local groundwater monitoring network in a metropolitan city, Korea. Hydrolog. Process. 19 (13), 2525–2537.

Lee, J.Y., Koo, M.H., 2007. A review of effects of land development and urbanization on groundwater development. J. Geolog. Soc. Korea 43 (4), 517–528, (in Korean with English abstract).

Lee, J.Y., Lee, K.K., 2000. Use of hydrologic time series data for identification of recharge mechanism in a fractured bedrock aquifer system. J. Hydrol. 229 (3–4), 190–201.

Lee, J.Y., Lee, K.K., 2004. A short note on investigation and remediation of contaminated groundwater and soil in Korea. J. Eng. Geol. 14, 123–130, (in Korean with English abstract).

Lee, J.Y., Lee, K.K., 2013. Remediation of contaminated groundwater: change of paradigm for sustainable use. J. KoSSGE 18 (6), 1–7.

Lee, J.Y., Yi, M.J., Lee, J.M., Ahn, K.H., Won, J.H., Moon, S.H., Cho, M., 2006. Parametric and non-parametric trend analysis of groundwater data obtained from national groundwater monitoring stations. J. KoSSGE 11 (2), 56–67, (in Korean with English abstract).

Lee, S.M., Min, K.D., Woo, N.C., Kim, Y.J., Ahn, C.H., 2003. Statistical models for the assessment of nitrate contamination in urban groundwater using GIS. Environ. Geol. 44, 210–221.

Lerner, D., 2002. Identifying and quantifying urban recharge: a review. Hydrogeol. J. 10, 143–152.

McGee, T., 2001. Urbanization takes on new dimensions in Asia's population giants. Population Today 29, 1–2.

Menberg, K., Blum, P., Schaffitel, A., Bayer, P., 2013. Long term evolution of anthropogenic heat fluxes into a subsurface urban heat island. Environ. Sci. Technol. 47, 9747–9755.

MoneyToday, 2014. Daily water use per capita in Seoul: 286 L. Available from: www.moneytoday.co.kr (in Korean).

Morris, B.L., Seddique, A.A., Ahmed, K.M., 2003. Response of the Dupi Tila aquifer to intensive pumping in Dhaka, Bangladesh. Hydrogeol. J. 11, 496–503.

Nazari, M.M., Burston, M.W., Bishop, P.K., Lerner, D.N., 1993. Urban groundwater pollution: a case study from Coventry, United Kingdom. Groundwater 31 (3), 417–424.

NGIC (National Groundwater Information Center), 2015. Groundwater statistics. Available from: www.gims.go.kr

OECD (Organization for Economic Cooperation and Development), 2012. Regional Population Density: Asia and Oceania, 2012: Inhabitants Per Square Kilometer, TL3 regions. OECD Regions at a Glance 2013. OECD Publishing, DOI: 10.1787/reg_glance-2013-graph37-en

Otto, B., Ransel, K., Todd, J., Lovaas, D., Stutzman, H., Bailey, J., 2002. Paving Our Way to Water Shortages: How Sprawl Aggravates the Effects of Drought. American Rivers, Natural Resources Defense Council and Smart Growth America, Washington DC.

Park, S.S., Kim, S.O., Yun, S.T., Chae, G.T., Yu, S.Y., Kim, S., Kim, Y., 2005. Effects of land use on the spatial distribution of trace metals and volatile organic compounds in urban groundwater, Seoul, Korea. Environ. Geol. 48, 1116–1131.

Park, Y.C., Jo, Y.J., Lee, J.Y., 2011. Trends of groundwater data from the Korean National Groundwater Monitoring Stations: indication of any change? Geosci. J. 15 (1), 105–114.

Phien-Wej, N., Giao, P.H., Nutalaya, P., 2006. Land subsidence in Bangkok, Thailand. Eng. Geol. 82 (4), 187–201.

Piancharoen, C., 1976. Ground water and land subsidence in Bangkok, Thailand. In: Proceedings of the Anaheim Symposium.

Poulsen, M.M., Kueper, B.H., 1992. A field experiment to study the behavior of tetrachloroethylene in unsaturated porous media. Environ. Sci. Technol. 26, 889–895.

SBS (Seoul Broadcasting System), 2015. Seoul–Tokyo, exchange of technologies on sinkhole mitigation. Available online at http://news.sbs.co.kr (accessed 15.02.2015.).

Seoul Economy, 2005. Groundwater seepage into subways, reuse to maintain the Cheonggyecheon. Available from: http://economy.hankooki.com (accessed 12.02.2015.).

Shah, T., Molden, D., Sakthivadivel, R., Seckler, D., 2001. The global groundwater situation: overview of opportunities and challenge. Economic Political Weekly 36 (43), 4142–4150.

Shah, T., Roy, A.D., Qureshi, A.S., Wang, J., 2003. Sustaining Asia's groundwater boom: an overview of issues and evidence. Nat. Resources Forum 27 (2), 130–141.

Slonecker, E.T., Jennings, D.B., Garofalo, D., 2001. Remote sensing of impervious surfaces: a review. Rem. Sens. Rev. 20, 227–255.

SMG (Seoul Metropolitan Government), 2006. Seoul Metropolitan Government Groundwater Plan. SMG, Seoul, 27 pp. (in Korean).

SMG (Seoul Metropolitan Government), 2015. Available from: http://seoul.go.kr (accessed 15.02.2015.).

Su, S., Jiang, Z., Zhang, Q., Zhang, Y., 2011. Transformation of agricultural landscapes under rapid urbanization: a threat to sustainability in Hang-Jia-Hu region, China. Appl. Geography 31, 439–449.

Sun, H., Grandstaff, D., Shagam, R., 1999. Land subsidence due to groundwater withdrawal: potential damage to subsidence and sea level rise in southern New Jersey, USA. Environ. Geol. 37 (4), 290–296.

Sung, I.H., Cho, B.W., Lee, B.J., Kim, T.K., Lee, B.D., 1997. Study on Protection and Reclamation for the Groundwater Resources in Seoul Area. Ministry of Science and Technology, Seoul, p. 418.

Taniguchi, M., Uemura, T., Jago-on, K., 2007. Combined effects on urbanization and global warming on subsurface temperature in four Asian cities. Vadose Zone J. 6 (3), 591–596.

Vázquez-Suñé, E., Sánchez-Vila, X., Carrera, J., 2005. Introductory review of specific factors influencing urban groundwater, an emerging branch of hydrogeology, with reference to Barcelona, Spain. Hydrogeol. J. 13 (3), 522–533.

Wiedemeier, T.H., Rifai, H.S., Newell, C.J., Wilson, J.T., 1999. Natural Attenuation of Fuels and Chlorinated Solvents in the Subsurface. John Wiley & Sons, New York, NY.

Won, J.S., Woo, N.C., Kim, Y.J., 2004. Analysis of influential factors on nitrate distribution in ground water in an urbanizing area using GIS. Econ. Environ. Geol. 37 (6), 647–655, (in Korean with English abstract).

Yang, H.S., Kim, I.S., Kang, S.H., Chang, Y.Y., Park, S.K., Ko, J.W., Kim, Y.J., Park CH, 2014. A study on remediation methods of contaminated soils at former military bases. Korean Chem. Eng. Res. 52 (5), 647–651, (in Korean with English abstract).

Yang, Y., Lerner, D.N., Barrett, M.H., Tellam, J.H., 1997. Quantification of groundwater recharge in the city of Nottingham, UK. Environ. Geol. 38, 183–198.

Yun, S.T., Choi, B.Y., Lee, P.K., 2000. Distribution of heavy metals (Cr, Cu, Zn, Pb, Cd, As) in roadside sediments, Seoul metropolitan city, Korea. Environ. Technol. 21, 989–1000.

Yun, S.W., Choi, H.M., Lee, J.Y., 2014. Comparison of groundwater levels and groundwater qualities in six megacities of Korea. J. Geol. Soc. Korea 50 (4), 517–528, (in Korean with English abstract).

Zhu, K., Bayer, P., Grathwohl, P., Blum, P., 2015. Groundwater temperature evolution in the subsurface urban heat island of Cologne, Germany. Hydrologic. Process. 29 (6), 965–978.

CHAPTER 19

Groundwater Environment in Tokyo, Japan

Binaya Raj Shivakoti* and Vishnu Prasad Pandey**,†
*Water Resources Management Team, Natural Resources and Ecosystem Area, Institute for Global Environmental Strategies (IGES), Japan
**Department of Civil Engineering, Asian Institute of Technology and Management (AITM), Nepal
†Water Engineering and Management, Asian Institute of Technology (AIT), Thailand

19.1 INTRODUCTION

Tokyo Metropolis (or Tokyo) is located in the southern part of the Kanto Plane, which is the largest plain in Japan (Figure 19.1). It is bordered by the Edogawa River and Chiba Prefecture in the east, mountainous areas of Yamanashi Prefecture in the west and Saitama Prefecture in the north, and the Tamagawa River, Kanagawa Prefecture, and the Tokyo Bay in the south. Tokyo's elevation ranges from sea level to above 2000 m moving from east to west. Mount Kumotori (2017 m) is the highest mountain, located at the boundary between Saitama, Yamanashi, and Tokyo. Formerly known as Edo, Tokyo has been a political and economic center of Japan for some 400 years. It became the capital city in 1968 during the Mejii Period. It is the largest city in Japan in terms of area, population, and contribution to GDP. Tokyo's total area of 2188 km^2 is divided into 23 special wards (622 km^2), Tama Area (1160 km^2) and a chain of small islands (406 km^2)[1]. Over 13 million people were living in Tokyo in 2013, making it one of the most densely populated cities (>5900 persons/km^2) in the world. Nearly 30% of the country's population is concentrated in Tokyo and its three adjacent prefectures, Saitama, Kanagawa, and Chiba, which are also known as Tokyo Megalopolis Region or Greater Tokyo Area.

After the Second World War (WWII), Tokyo underwent rapid urbanization and became one of the earliest megacities in 1963, when its population passed 10 million. Urbanization has resulted in the rapid conversion of land cover into residential areas, industrial areas and for infrastructure projects such as roads, railways, and business complexes. The number of industries steadily increased from 1950, reaching its peak in 1975 (IGES, 2007). This

---

[1] This chapter does not include an assessment of these small islands.

Groundwater Environment in Asian Cities
http://dx.doi.org/10.1016/B978-0-12-803166-7.00019-2
Copyright © 2016 Elsevier Inc.
All rights reserved.
451

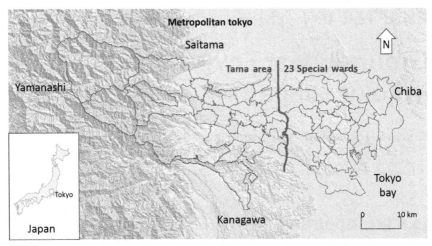

Figure 19.1 *Location Map of Metropolitan Tokyo. Source: Geology GIS data, Metropolitan Tokyo, http://doboku.metro.tokyo.jp/.*

increase in population and industrialization has led to an increase in demand for water and the emergence of environmental problems. Rivers flowing through Tokyo were heavily polluted at that time due to the discharge of domestic and industrial effluents. Groundwater was heavily exploited to cope with the increasing demand for water and the decreasing availability of surface water sources. Groundwater used to be one of the major sources of water supply before 1970 (Onodera, 2011). The overuse of groundwater soon resulted in the lowering of the water table, progression of land subsidence, and salt-water intrusion. However, the subsequent introduction of regulatory acts and their strict implementation was successful in controlling groundwater problems. Strict control over groundwater uses, especially for industry, through rationing and improvement in the surface water supply system drastically reduced the demand for groundwater and prompted rapid recovery of the groundwater table. Although a rise in the level of the groundwater table was a positive outcome, a new and unpredicted problem soon emerged. Increasing groundwater levels exerted a buoyant pressure on the underground groundwater infrastructure and seepage occurred, such as was seen in the subway system (Hayashi et al., 2009).

The historic problems such as groundwater depletion, land subsidence, and pollution of groundwater are more or less in a controlled state; however, other more modern issues such as the impacts of climate change, groundwater seepage to underground infrastructure, and micropollutant contamination have recently emerged. The case of Tokyo could be a valuable lesson for other Asian cities to not only help solve their immediate problems, but also to help in promoting the sustainable use of groundwater and in considering natural recharge, discharge, and water quality problems.

19.2 GEOLOGY, GROUNDWATER AQUIFERS, AND CLIMATE

The geology of Tokyo (Kanto Plain) is affected by different natural processes such as tectonic movements, volcanic eruptions, erosion and sedimentation, and weathering (Hayashi et al., 2009). The lowland alluvial plains are mainly composed of deep quaternary sediments (from the Holocene and Pleistocene). The mountains in the northern and western parts are composed of consolidated rocks (Mesozoic, Paleozoic, Miocene, Pliocene) and Pleistocene sediments, which also form the basement of the plain (Figure 19.2). Major rivers originating in the surrounding mountainous such as the Tone River, Ara River, and the Tama River transport and deposit the sediments into the lowland plains and Tokyo Bay. Intermittent tectonic movements and volcano deposits of ash during these different periods have also affected the present day geological composition of the Kanto Plain.

Confined and unconfined aquifers are distributed along different depths in the Kanto Plain. There are shallow unconfined aquifers (up to about

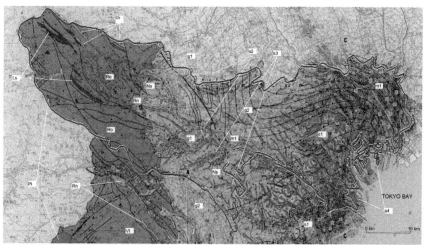

s1–s3: sand rich sediments; m1, s4: sand and mud sediments; g1–g3: gravel rich sediments; Ns: silt, sand and gravel; Rn: consolidated rock; Ro: consolidated rock (excluding chert, limestone and plutonic rocks); Ch: chert; Ls: limestone; Pi: plutonic rock; Vl: volcanic rocks; ▣ Observation wells for groundwater level and land subsidence ● Observation wells for groundwater level; ─── Administrative boundary

Figure 19.2 *Geological Map of Tokyo and Distribution of Observation Wells for Groundwater Level and Land Subsidence. Source: Groundwater Map of Tokyo, Chiba, Kanagawa Prefecture, Ministry of Land, Infrastructure and Transport Japan.*

70 m below Tokyo sea level) in the alluvial layer, which mainly consists of loose sand and gravel above. The lower part of the alluvial layer consists of soft clay followed by alternating layers of sand, clay, and gravels. Below the alluvial layers is the confined aquifer system. The confined aquifer below the alluvial layer is composed of alternating gravel in the upper part and sand and gravel in the lower parts. Depending on the location, the base of the unconfined aquifers ranges from 100 m to 600 m deep. Below the confined aquifer, Pliocene sediments extend more than 2000 m thick and mainly consist of alternating sandstones and mudstones. There is also a reservoir system of methane gas dissolved in water. Most groundwater abstraction takes place from aquifers less than 400 m deep (Hayashi et al., 2009).

Tokyo has a warm humid climate and experiences four distinct seasons consisting of a hot summer, cold winter, and mild autumns and springs. On hot days the temperature can reach over 35°C. and in the winter can drop below freezing (Figure 19.3).

Like other parts of Japan, Tokyo also receives frequent precipitation with an annual average of 1509 mm during 1950–2014 (Figure 19.4). Most of its rainfall is concentrated in the months of July–October and can be divided into two parts. First, the June–July period is a rainy season when winds from the Pacific Ocean bring rainfall. Second, tropical typhoons, which occur

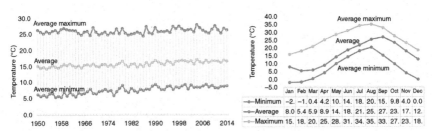

Figure 19.3 *Annual and Seasonal Variation of Temperature at Tokyo Station.* Source: *Japan Meteorological Agency, http://www.data.jma.go.jp/gmd/risk/obsdl/index.php.*

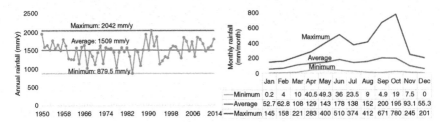

Figure 19.4 *Annual and Seasonal Variation in Rainfall at Tokyo Station.* Source: *Japan Meteorological Agency, http://www.data.jma.go.jp/gmd/risk/obsdl/index.php.*

during summer and autumn, cause high-intensity rainfall and often induce flash floods. During winter, cold northern winds from the Sea of Japan also bring rains and snowfall.

19.3 DRIVERS

Rapid economic development has affected Tokyo's groundwater environment significantly in the past (Table 19.1). Rising economic opportunities attracted an influx of people from various parts of Japan. Tokyo underwent a massive urbanization to accommodate industries, business and residents, and to provide social services to its inhabitants. Population growth and urbanization are the two main drivers affecting the groundwater resources in Tokyo. More recently, the potential impacts of climate change on the groundwater environment have also become apparent.

19.3.1 Population Growth

Tokyo has been attracting people from all over Japan for a long time. Figure 19.5 shows the population trend between 1920 and 2013. The population of Tokyo was already over 6 million in 1935. Due to WWII casualties, the population growth was briefly halted and actually decreased to 3.5 million in 1945. After WWII, Tokyo saw rapid population growth, passing 10 million (i.e., a more than threefold increase) by 1965. Since then population growth has slowed and had increased by only 22% at the end of 2013 when it reached 13.3 million. A similar growth pattern can also be seen around the Greater Tokyo area, where the population increased rapidly by almost 20 million between 1945 and 1980 and then grew modestly before reaching 36.6 million in 2013. People living in Greater Tokyo travel to Tokyo

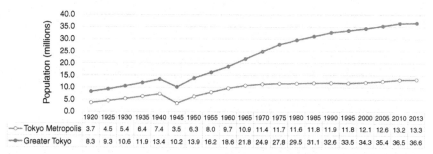

	1920	1925	1930	1935	1940	1945	1950	1955	1960	1965	1970	1975	1980	1985	1990	1995	2000	2005	2010	2013
Tokyo Metropolis	3.7	4.5	5.4	6.4	7.4	3.5	6.3	8.0	9.7	10.9	11.4	11.7	11.6	11.8	11.9	11.8	12.1	12.6	13.2	13.3
Greater Tokyo	8.3	9.3	10.6	11.9	13.4	10.2	13.9	16.2	18.6	21.8	24.9	27.8	29.5	31.1	32.6	33.5	34.3	35.4	36.5	36.6

Figure 19.5 *Population Trend during 1920–2013 in Tokyo Metropolis and Greater Tokyo Area. Source: Statistical Survey Department, Statistics Bureau, Ministry of Internal Affairs and Communications.*

regularly for work, business, and other purposes. It has been estimated that over 2.4 million "day-time citizens" commute every day to Tokyo.

19.3.2 Urbanization

Urbanization in Tokyo is characterized by the massive land cover changes that occurred in the post-WWII period. Economic development and population growth were the two main drivers for the rapid urban transformation of Tokyo. Agricultural plains and natural land were converted into residential, industrial, or commercial areas. High economic growth and resultant opportunity has led to the development of industries, growth of the service sector, and the establishment of business centers. High-rise buildings, residential areas, and transport networks (roads and railway lines) were developed to accommodate the high population density and support economic growth.

Currently, the Special 23 Wards are the most urbanized area in Tokyo with over 90% of its land occupied by buildings (58% including residential and commercial use) and open spaces (35% including unused land, public space, and roads) (Figure 19.6). The remaining area (~7%) is occupied by water bodies, agriculture, forest, and natural landscape. With the increase of population, urbanization also spread to Tama Area and adjacent prefectures such as Kanagawa, Chiba, and Saitama. In Tama Area, nearly half of the land cover is composed of forest and natural landscape in the western and

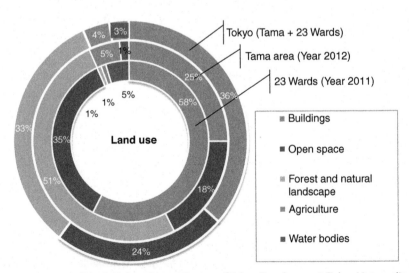

Figure 19.6 *Land Use in Tokyo. Source: Bureau of Urban Development, Tokyo Metropolitan Government, http://www.toshiseibi.metro.tokyo.jp.*

northern hilly areas. Toward the east and south of Tama Area are built-up areas (25.1%) and open spaces (17.6%). Agriculture (5.1%) and water bodies (1.5%) are the other land types.

19.3.3 Temperature Change

Climate change and urban land use are being considered as a new driver giving rise to a "heat island" effect. Between 1950 and 2013, the average minimum temperature increased by 1.3°C, while the average and average maximum showed less than 1°C change. A decrease in vegetation and natural areas and an increase of constructed areas, such as pavements, roads, and buildings, are responsible for the heat-island effect.

19.4 PRESSURES

19.4.1 Demand for Water

There was a high demand for water during the period of rapid economic development in the post-WWII period. Water supply wells were installed by public water supply authorities, industries and businesses, and individual households. Groundwater abstraction reached its peak during the 1970s. After that period, groundwater abstraction saw a rapid decline (Figure 19.7) following the introduction of legal and regulatory restrictions on groundwater abstraction, provisioning of an alternative water supply from surface

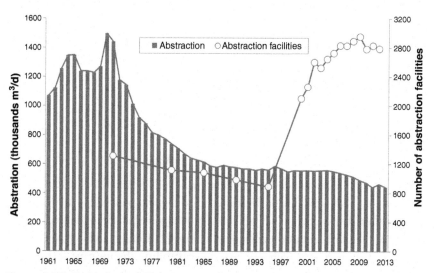

Figure 19.7 *Groundwater Abstraction and Number of Abstraction Facilities (Each Facility May Consist of One or More Wells). Source: TMG (2014).*

water sources and improvement in water use efficiency, including rationing, recycling, and reuse, by industry. At present, the majority of Tokyo's water supply is sourced from surface water. In 2013, groundwater abstraction was a mere 443,000 m^3/day, which is about 30% of peak abstraction in 1971 and only a small fraction (~6.5%) of the current 6.8 million cubic meters (MCM)/d supply capacity of the Bureau of Waterworks under Tokyo Metropolitan Government.

19.4.2 Discharge of Pollutants

There was widespread pollution during the early period of rapid economic development. At that time, untreated effluent discharges from industries and residential areas led to the deterioration of water quality. As polluted water was not fit for direct use, this too exerted pressure on groundwater. However, subsequent control of industrial discharges and improvement in sewer wastewater treatment systems helped reverse the state of water quality after the early 1970s. While quality-related problems are not severe and are considered to be at normal levels, there are still risks of groundwater contamination from industrial discharges such as volatile organic compounds (VOCs), nonpoint sources (such as nitrite and nitrate). and other kinds of micropollutants (IGES, 2007).

19.4.3 Land Use Change

Land use change has transformed much of Tokyo into a highly urbanized area – more than 90% in the Special 23 Wards area and over 50% in Tama Area. Loss of natural areas and an increase in impervious built-up areas are affecting groundwater recharge and causing an increase in surface temperature. Tokyo's temperature has seen an upward trend (Figure 19.3), which has induced a heat island effect (Taniguchi, 2011).

19.5 STATE

19.5.1 Groundwater Abstraction and Use

At its peak, groundwater abstraction reached over 1.4 MCM/d in 1970 (Figure 19.7). At that time, groundwater used to be one of Tokyo's major water sources due to the pollution of surface water bodies and a lack of adequate alternative sources. Subsequent legislative measures, rationing, and a gradual shift to surface water sources were effective in controlling the upward abstraction trend. The number of abstraction facilities also showed a decline from 1308 facilities in 1970 to 888 facilities in 1995 (Figure 19.7). However, after that period, groundwater abstraction facilities increased and reached 2440 units in 2012. This increase in abstraction facilities indicates a preference for groundwater for specific purposes, but not in terms of volume of use. The majority of abstracted groundwater (>70%) is used for drinking water. Other uses include manufacturing, heating and cooling, emergency use, washing and cleaning, public use (such as in parks), public baths, groundwater remediation, environmental flows, and ponds. More than 95% of these abstraction facilities are small scale and account for less than 10% of the total groundwater abstraction (Figure 19.8). More than 95% of groundwater facilities are concentrated in Tama Area, including 34 large stations (with a capacity of 5000 m³/d and over), which can pump over three quarters of total groundwater abstraction.

19.5.2 Groundwater Quality

Regular groundwater monitoring was started by the local government in 1982. Currently, 28 water quality indices for regular monitoring are listed in the groundwater quality standards for human health. They include VOCs, toxic substances (such as heavy metals, PCB, cyanide, benthiocarb, simazine, thiuram, and selenium), and nitrogen (nitrate and nitrite). At present,

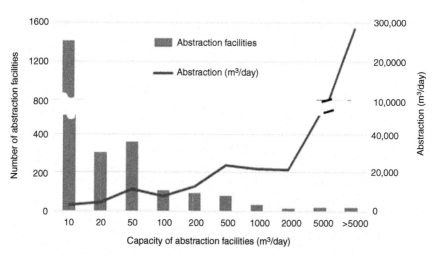

Figure 19.8 *Capacity of Groundwater Abstraction Facilities and Volume of Abstraction.* Source: TMG (2014).

the quality of groundwater in Tokyo is considered good. According to the Bureau of Environment, Tokyo Metropolitan Government (TMG) results of baseline groundwater quality monitoring, all these items are still detectable in the majority of observation wells, but their concentrations are usually below the standard value (Figure 19.9). There are only limited cases where concentrations exceed the minimum standard. According to results from continuous monitoring, lead, fluorine, and boron were detected at a higher concentration than the standard value in less than 5% of observation wells during 2006–2013 (Figure 19.10). Similarly, a higher than standard value of arsenic was found in 3–6% of wells every year between 2006 and 2016. In the case of nitrate and nitrite, higher than standard values were detected in 10–20% of observation wells. Nonpoint sources rather than point sources could be responsible for that. Among VOCs, tetrachloroethylene has been detected in a high concentration in 20–30% of the wells each year. These detected cases confirm that these chemicals still find their way into the aquifer. Stricter control on the release of these compounds means that they could be residuals of industrial discharge during the 1970s, 1980s, and 1990s. At that time more than 50,000 t of VOCs were produced

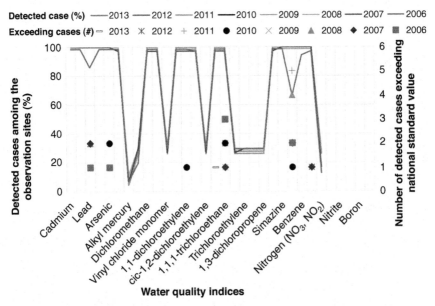

Figure 19.9 *Detected Cases of Water Quality Parameters under the Baseline Survey in Different Years at Observation Wells. Source: Bureau of Environment, Tokyo Metropolitan Government, http://www.kankyo.metro.tokyo.jp/.*

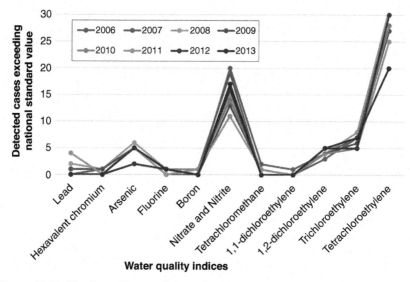

Figure 19.10 *Number of Detected Cases for Water Quality Parameters Exceeding the Minimum Standard Value during Continuous Monitoring. Source: Bureau of Environment, Tokyo Metropolitan Government, http://www.kankyo.metro.tokyo.jp/.*

annually in Japan (IGES, 2007). Besides these substances, there are also emerging issues about new kinds of substances such as pharmaceutical and personal care products (PPCPs), and perfluorinated compounds (PFCs), which have been found by some studies (Kuroda et al., 2012; Kuroda et al., 2014). Although these compounds do not pose an immediate threat, they could have long-term impacts on human health due to their potential to bioaccumulate and induce chronic health problems.

19.5.3 Land Subsidence

Land subsidence has a long history in Tokyo, and was first monitored as early as 1882 (Taniguchi, 2011). Cumulative land subsidence of 1 m was observed between 1882 and the 1930s (TMG, 2011). After that period, excessive groundwater use and the resultant loss in hydraulic head intensified land subsidence to more than 4 m by around 1970 at some observation sites (Figure 19.11). At that time, a maximum land subsidence of more than 15 cm/y was observed at some observation sites. Since 1979, land subsidence became more or less stable and maximum land subsidence has been below 5 cm/y (Figure 19.12). Due to the decreasing use of groundwater and natural aquifer recharge, maximum land subsidence has been limited to less than 1.5 cm/y since 2000.

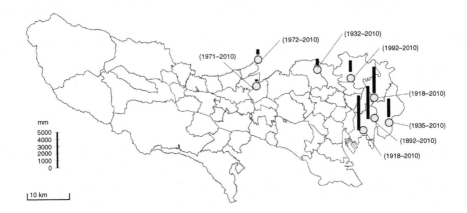

Figure 19.11 *Cumulative Land Subsidence Observed at Different Observation Sites in Tokyo.* Source: *Bureau of Environment, Tokyo Metropolitan Government, http://www.kankyo.metro.tokyo.jp/.*

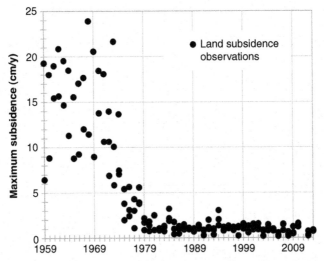

Figure 19.12 *Observed Maximum Land Subsidence in Different Years.* Source: *Bureau of Environment, Tokyo Metropolitan Government, http://www.kankyo.metro.tokyo.jp/.*

19.6 IMPACTS

19.6.1 Decline in Groundwater Level

Groundwater levels have been fluctuating greatly over time. In the past, excessive pressure on groundwater from urbanization and industrialization has depleted the aquifer storage significantly, especially prior to the 1970s. At that time, groundwater rather than surface water was the main source for water supply. The groundwater level recovered rapidly after the introduction of restrictive policies on resource use (Figure 19.13). By 2005, the groundwater level was already more than 10 m below Tokyo mean sea level. Since this period, a new issue, related to leakage and increased pressure from infrastructure, has been observed (Hayashi et al., 2009). In order to tackle this problem, a new strategy to balance abstraction and recharge has to be established.

19.6.2 Groundwater Pollution

During its peak economic development, both the major rivers and the Tokyo Bay area were heavily polluted and became an issue of public concern. Groundwater quality did not receive the same level of attention as a result of several factors (IGES, 2007): higher priority given to controlling problems such as land subsidence and saline water intrusion than to quality issues, urgency to control worsening pollution of surface water sources, reduced abstraction by the main users (i.e., industries), less concern about quality from the majority of domestic users who receive groundwater only after treatment, and the introduction of wastewater treatment technologies which contributed to reducing groundwater contamination.

Discharge of pollutants from households and industries also threatened groundwater quality, although no serious problems were found by the user

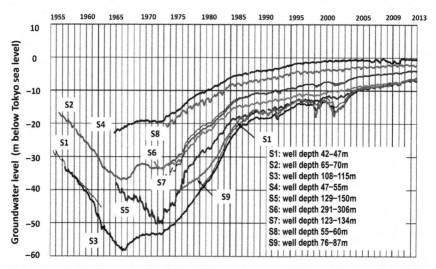

Figure 19.13 *Recovery Trend of Groundwater Level at Different Observation Wells.*
Source: TMG (2014).

(IGES, 2007). However, the release of new types of pollutants like VOCs by industry was found as a potential source. Similarly, nonpoint sources such as runoff of nitrogen from agriculture also pose threats to groundwater quality, even today.

19.6.3 Change in Subsurface Temperature

Anomalies in subsurface temperature is a new issue in lower lying areas of Tokyo, as well as the Kanto Plain, which is caused by global warming and heating of impervious urban objects. Studies have revealed that heat transferred to the subsurface environment has been accumulating and has extended up to 140 m below groundwater level (Taniguchi, 2011). Such changes could be a useful indicator to help assess the rate of urbanization or to understand subsurface flow. However, accumulated heat could also affect groundwater quality through geochemical processes and the microbial environment.

19.7 RESPONSES

Tokyo faced serious groundwater-related problems such as groundwater depletion, land subsidence, saline water intrusion, and pollution earlier than other groundwater-dependent cities in Asia. A number of measures have been implemented by the government to overcome these groundwater problems.

19.7.1 Legal and Regulatory Measures

Although Japan lacks any distinct law on groundwater, problems related to groundwater are addressed through individual laws and acts (IGES, 2007; Sato et al., 2011). Specific laws and acts relevant to the control of groundwater abstraction and related negative impacts such as land subsidence, include Industrial Rational Use of Groundwater 1970, Water Law 1956, Building Water Law 1962, and Japan Water Cycle Law 2014 (Table 19.1). As industry was the primary user of groundwater, most of the laws, acts, or ordinances were related to minimizing industrial dependence on groundwater. Other categories of groundwater related laws are related to protection of groundwater quality, prevention of groundwater contamination/pollution, regulations on hot springs and natural gas exploitation, protection of source quality for drinking water, and prevention of harmful impacts on groundwater environment (Sato et al., 2011).

Table 19.1 Status of DPSIR indicators in Tokyo, Japan

	Indicators	Past	Recent
Drivers	Population growth	3.7 million (1920)	13.3 million (2013)
	Urbanization		Urban areas = > 60% (2012)
	Temperature change	Average minimum temperature increased by less than 1°C during 1950–2013	Average minimum temperature increased by 1.3°C, while average and average maximum
Pressures	Demand for water	Industry, domestic uses	Industry, domestic uses
	Discharge of pollutants	Industry, domestic wastewater, agriculture runoff (serious)	Industry, domestic wastewater, agriculture runoff (less serious)
	Land use change	Conversion to built-up areas	Mostly built-up area: 90% in Special 23 Wards and over 50% in Tama Area
State	Groundwater abstraction and use	1.4 MCM/d	<0.5 MCM/d
	Groundwater quality	Contaminated	Relatively normal
	Land subsidence	>5–23 cm/y (before 1975)	<5 cm/y (after 1975)
Impacts	Decline in groundwater level	>50 m at some observation sites (before 1970)	<10 m (after 1995)
	Groundwater pollution	VOCs, nitrite and nitrate, heavy metals (1982)	VOCs, nitrite and nitrate, heavy metals (2013), limited cases only
	Change in subsurface temperature	Not known	Recent phenomenon: heat transferred to subsurface environment has been accumulating and has extended up to 140 m below groundwater level
Responses	Legal and regulatory measures	Industrial Water Law 1956 Building Water Law 1962 Rational Use of Groundwater 1970	Japan Water Cycle Law 2014
	Restrictions on pumping Alternative water supply Stricter control on water pollution	Water Pollution Prevention Law 1970	6.8 MCM/d (2013), capacity of the Bureau of Waterworks Soil Contamination Countermeasures Law 2003

19.7.2 Restrictions on Pumping

Restrictions on pumping were aggressively implemented in Tokyo. Since groundwater in Tokyo was also used for the exploration of natural gas, local government purchased the rights from industry to abstract the gas so that it could stop further exploration (IGES, 2007). Unlike in Osaka, an alternative source of water was not readily available, so the governor introduced a policy of rational water use under which users were required to report their abstraction regularly.

19.7.3 Alternative Water Supply

For the effective implementation of the regulatory measures and restrictions on the use of groundwater, an alternative water supply was necessary to substitute for water demand. For that purpose the Bureau of Waterworks started to increase its supply capacity by augmenting from various surface water sources. These included the construction of multiple dams at upstream sites, control of leakage, and improvement of tap water quality by introducing advanced treatment processes.

19.7.4 Stricter Control on Water Pollution

In order to address groundwater pollution control at the national level, the "Water Pollution Prevention Law" was amended in 1989 and set regulations on groundwater quality management (IGES, 2007). Four elements of groundwater quality management were determined in the law: (i) implementation of regular water quality monitoring by provincial government; (ii) prohibition of discharging hazardous wastewater into the ground by industry; (iii) mandatory notification for the establishment of facilities treating hazardous materials (notification by industry and examination by the governor), and (iv) implementation of emergent measures for accidental groundwater pollution (notification by industry and examination by the governor). Of the four policy measures, three are for the prevention of pollution, while the other is for mitigation. In addition to those measures, in order to support these groundwater pollution policies, the Environmental Standard for Groundwater Contamination was established in 1997. In order to effectively conduct regular groundwater quality monitoring within the limited budget, there are three types of survey with different purposes: (i) a baseline survey of 240 wells in a 4-year period (survey of a quarter of wells per year); (ii) a survey of wells surrounding contamination points to detect the source and spread of contamination; and (iii) a periodical survey to monitor annual variation of the contaminants in detected cases.

19.8 SUMMARY

Groundwater has been a traditional source of water supply in Japan for a long period of time. After the country entered a rapid phase of industrialization and urbanization, there was an obvious increase in the pressures on the groundwater environment. Uncontrolled abstraction of groundwater soon led to the emergence of problems such as drawdown of water tables, saline water intrusion near the Tokyo Bay area, and land subsidence. The Tokyo government managed to overcome all of these problems by introducing hard and soft measures. Excessive control on groundwater abstraction, which is still prevalent, became a new problem when the groundwater level recovered rapidly to a point where it became a problem for the stability of the extensive underground infrastructure of Tokyo. This provided a new lesson that management intervention should not only consider the preservation of resources, but also promote sustainable abstraction. It is also important to manage groundwater with a broader perspective as envisioned in the Japan Water Cycle Law 2014.

ACKNOWLEDGMENTS

The authors would like to acknowledge SEA-EU-NET II Project (http://www.sea-eu.net/about/aims_results) for supporting the development of this chapter. The authors acknowledges the Kakenhi Project (No. 143010435206-0001) for supporting this research.

REFERENCES

Hayashi, T., Tokunaga, T., Aichi, M., Shimada, J., Taniguchi, M., 2009. Effects of human activities and urbanization on groundwater environments: an example from the aquifer system of Tokyo and the surrounding area. Sci. Total Environ. 407, 3165–3172.

IGES, 2007. Sustainable Groundwater Management in Asian Cities. Institute for Global Environmental Strategies (IGES), Hayama, Japan.

Kuroda, K., Murakami, M., Oguma, K., Takada, H., Takizawa, S., 2014. Investigating sources and pathways of perfluoroalkyl acids (PFAAs) in aquifers in Tokyo using multiple tracers. Sci. Total Environ. 488–489, 51–60.

Kuroda, K., Murakami, M., Oguma, K., Muramatsu, Y., Takada, H., Takizawa, S., 2012. Assessment of groundwater pollution in Tokyo using PPCPs as sewage markers. Environ. Sci. Technol. 46 (3), 1455–1464.

Onodera, S., 2011. Subsurface pollution in Asian megacities. In: Groundwater and Subsurface Environments: Human Impacts in Asian Coastal Cities, Taniguchi, M. (Ed.), Springer, Japan, Tokyo.

Sato, K., Shichinohe, K., Ueno, T., 2011. Groundwater related laws in Japan. In: Findikakis, A.N., Sato, K. (Eds.), Groundwater Management Practices. CRC Press/Balkema, The Netherlands.

Taniguchi, M., 2011. Groudwater and Subsurface Environment: Human Impacts in Asian Coastal Cities. Springer, Japan, Tokyo.

TMG., 2011. Assessment of Land Subsidence and Status of Groundwater in Tokyo: Summary of 2010 Groundwater Exploratory Survey (in Japanese). Environment Bureau, Tokyo Municipial Government (TMG), Tokyo.

TMG, 2014. Status Report on Groundwater in Tokyo 2012 (in Japanese). Tokyo Metropolitian Government (TMG), Tokyo.

ANNEXURE 1

About the Editors

Dr Sangam Shrestha
(Email: sangam@ait.asia)

Dr Shrestha is an Associate Professor of Water Engineering and Management at the Asian Institute of Technology (AIT), Thailand. He is also a Visiting Faculty of the University of Yamanashi, Japan, and Research Fellow of the Institute for Global Environmental Strategies (IGES), Japan. His research interests are within the field of hydrology and water resources including, climate change impact assessment and adaptation in the water sector, integrated water resources management, and groundwater assessment and management. Dr Shrestha has published several papers in peer-reviewed international journals and presented more than three dozen conference papers ranging from hydrological modeling to climate change impacts and adaptation in the water sector. His recent publications include "Climate Change and Water Resources" and "Kathmandu Valley Groundwater Outlook." His present work responsibilities at AIT include delivering lectures at the postgraduate and undergraduate levels, supervising research to postgraduate students (Masters and Doctoral), and providing consulting services on water-related issues to government and donor agencies and research institutions. He has conducted several projects relating to water resources management, climate change impacts, and adaptation with awards from International organizations, such as APN, CIDA, EU, FAO, IFS, IGES, UNEP, UNESCO.

Groundwater Environment in Asian Cities
http://dx.doi.org/10.1016/B978-0-12-803166-7.00028-3

Copyright © 2016 Elsevier Inc.
All rights reserved.

Dr Vishnu Prasad Pandey
(Email: vishnu.pandey@gmail.com)

Dr Pandey is a Research Fellow and Affiliated Faculty at Asian Institute of Technology (AIT), Thailand. He has worked as a postdoctoral researcher for nearly 3 years at the University of Yamanashi in Japan and Research Faculty for nearly 2 years with Asian Institute of Technology and Management (AITM) in Nepal before joining AIT. Dr Pandey has received several awards and fellowships and published over two dozen of peer-reviewed journal papers, several book chapters and conference papers. His recent books include "Climate Change and Water Resources" and "Kathmandu Valley Groundwater Outlook." His core area of research interest and expertise is related to various aspects of groundwater characterization, assessment, and management. His publications related to groundwater include development of groundwater sustainability infrastructure index, evaluation of groundwater environment using DPSIR framework, analyzing hydrogeologic characteristics, estimating groundwater storage potential, and analyzing groundwater markets, among others. Dr Pandey received BEng (Civil) from Tribhuvan University (Nepal), MEng (Water Engineering and Management) from AIT (Thailand) and PhD (Groundwater Management) from University of Yamanashi (Japan).

Dr Binaya Raj Shivakoti
(Email: shivakoti@iges.or.jp)

Dr Shivakoti has been working at IGES as a water resources specialist since 2010. He has over 10 years of experience in water resources management related projects and interdisciplinary research activities. He has accumulated diverse expertise in different aspects of water including water and sanitation, modeling and water quality assessment, GIS and remote sensing, wastewater management, climate change adaptation, policy assessments, and groundwater management. He has been involved in various projects, knowledge sharing and networking on groundwater thorough the Asia Pacific Water Forum Regional Hub for Groundwater Management which IGES has been mobilizing. He obtained PhD from Kyoto University in 2007 and MSc from Asian Institute of Technology (AIT) in 2004.

Dr Shashidhar Thatikonda

(Email: mrshashidhar@gmail.com)

Dr Shashidhar is an Assistant Professor of Department of Civil Engineering at Indian Institute of Technology, Hyderabad. He obtained PhD from Indian Institute of Technology, Madras in the year 2006. He was a visiting faculty at Asian Institute of Technology, Bangkok; University of Illinois, Urbana Champaign; and University of Santiago De Compostela, Spain. He has been conducting research and teaching courses related to water resources and environmental engineering since 2001. His research interest includes *in situ* bioremediation of contaminated aquifers, contaminant transport modeling, isotope hydrology, remote sensing and GIS applications in water resources, and hydroclimate. He has several publications in peer-reviewed international journals. Presently working on various research and consultancy projects and studies related to hydrogeology and aquifer mapping, conjunctive use of surface and groundwater, bioremediation of contaminated aquifers, lake studies, floods, hydraulic transients in pipe networks, etc. He is a member of state pollution control board and member of board of studies for various universities at India. He received awards such as "Young Engineer" and "Young Engineer of the Year – 2010" from Institute of Engineers India.

ANNEXURE 2

About the Authors

Mr Aditya Sarkar
(Email: adi_sarkar.com@rediffmail.com)

Mr Sarkar is a PhD research scholar in the Department of Geology, University of Delhi, India. He earned his Master's degree in Geology from the same university in 2011. He has published research articles in referred journals and has been associated with oral/poster presentation in conferences. He had also cleared the CSIR-UGC joint NET-JRF test and is currently receiving Senior Research Fellowship from CSIR. His research areas include Hydrogeology, Geochemistry, Sedimentology, and Environmental Geology.

Mr Anirut Ladawadee
(Email: anirut.l@dgr.mail.go.th)

Mr Anirut is a Hydrogeologist with Bureau of Groundwater Conservation and Restoration, Department of Groundwater Resource, Thailand. He has been conducting study projects related to land subsidence, groundwater monitoring, and database management. Recently, he has started to work in the field of Geographical Based Management Information System for Groundwater Management in the Groundwater Critical Zone, Groundwater Quality Conservation and Protection, Groundwater Quality Assessment, Contamination Monitoring System in Thailand.

Groundwater Environment in Asian Cities
http://dx.doi.org/10.1016/B978-0-12-803166-7.00029-5

Copyright © 2016 Elsevier Inc.
All rights reserved.

473

Dr Binaya Raj Shivakoti

(Email: shivakoti@iges.or.jp)

Dr Shivakoti has been working at IGES as a water resources specialist since 2010. He has over 10 years of experience in water resources management related projects and interdisciplinary research activities. He has accumulated diverse expertise in different aspects of water including water and sanitation, modeling and water quality assessment, GIS and remote sensing, wastewater management, climate change adaptation, policy assessments, and groundwater management. He has been involved in various projects, knowledge sharing and networking on groundwater thorough the Asia Pacific Water Forum Regional Hub for Groundwater Management which IGES has been mobilizing. He obtained PhD from Kyoto University in 2007 and MSc from Asian Institute of Technology (AIT) in 2004.

Dr Bui Tran Vuong

(Email: buitranvuong@gmail.com)

Dr Vuong is the Deputy Director General of the Division of Water Resources Planning and Investigation for the South of Vietnam (DWRPIS), Ministry of Natural Resources and Environment of Vietnam. He is now responsible for assessment of impacts of groundwater abstraction and climate changes and sea level rises on groundwater resources for Mekong Delta, Vietnam. He has experiences in hydrogeological mapping, assessment of groundwater quality and quantity, groundwater monitoring, groundwater artificial recharge, and groundwater flow modeling. His recent published books consists of Chapter 7 of The Mekong Delta System; Exploitation and Protection of Groundwater Resources; Investigation, Exploitation and Treatment of Water for domestic use. He is also an author of several articles such as groundwater flow model at Le Hong Phong area; the formulation of chemical components of groundwater in coastal plains of Nam Trung Bo; Groundwater Artificial Recharge and its application in Mekong Delta and Ho Chi Minh City. Dr Bui completed his MSc (Hydrology and Water Resources) in 2000 from UNESCO-IHE, The Netherlands and PhD (Utility and Protection of Natural Resources and Environment) in 2009 from Institute of Natural Resources and Environment, Ho Chi Minh City, Vietnam.

Mr Domingos Pinto

(Email: ingobet@gmail.com)

Mr Pinto is a PhD scholar of Water Engineering and Management at Asian Institute of Technology (AIT), Thailand. After completing his Civil Engineering in 2003 and Master of Engineering in 2009, Mr Pinto has worked as National Engineer Consultant for several international agencies (UNDP, GIZ, FAO, AECID, and Care Australia) to provide service as project construction manager, designer, prepare bill of quantity, and assist the project implementation. He is also working for the Government of East Timor, Ministry of Finance in cooperation with Ministry of Public Work to assist project design, evaluate the project financial proposal before proceed to National Procurement for the Biding Process. In August 2012, he continued his PhD at Asian Institute of Technology, Thailand. Currently, he is conducting a research with the topic *"Assessment of Groundwater Potential Zone and Its Sustainable Yield in Dili Alluvial Plain, Timor Leste,"* using following tools: GIS and Remote Sensing, Wetspass and ModelMuse (Modflow 2005).

Dr Ekaterina I. Nemaltseva

(Email: kniiir@mail.ru)

Dr Nemaltseva is a leading researcher of the Laboratory Rational Ground Water Use (Institute of Water Problems and Hydropower, National Academy of Science, Kyrgyz Republic). His field of PhD dissertation is hydrogeology and water resources. In the last decade he was an investigator of three international projects (the main directions are groundwater protection and management, protection from groundwater flood, assessment of groundwater vulnerability). She was also an investigator of few projects in Kyrgyzstan. Dr Nemaltseva is the author of more than 25 scientific papers. Experience includes: (1) calculation and forecast of water and salt balances of irrigated areas; (2) assessment of groundwater vulnerability; (3) creation of the database for keeping and analyses of hydrogeological and water economy information.

Dr Gennady M. Tolstikhin
(*Email: G. Tolstihin@mail.ru*)

Dr Tolstikhin is the Chief Hydrogeologist of the Kyrgyz Hydrogeological Survey. He took an active part in exploration and assessment of main groundwater deposits in the Chu Valley in Kyrgyzstan. Dr Tolstikhin is also in charge of the Laboratory of Ecology of Water Resources and water use at the Institute of Water Problems of the Academy of Sciences of Kyrgyzstan. Also, he is the author of 25 scientific papers in the following domains: environmental control and rational use of groundwater; ecologically based management of groundwater resources for drinking water supply; and radiation security of the environment. Dr Tolstikhin has participated in four projects of the International Research and Development Center, among them: (1) KR-850 "Joint international nuclear pollution studies of Transboundary Rivers in the Central Asia with a view of its non-dissemination, 2003–2006"; (2) KR-1430 "Formation factors studies and estimation of the Lower-Naryn HPS cascade water basins influence on water resources quality of the river Naryn water-collecting area, with the use of isotopic methods." Dr Tolstikhin holds a course in "The methodology of groundwater studies," with the Razzakov Technical University, Bishkek.

Mr Haryadi Tirtomihardjo
(*Email: haryadi_tirtomihardjo@yahoo.com*)

Mr Tirtomihardjo is a senior hydrogeologist at the Central of Groundwater Resources and Environmental Geology (CGREG), Ministry of Energy and Mineral Resources (MEMR). He has been working for MEMR since 1981 and has experiences on hydrogeological mapping and groundwater development for rural water supply in Java areas.

After completing his postgraduate study in International Institute of Hydraulics and Environmental Engineering (IHE) Delft, The Netherlands in 1984, Haryadi Tirtomihardjo conducted various groundwater studies in Indonesia, mainly on assessment of groundwater resource potential within groundwater basins, groundwater exploration and development in areas of the water shortage, monitoring and groundwater conservation in urban groundwater basins (e.g., Jakarta GB, Bandung GB, Semarang GB, Denpasar GB, etc.), and groundwater flow simulation models as well. Since 2000, i.e., era of decentralization in Indonesia, Mr Haryadi Tirtomihardjo has been intensively taking part in the preparation of regulation on groundwater management in Indonesia, e.g., Ministerial Decree of EMR No. 1451K/10/MEM//2000 on Technical Guidelines of Groundwater Management, Governmental Decree No. 43/2008 on Groundwater, Presidential Decree No. 26/2011 on Groundwater Basin of Indonesia, and Technical Guidelines on Establishment and Decision of Groundwater Conservation Zones (will be issued as Ministerial Decree of EMR). Since 2012, Haryadi and colleagues have been involved in project on Comparative Research of Groundwater Management in the Coastal Areas of Southeast Asia funded by ADB and Unesco-IHE, which selected case study is Jakarta Groundwater Basin.

Ms Heejung Kim
(Email: re503@snu.ac.kr)

Dr Kim is a senior researcher at the Research Institute of Basic Sciences in Seoul National University, Korea. She received her PhD and MS degrees from School of Earth and Environmental Sciences, Seoul National University, Korea. She earned her BS degree of Environmental Science and Engineering at Ewha Womans University, Korea. She was awarded the Best Paper Awards from Korean Wetland Society and Korean Society of Soil and Groundwater Environment. She was selected as a recipient of Frontier Scholarship from Ewha Womans University. Her research interests include groundwater-stream water interaction, delineation study of hyporheic zone, and microbial diversity in soil and groundwater.

Dr Jingli Shao

(Email: jshao209@163.com)

Dr Jingli Shao, got his PhD degree from China University of Geosciences (Beijing). He is now a full professor of School of Water Resources and Environment, China University of Geosciences, Beijing. His research interests focus on groundwater numerical simulation, local hydrogeology, assessment and management of groundwater resources, groundwater environmental issues. And he has published more than 100 research papers. He won the second prize of national scientific and technological achievements in 2009, Dayu Reward of MWR in 2008 and five times second prize of scientific and technological achievements of MLR.

Dr Jin-Yong Lee

(Email: hydrolee@kangwon.ac.kr)

Dr Lee is an Associate Professor of Groundwater and Soil Environment at Kangwon National University, Korea. He earned his PhD degree from Seoul National University. He has published over 180 research papers in international and Korean journals. He was awarded Young Geologist Award, Commendation Award from Environment Minister, Academic Award and Best Paper Award from The Geological Society of Korea, Academic Award from Korea Federation of Water Science and Engineering Societies, Academic and Contribution Awards from Korean Society of Soil and Groundwater Environment. In 2014, he was selected as a Hanrim Leading Scientist by The Korean Academy of Science and Technology. He has served as Deputy Editor in Chief of Journal of the Geological Society of Korea, Journal of Soil and Groundwater Environment, and Associate Editor of Geosciences Journal. Dr Jin-Yong Lee is visiting professor (2014–2015) Department of Geological Sciences, University of Colorado, Boulder, CO, USA.

Dr Jiurong Liu

(Email: jiurong@263.net)

Dr Liu is the general engineer of Beijing Institute of Hydrogeology and Engineering Geology, China. He majors in hydrogeology and environmental geology, geothermal resources and development. He got his PhD degree from China University of Geosciences, Beijing, and has published over 40 research papers in international and Chinese journals. He has won Beijing Science and Technology Award and Beijing Geological Science and Technology Award, Awards of Science and Technology of Land and Resources, etc. He participated in UNU Geothermal Training Programme in 1999.

Ms Kabita Karki

(Email: kabita.geo@gmail.com)

Ms Karki is working as a Geologist at Department of Mines and Geology, Nepal. She completed her Master degree on Geology from Tribhuwan University in 2013 AD. She has been involved in different hydrogeological research like study of groundwater extraction status and well inventory in the northern and southern groundwater district of Kathmandu Valley. Hydrogeological study on the formation of the Bis Hazari Tal, a wetland in Chitwan District, Central Nepal, Using Lichenometry to assess long-term GLOF and landslide frequency in the Nepal Himalaya.

Dr Kang-Kun Lee

(Email: kklee@snu.ac.kr)

Dr Lee is a full professor of Hydrogeology and an associate dean of the College of Science at Seoul National University, Korea. He received his PhD from Purdue University, West Lafayette, IN, USA. He has been studying groundwater in Korea since he joined Seoul National University in 1993. He was awarded a medal from the Korean Government at the national ceremony for World Water Day in 2011. He was the president of the Korean Society of Soil and Groundwater Environment during 2011–2012 and is currently serving as a Vice President of the Geological Society of Korea. He has served as an editor for Geosciences Journal, associate editors for Journal of Hydrology and Hydrogeology Journal, and national representative of the International Association of Hydrological Sciences. Dr Lee has published 122 peer-reviewed international journal articles that have received about 1300 ISI Web of Science citations.

Dr Khin Kay Khaing

(Email: khine.khinkay@gmail.com)

Dr Khin Kay Khaing is a lecturer at the Department of Geography, University of Yangon, Myanmar. She obtained Master of Arts degree in 1999 and a doctorate degree in 2006 at the University of Yangon. She is currently teaching undergraduate as well as postgraduate courses for over 18 years. She has been conducting research on groundwater resources and sustainable use, climate change, and climatic conditions of Myanmar, urban environment and water supply. Her research interest includes climate change impact, vulnerability to hazards, adaptation and emergency preparedness, water resource management and sustainability, and environmental geography. She has contributed to some publications in local and international research journals. She also has a responsibility on supervision of Master of Research theses and PhD dissertations at the University of Yangon, Myanmar.

Mr Le Hoai Nam

(Email: lenamdhbk04@yahoo.com)

Mr Nam is an Engineer in Department of Groundwater Planning and Investigation, Division of Water Resources Planning and Investigation for the South of Vietnam, Ministry of Natural Resources and Environment, Vietnam. He has completed a BEng (Geotechnics) degree from Ho Chi Minh City University of Technology, Vietnam. He has experiences in engineering–geological and hydrogeological mapping, groundwater flow and solute transportation modeling. He is now responsible for building a groundwater flow model as well as a solute transportation model in the project on "Protecting Groundwater in Ho Chi Minh City."

Dr Liya Wang

(Email: liyawang1979@126.com)

Dr Liya-wang is a senior engineer, who works for Beijing Institute of Hydrogeology and Engineering Geology, China. She majors in hydrogeology and environmental geology and earned her PhD degree from China University of Geosciences, Beijing. She has published over 10 research papers in international and Chinese journals.

Dr Muhammad Basharat

(Email: basharatm@hotmail.com)

Dr Basharat is an Additional Director at International Waterlogging and Salinity Research Institute (IWASRI) of Pakistan Water and Power Development Authority (WAPDA), Lahore, Pakistan. He completed his PhD (Engineering Hydrology) in 2012 from Engineering University, Lahore and is now a leading researcher for Indus Basin Irrigation

System. The major goal of his research is to quantify the interaction of surface and groundwater in various on-ground environments and varying climatic conditions, using modern tools of RS/GIS and modeling for assessment and performance improvement of irrigation systems, with ultimate objectives of sustainable groundwater management. His research papers in national and international impact journals promote the rationalization of surface water allocations, with the objectives to provide relief to groundwater-stressed areas. Groundwater movement and saline intrusion of fresh groundwater from adjoining saline pockets is of primary interest to him. He is a resource person for WAPDA Engineering Academy and many other institutes, and offering part-time consultancy services on various aspects of water management. He is also a coauthor of the book "Half The Water: Groundwater and its Management in Pakistan."

Dr Oranuj Lorphensri
(*Email: oranuj.l@dgr.mail.go.th*)

Dr Oranuj is Director of Groundwater Control Bureau, Department of Groundwater Resource, Thailand. She has been conducting study projects related to land subsidence groundwater monitoring, database management. Recently, she has started to work in the field of groundwater management, which included diverse aspect in implementing Groundwater Law and related regulations, as well as decentralization of groundwater management task to local administrative level. In addition, she has the position of Board Secretary of Groundwater Development Fund. This board has been major funding machine providing essential support to several study and groundwater management projects in Thailand. Dr Oranuj has several publications in Water Research Journal, and published *Review of Groundwater Management and Land Subsidence in Bangkok, Thailand* in Groundwater and Subsurface Environments, Springer.

Mr Phan Nam Long
(*Email: phannamlong89@gmail.com*)

Mr Long is an Engineer in the Division of Water Resources Planning and Investigation for the South of Vietnam, Ministry of Natural Resources and Environment, Vietnam. He has earned an MEng Degree in Hydrogeology and Environmental Geology from Gadja Mada University, Indonesia. He has experience in hydrogeological mapping, groundwater flow, and solute transportation modeling. He is now working on "Protecting Groundwater in Ho Chi Minh City."

Mr Qing Yang
(*Email: yq@bjswd.com*)

Mr Yang is a Senior Engineer, who works for Beijing Institute of Hydrogeology and Engineering Geology, China. He is then Vice Director of Geological Environmental Research Institute. He majors in hydrogeology and environmental geology and obtained a Master's Degree from China University of Geosciences. He has published over 30 papers on international and Chinese journals. He has won Beijing Science and Technology Award and Beijing Geological Science and Technology Award, etc. He is a member of Geological Society of China.

Dr Qiulan Zhang

(Email: qlzhang919@163.com)

Dr Qiulan Zhang got her PhD from Utrecht University, the Netherlands in 2013. From 2014, she works as an Assistant Professor in School of Water Resources and Environment, China University of Geosciences, Beijing. Her research interests focus on solute and colloid transport in porous media, assessment of water resources. She has published five SCI journals in Water Resources Research, Vadose Zone Journal, etc.

Dr Rabin Malla

(Email: rabin@creew.org.np)

Dr Malla is working as an Executive Director at Center of Research for Environment, Energy and Water (CREEW), Nepal. He completed his PhD degree on Environmental Engineering from University of Yamanashi, Japan in 2010. He has been involved in research activities relating to groundwater of Kathmandu valley that includes groundwater and river water quality (physio-chemical and microbiological) assessment; groundwater treatment for reducing ammonia/nitrate and iron; analysis of heavy metals in river-sediment, river water and plants; and biomass modeling and nutrient uptake modeling by plants. There are 8 peer-reviewed journal and book chapters authored and co-authored by him and several publications in the conference and symposium proceedings.

Dr Rafael Litvak

(Email: lit1@mail.kg)

Dr Litvak is a Head of Ground Water Modeling laboratory of Kyrgyz Scientific and Research Institute of Irrigation, Kyrgyz Republic. In the last decade he was Principal Investigator of four International Projects (the main directions are groundwater protection and management, protection from groundwater flood, assessment of groundwater vulnerability). He was the main investigator of dozen projects in Kyrgyzstan. The basic research instruments are groundwater modeling and mathematical approach to hydrogeological and hydrologic problems. Dr Litvak is the author of more than 50 scientific papers. Experience includes: (1) problems of management and protection of water resources (groundwater and surface water); (2) groundwater flood protection; (3) groundwater modeling, mathematical assessment of hydrogeological processes; (4) problems of mitigation negative sequels of climate change for Central Asia intermountain basins. Pedagogical experience: (1) leaderships of Master's and PhD thesis; (2) teaching of the groundwater dynamics (Russian–Kyrgyz Slavonic University, Kyrgyzstan).

Ms Rong Wang

(Email: happywangrong@126.com)

Ms Rong Wang is a Senior Engineer, who works in Land Subsidence Research Institute of Beijing Institute of Hydrogeology and Engineering Geology, China. She majors in monitoring land subsidence and ground fissure and obtained a Master's degree from China University of Geosciences. She has published over 20 papers on international and Chinese journals. She is a member of Geological Society of China.

Dr S.V.N. Rao

(*Email: shedimbi@yahoo.com*)

Dr Rao is presently working as a Project Director, with WAPCOS, Regional Office, Hyderabad, India (http://wapcos.gov.in/). He has an advanced understanding of hydrology and water resources problems. Presently working as Team Leader for Floodplain Groundwater Management project in River Yamuna near Delhi. Formerly Dr Rao was working as a Principal Engineer/Senior Hydrologist and Modeler with DHI India in several projects related to groundwater and surface water. Before joining DHI India he was working as a Senior Scientist F&Head in various capacities with National Institute of Hydrology at Roorkee under the Ministry of Water Resources, Govt. of India and its regional offices at Jammu and Kakinada. Dr Rao holds a PhD in Civil Engineering with specialization in Water Resources from IIT Madras. His large experience of over 25 years spans in the Government and Private sectors as a Consultant, Scientist-Hydrologist, Modeler, and Trainer in a multicultural environment. He has worked with various international agencies (UN, UNESCO, UNDP, World Bank) on many diverse hydrological problems. Dr Rao has more than 5 years teaching experience in Civil Engineering as a Lecturer. He is a specialist in groundwater assessment and management modeling, and has a large number of peer-reviewed publications in hydrology and water resources.

Dr Sangam Shrestha

(*Email: sangam@ait.asia*)

Dr Shrestha is an Associate Professor of Water Engineering and Management at the Asian Institute of Technology (AIT), Thailand. He is also a Visiting Faculty of the University of Yamanashi, Japan, and Research Fellow of the Institute for Global Environmental Strategies (IGES), Japan. His research interests are within the field of hydrology and water resources including, climate change impact assessment and adaptation in the water sector, integrated water resources management, and groundwater assessment and management. Dr Shrestha has published several papers in peer-reviewed international journals and presented more than three dozen conference papers ranging from hydrological modeling to climate change impacts and adaptation in the water sector. His recent publications include "Climate Change and Water Resources" and "Kathmandu Valley Groundwater Outlook." His present work responsibilities at AIT include delivering lectures at the postgraduate and undergraduate levels, supervising research to postgraduate students (Masters and Doctoral), and providing consulting services on water-related issues to government and donor agencies and research institutions. He has conducted several projects relating to water resources management, climate change impacts, and adaptation with awards from International organizations, such as APN, CIDA, EU, FAO, IFS, IGES, UNEP, UNESCO.

Mr Shakir Ali
(Email: shakiriitb@gmail.com)

Mr Ali is an MPhil scholar in Department of Geology, University of Delhi, India. He earned MTech degree from Indian Institute of Technology, Bombay. He has published one international paper and two abstracts in national conferences. His research area includes Hydrogeochemistry, Geochemistry, etc.

Dr Shashank Shekhar
(Email: shashankshekhar01@gmail.com)

Dr Shekhar is an Assistant Professor at University of Delhi, India. He did his graduation, postgraduation and PhD from Department of Geology, University of Delhi. He started his career as a teacher in the same department in 1998, moved to Central Ground Water Board, Government of India as Scientist (2001–2009) and again joined back the same University in 2009. His current research interest includes Integrated Groundwater Management Study (using analytical approach and numerical modeling approach), Hydrogeological Mapping and Investigations, Well Hydraulics, Tube Well (production well) Design and Construction, Stream Aquifer Dynamics, etc. As a Scientist at Central Ground Water Board, he was associated with design, construction, and pumping test on about 80 tube-wells, designing of rainwater harvesting structures, integrated groundwater management studies, and numerical modeling studies. He has authored about 17 articles in peer-reviewed journals and 8 articles in edited volumes/conference proceedings, etc. He has been associated as author with about 26 papers/posters presented in national and international seminar/workshops. Besides normal teaching he has delivered about 19 invited lectures.

Dr Shashidhar Thatikonda

(Email: mrshashidhar@gmail.com)

Dr Shashidhar is an Assistant Professor of Department of Civil Engineering at Indian Institute of Technology, Hyderabad. He obtained PhD from Indian Institute of Technology, Madras in the year 2006. He was a visiting faculty at Asian Institute of Technology, Bangkok; University of Illinois, Urbana Champaign; and University of Santiago De Compostela, Spain. He has been conducting research and teaching courses related to water resources and environmental engineering since 2001. His research interest includes *in situ* bioremediation of contaminated aquifers, contaminant transport modeling, isotope hydrology, remote sensing and GIS applications in water resources, and hydroclimate. He has several publications in peer-reviewed international journals. Presently working on various research and consultancy projects and studies related to hydrogeology and aquifer mapping, conjunctive use of surface and groundwater, bioremediation of contaminated aquifers, lake studies, floods, hydraulic transients in pipe networks, etc. He is a member of state pollution control board and member of board of studies for various universities at India. He received awards such as "Young Engineer" and "Young Engineer of the Year – 2010" from Institute of Engineers India.

Mr Suman Kumar

(Email: geology.suman@gmail.com)

Mr Kumar is Senior Research Scholar in Department of Geology, University of Delhi, India. He completed his graduation and postgraduation in Geology from University of Delhi, Delhi. He has qualified CSIR-NET in 2011 and since then is working as a PhD Research Scholar in the same department. His areas of interest include hydrogeology, hydrology, and allied fields.

Dr Tussanee Nettasana

(Email: tussanee.n@dgr.mail.go.th)

Dr Tussanee is a senior hydrogeologist of Groundwater Conservation and Restoration Bureau, Department of Groundwater Resource, Thailand. She has been conducting study projects related to groundwater exploration and assessment, groundwater modeling, and groundwater management. In addition, she is the Director of Groundwater Impact Assessment in the Critical Area and the Lower Chao Phraya Basin Project and Director of Groundwater Coordinating Center for supporting the Royal Initiative Project. Dr Tussanee has several publications on hydrogeological mapping, groundwater assessment, and conceptual model uncertainty.

Dr Vishnu Prasad Pandey

(Email: vishnu.pandey@gmail.com)

Dr Pandey is a Research Fellow and Affiliated Faculty at Asian Institute of Technology (AIT), Thailand. He has worked as a postdoctoral researcher for nearly 3 years at the University of Yamanashi in Japan and Research Faculty for nearly 2 years with Asian Institute of Technology and Management (AITM) in Nepal before joining AIT. Dr Pandey has received several awards and fellowships and published over two dozen of peer-reviewed journal papers, several book chapters, and conference papers. His recent books include "Climate Change and Water Resources" and "Kathmandu Valley Groundwater Outlook." His core area of research interest and expertise is related to various aspects of groundwater characterization, assessment, and management. His publications related to groundwater include development of groundwater sustainability infrastructure index, evaluation of groundwater environment using DPSIR framework, analyzing hydrogeologic characteristics, estimating groundwater storage potential, and analyzing groundwater markets, among others. Dr Pandey received BEng (Civil) from Tribhuvan University (Nepal), MEng (Water Engineering and Management) from AIT (Thailand), and PhD (Groundwater Management) from University of Yamanashi (Japan).

Mr Woo-Hyun Jeon

(Email: simple950@hanmail.net)

Mr Jeon is a PhD candidate of Groundwater and Soil Environment at Kangwon National University, Korea. He has published one research paper in international journal and six research papers in Korean journals. He has presented three Korean patents and was awarded The Best Poster from The Geological Society of Korea. He has presented 10 posters in international conferences. He received a completion certification of groundwater modeling short course from National Ground Water Association. His research interests include remediation of TCE-contaminated groundwater, spatial and temporal variations of soil temperatures in the Antarctic, and groundwater–stream water interaction in the hyporheic zone.

Ms Zhiping Li
(Email: lzp2997@126.com)

Ms Zhiping Li is a Senior Engineer, who works for Beijing Institute of Hydrogeology and Engineering Geology, China. She majors in Hydrogeology and Environmental Geology and obtained a Master's degree from Ji Lin University. She has published 26 papers on international and Chinese journals and 3 books as a collaborator. She won Beijing Science and Technology Award and Beijing Geological Science and Technology Award, etc. She is engaged as an environmental impact assessment expert by Beijing Municipal Environmental Protection Bureau.

INDEX